Stem Cell Biology and Regenerative Medicine

Volume 69

Series Editor

Kursad Turksen, Ottawa Hospital Research Institute, Ottawa, ON, Canada

Our understanding of stem cells has grown rapidly over the last decade. While the apparently tremendous therapeutic potential of stem cells has not yet been realized, their routine use in regeneration and restoration of tissue and organ function is greatly anticipated. To this end, many investigators continue to push the boundaries in areas such as the reprogramming, the stem cell niche, nanotechnology, biomimetics and 3D bioprinting, to name just a few. The objective of the volumes in the Stem Cell Biology and Regenerative Medicine series is to capture and consolidate these developments in a timely way. Each volume is thought-provoking in identifying problems, offering solutions, and providing ideas to excite further innovation in the stem cell and regenerative medicine fields.

More information about this series at http://www.springer.com/series/7896

Firdos Alam Khan
Editor

Advances in Application of Stem Cells: From Bench to Clinics

Humana Press

Editor
Firdos Alam Khan
Department of Stem Cell Research
Institute for Research and Medical Consultations
Imam Abdulrahman Bin Faisal University
Dammam, Saudi Arabia

ISSN 2196-8985 ISSN 2196-8993 (electronic)
Stem Cell Biology and Regenerative Medicine
ISBN 978-3-030-78100-2 ISBN 978-3-030-78101-9 (eBook)
https://doi.org/10.1007/978-3-030-78101-9

This Humana imprint is published by the registered company Springer Nature Switzerland AG
The registered company address is: Gewerbestrasse 11, 6330 Cham, Switzerland

Preface

The field of stem cell biology is expanding with a continuous surge of new information related to its applications. Over the past few years, stem cells have been extensively used in various biological applications, especially in cell therapy, tissue engineering, in vitro drug testing. There is no single book available that comprehensively describes the significance of stem cells' various applications derived from embryonic and adult sources from laboratory to clinics point of view. Most of the books are either written about the basics of stem cells or purely commercialized aspects separately. This book discusses the basics and advance topics of stem cells that help the researchers, students, and professionals a single source of updated information about stem cells and in various applications.

This book is divided into 12 chapters and covers topics such as in vitro cell culture, 3D cell culture, cell therapy, tissue engineering, cell factory, cell functionality, in vitro drug testing, organ development, autologous transplantation, allogeneic transplantation, adult stem cells, multipotent stem cells, induced pluripotent stem cells, a pluripotent, and embryonic stem cell. We will also discuss various stem cell-based products, commercialization, IPR, and market of stem cell-based products, challenges of stem cell therapy, current research trends, and career. As stem cell technology is expanding, we thought it is time to overview the stem cell field and write a complete book dedicated to advances in the application of stem cells from bench to clinics. This book provides comprehensive and updated information on all aspects of stem cell research. There are many books written on stem cell biology and research on different topics. Still, there is no single book available that discusses all advanced topics of application of stem cells from bench to clinics.

This book has tried to include all the topics directly or indirectly related to the stem cell field. The primary objective is to provide the students, researchers, and professionals a single source of information about stem cells' applications. There are 12 chapters in this book, and each chapter contains the updated information with beautiful illustrations. We have discussed the various topics basics of stem cell biology, types and classifications of stem cells, and the method of isolation and characterization of stem cells, differentiation of stem cells into neuronal, cardiomyocyte, and hepatocyte pancreatic lineage, and differentiation into other cell types. The book also discusses topics such as in vivo transplantation in animal models, stem cells in regenerative medicine, clinical trials, stem cell production, and stem cell-based products in the market, and commercialization of stem cell products.

Dammam, Saudi Arabia

Firdos Alam Khan, Ph.D.
Editor

Acknowledgements

First, I am grateful to the Almighty Allah, who gave me the strength to complete this book's edition on the stipulated time. I am thankful to many people, especially to Dr. Kursad Turksen, Editor-in-Chief, Stem Cell Biology and Regenerative Medicine Series, for giving me an opportunity to edit this book. I am also thankful to Dr. Gonzalo Cordova, Associate Editor, Biomedicine, Cell Biology, Molecular Biology, Biochemistry, and Biophysics, Springer Nature, who supported me to complete the task. I am thankful to all Springer Nature production team members for their support and cooperation.

I am grateful to all the authors and especially to all corresponding authors for their immense contributions and timely completion of the work. I want to thank the entire management team of the Institute for Research & Medical Consultations (IRMC), Imam Abdulrahman Bin Faisal University, Dammam, Saudi Arabia, for their support, especially to Prof. Ebtesam Al-Suhaimi, Dean, IRMC, Imam Abdulrahman Bin Faisal University, Dammam, Saudi Arabia, for her constant encouragement.

I am thankful to all my teachers and mentors, especially Prof. Nishikant Subhedar and Late Prof. Obaid Siddiqi FRS, for their immense contributions in shaping my research career. I am also thankful to all my friends, well-wishers, and colleagues for their support and cooperation.

I am grateful to my entire family members, especially to my father Late Nayeemuddin Khan and mother Sarwari Begum, my brothers (Aftab Alam Khan, Javed Alam Khan, Intekhab Alam Khan, Sarfaraz Alam Khan), my sisters (Sayeeda Khanum, Faheemida Khanum, Kahkashan Khanum, Ayesha Khanum), my wife Samina Khan, and my sons (Zuhayr Ahmad Khan, Zaid Ahmad Khan, and Zahid Ahmad Khan) and my daughter (Azraa Khan), my father-in-law (Abdul Qayyum Siddiqi) and mother-in-law (Uzma Siddiqi). All of them, in their ways, supported me.

Enjoy reading!

Firdos Alam Khan, Ph.D.
Professor and Chairman
fakhan@iau.edu.sa

Contents

About the Editor

Prof. Firdos Alam Khan is Professor and Chairman, Department of Stem Cell Biology, Institute for Research and Medical Consultations (IRMC), Imam Abdulrahman Bin Faisal University, Dammam, Saudi Arabia. In addition to his research work, he has been teaching (Cell Physiology and Biochemistry and Stem Cell Biology and Regenerative Medicine) courses to M.Sc. students. He has done his Ph.D. degree in Zoology with a Neuroscience specialization from Nagpur University, India. Over the past 25 years, he has been involved in teaching various courses such as cell biology, pharmacology, business of biotechnology, biomedicine, cell and tissue engineering, bioethics and IPR to undergraduate and postgraduate students. He was associated with Manipal University (now Manipal Academy of Higher Education) Dubai Campus, United Arab Emirates. He was Professor and Chairperson, School of Life Sciences, and served as Chairman of the Research & Development Program. He did his first postdoctoral fellowship from the National Centre for Biological Sciences (NCBS), Bangalore, India, and a second postdoctoral fellowship from Massachusetts Institute of Technology (MIT), Cambridge, USA. He has also worked with the Reliance Life Sciences as a Research Scientist in the Stem Cell Research Laboratory. He has been granted two US patents in stem cell biology and published three US patents on the use of biomaterials in cancer treatment. He has authored two books, namely "Biotechnology Fundamentals (textbook)" (1st edition 2009, 2nd edition 2014, 3rd edition 2020) and "Biotechnology in Medical Science" published by CRC Press, Taylor & Francis Group. He has also edited one book, "Application of Nanomaterials in Human Health," in 2020 published by Springer Nature. He was Associate Editor of research journal 3Biotech from 2010 to 2020. He has been a recognized reviewer for the scientific journals, Nanomedicine, International Journal of Nanomedicine, Scientific Reports, IET Nanobiotechnology, Drug Design, Development, and Therapy, Journal of Nanostructure in Chemistry, Aging, Current Pharmaceutical Biotechnology, Tissue Engineering, Nano Express. He has published more than 70 research papers in peer-reviewed journals. He is currently studying the impact of different biomolecules, biomaterials, and nanomaterials on stem cell differentiation and cancer treatment.

Chapter 1
Basics of Stem Cells

Dhvani H. Kuntawala and Glen J. P. McCann

Abstract Stem cells have been researched for over 100 years. It all started in 1908, when a histologist Alexander Maksimov coined the name stem cells. Many key scientists have noted the potential in researching stem cells in the following the years after. These cells have the ability to renew and differentiate themselves into a wide range of cell types. Stem cells have two classifications accordingly to their properties—pluripotent and multipotent. Pluripotent cells are able to differentiate into three germ layers while multipotent cells can differentiate into only a few limited types of cells. Hence, they are important to the repair, development, preservation and growth of many organs from the earliest stages of life. Stem cells are also obtained from many sources and found throughout the life cycle from embryos to adults. Research has also helped scientists understand stem cells in different species (animals and humans) for many years. Recognition of the value of the field has seen scientists awarded Nobel prizes on discoveries regarding stem cells. This chapter describes the basics of stem cells: their early discovery, structure, morphology, characteristics, differences, location, function, roles, sources and Nobel Prize research carried out.

Keywords Stem cells · Adult stem cells · Embryonic stem cell · Cancer stem cells · Induced pluripotent stem cells · Pluripotent cells · Multipotent cells

Abbreviations

ASCs	Adult Stem Cells
ESC	Embryonic Stem Cell
hESC	Human Embryonic Stem Cell

D. H. Kuntawala · G. J. P. McCann (✉)
Leicester School of Pharmacy, De Montfort University, The Gateway, Leicester, UK
e-mail: gmccann@dmu.ac.uk

F. A. Khan (ed.), *Advances in Application of Stem Cells: From Bench to Clinics*,
Stem Cell Biology and Regenerative Medicine 69,
https://doi.org/10.1007/978-3-030-78101-9_1

iPSC Induced Pluripotent Stem Cell
MSCs Mesenchymal Stem Cells.

Definition of Stem Cells

Stem cells are a different class of cells present in tissues and organs of the body at all stages of development, from an early sequential series of potent processes to adult life. They can be explained as the cellular building blocks of the body that have the potential to self-renew (make copies of themselves) and differentiate (mature into more specialized cells) over a period of time. Differentiation is essential for the growth of an adult organism. Studies have shown different forms, obtained from the embryo and the adult life (Carpenedo & McDevitt, 2013; Snoeckx et al., 2009) as these cells play specific roles, such as becoming blood, brain, bone or skin cells (Morrison & Kimble, 2006). Therefore, stem cells can replace damaged cells by acting like a repair system in organisms (Łos et al., 2019).

Stem cells have been distinguished from four sources: embryonic, germinal, those extracted from carcinomas, and somatic stem cells. They are known as adult stem cells once located in the postnatal tissues (Gonzalez & Bernad, 2012). There are four basic types of stem cells that occur at different periods of life: Adult, embryonic, induced pluripotent and diseased cells in cancers that may also exhibit quite a few stem cell properties. (Alvarez et al., 2012). Embryonic stem cells have the ability to differentiate into three embryonic germ layers (endoderm, mesoderm, and ectoderm). Furthermore, they can differentiate within the embryonic extension that leads to obtaining any of the 220 kinds of cells in an individual (Gucciardo et al., 2008).

Historical Background of Stem Cell Research

In 1908, the histologist Alexander Maksimov suggested the expression "stem cells." He developed and introduced a theory of blood and cell origin and differentiation. Many papers were published by other researchers regarding hematopoietic stem cells after Alexander's discovery. Evidence demonstrated functional cells could be non-specialized and were termed hematopoietic (immature cells that can grow into red, white blood cells, and platelets). These stem cells were found in the marrow in 1932 by Florence Sabin. In 1950, E. Donnall Thomas initiated his work on bone marrow transplantation to support hematopoietic stem cell actuality. He executed the first bone marrow transfer to treat leukemia. This transplant comprised of similar twins, of which one of them had leukemia. In 1957, Thomas also performed the first allogeneic transplantation. Later, evidence was building of neurons in adults being able to be produced from the neural stem cells (adult neurogenesis), and knowledge of stem cell activity in the brain was also contributed to in 1960. Ernest McCulloch and his research team illustrated the existence of revival cells in the bone marrow of a mouse during 1963. The first HLA matched human bone marrow transfer was carried out in 1968 by Robert Good. In 1978, hematopoietic stem cells were located in human placental cord blood. Derivation of mouse embryonic stem cells from the pluriblast was achieved by Gail Martin in 1981, she invented the expression "embryonic stem cell". Bonnet and Dick located cancer stem cells in blood-forming tissue or cells, also known as hematological cancer, in 1997. Cancer stem cells usually have a normal stem cell trait, and this gives them the ability to produce many different cell types in a particular cancer sample.

James Thomson and co-workers derived the first human ESC cell line from the pluriblast (inner cell mass) of early embryos that generally go on to give rise to a foetus in 1998. A year later, adult mouse tissues were altered to give rise to various cell kinds, and this indicated that liver cells could be produced from cells obtained from the bone marrow. The first human embryo was cloned at an early stage of about 4–6 cells by researchers to generate ESCs in 2000, thereafter many articles on plasticity in adult stem cells were issued. In 2002, the first human embryonic stem cell trial took place in the USA. Government funding was banned by George W. Bush for embryonic stem cell research because of ethical (social) issues concerning the disturbance of the embryo. This ban was lifted in 2009, and the research was allowed to continue in the USA. Then biotechnology company Geron Corporation carried out a trial, hoping to trigger nerve growth in patients that have spinal cord injury by using GRNOPC1, a human embryonic stem cell-derived oligodendrocyte progenitor cells (OPCs) therapy. No formal outcomes from this trial were issued as such; preparatory conclusions of the study were addressed in October 2011 at the American Congress of Rehabilitation Medicine (ACRM) conference.

From 2004 to 2005, Hwang Woo-suk was alleged to have made many human ESC lines from human oocytes that had not been fertilized. Some of this work was later shown to have been forged. In 2006, Kazutoshi Takahashi and Shinya Yamanaka published their rat induced pluripotent stem cell work. They found the potential to encourage cellular pluripotency, and it transformed the perspective of stem cell research. During 2007, various groups reported normal skin cells being reprogrammed in mice back to an embryonic state. Mario Capecchi, Martin Evans, and Oliver Smithies in Physiology and Medicine by the end of 2007, a Nobel Prize was granted for their gene research on ESCs from mice. The researchers published the induction of pluripotent stem cells. Raymond Wong et al. resolved the part of intercellular communication connections via gap junctions in both somatic and embryonic stem cells in 2008. Vanessa Hall discovered ESCs in porcine as an origin for human cell replacement treatment in 2009, and also the genomic profiling of mesenchymal stem cells was carried out in the same year by Danijela Menicanin et al. In 2010, Yue Xu et al. revealed the importance of cell adhesion and growth factor signalling regulatory techniques for pluripotent stem cell survival and self-renewal.

In 2012, a Nobel Prize was granted to Shinya Yamanaka and Sir John Gurdon to discover developed cells being reprogrammable to give rise to many different cell types (pluripotent). The first derivation of human embryonic stem cells (hESCs) by a therapeutic (designing a cloned embryo to produce embryonic stem cells with identical DNA as the donor cell) cloning was in 2013 by Tachibana et al. The first clinical trial with human induced pluripotent stem cells (hiPSCs) was initiated in 2014. Liao et al. in 2015 reported CRISPR/cas9 technology being applied to human embryonic stem cell (hESC) gene editing; later during that year, other research was carried out by Takasato et al. on the generation of organoids from hESC for modelling foetal organ morphogenesis. During the year 2017, Jun Wu published his finding on a chimeric pig embryo populated with hIPSCs (Eguizabal et al., 2019; Kumar et al., 2010; Shihadeh, 2015).

Structure and Morphology of Stem Cells

From 2004 to 2005, Hwang Woo-suk was alleged to have made many human ESC lines from human oocytes that had not been fertilized. Some of this work was later shown to have been forged. In 2006, Kazutoshi Takahashi and Shinya Yamanaka published their rat induced pluripotent stem cell work. They found the potential to encourage cellular pluripotency, and it transformed the perspective of stem cell research. During 2007, various groups reported normal skin cells being reprogrammed in mice back to an embryonic state. Mario Capecchi, Martin Evans, and Oliver Smithies in Physiology and Medicine by the end of 2007, a Nobel Prize was granted for their gene research on ESCs from mice. The researchers published the induction of pluripotent stem cells. Raymond Wong et al. resolved the part of intercellular communication connections via gap junctions in both somatic and embryonic stem cells in 2008. Vanessa Hall discovered ESCs in porcine as an origin for human cell replacement treatment in 2009, and also the genomic profiling of mesenchymal stem cells was carried out in the same year by Danijela Menicanin et al. In 2010, Yue Xu et al. revealed the importance of cell adhesion and growth factor signalling regulatory techniques for pluripotent stem cell survival and self-renewal.

In 2012, a Nobel Prize was granted to Shinya Yamanaka and Sir John Gurdon to discover developed cells being reprogrammable to give rise to many different cell types (pluripotent). The first derivation of human embryonic stem cells (hESCs) by a therapeutic (designing a cloned embryo to produce embryonic stem cells with identical DNA as the donor cell) cloning was in 2013 by Tachibana et al. The first clinical trial with human induced pluripotent stem cells (hiPSCs) was initiated in 2014. Liao et al. in 2015 reported CRISPR/cas9 technology being applied to human embryonic stem cell (hESC) gene editing; later during that year, other research was carried out by Takasato et al. on the generation of organoids from hESC for modelling foetal organ morphogenesis. During the year 2017, Jun Wu published his finding on a chimeric pig embryo populated with hIPSCs (Eguizabal et al., 2019; Kumar et al., 2010; Shihadeh, 2015).

Detailed morphology of embryonic stem cells was described by results that showed high nuclear to cytoplasmic proportion and compact colonies. Cells with prominent nucleoli and round colonies have also been seen. Additionally, these cells also have clear and smooth colony edges. When culturing hPSCs, they are grown

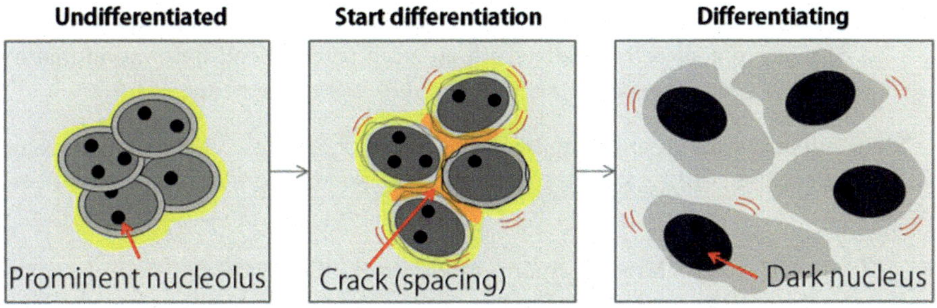

Fig. 1.1 Morphological changes in human pluripotent stem cells (hPSCs) (Wakui, 2017). Moving from left to right, as the cells specialize the nucleolus becomes harder to visualize within a dark nucleus. Cells also proceed to space themselves apart as their morphology becomes more specific

while maintaining morphological features (round, flat, and smooth edged) intact. In an undifferentiated state after reprogramming. Variations of cells usually occur in the culture due to impurities, with some cells that are not successful at reprogramming. Hence, during culture, they are selected to keep the cells that were reprogrammed and remove the unsuccessful cells. Once cells diverge from pluripotency to a differentiated condition, they have a white space that appears like a split space intercellularly. As seen in Fig. 1.1, the distance connecting the cells expands, as well as these cells develop an appearance similar to the differentiated cells. These cells slowly develop a dark and flat appearance. The differentiated cells within their nuclei have less relaxed chromatin when compared to the undifferentiated hPSCs, as chromatin undergoes an alteration in structure to heterochromatin. This will lead to a loss of translucency in the nuclei, as the nucleoli will become invisible in the course of the differentiated process under staged contrast microscopy.

Characteristics of Stem Cells

Stem cells vary in their developmental versatility or degree of plasticity. So, they can also be described according to their characteristic of self-renewal and plasticity; this is important for the renewal and recovery of the body. Other properties that allow stem cells to show their ability to exert the structure of organs and tissues are rapid proliferation and differentiation (Pavlović & Radotić, 2017).

There four main types of stem cells:

I. Embryonic stem cells—they are pluripotent stem cells obtained from the inner cell mass (ICM) of an early stage embryo called blastocyst.
II. Adult stem cells—they are any cells taken from a fully grown tissue. These include endodermal, mesodermal and ectodermal origins.
III. Cancer stem cells—these have been seen associated with nearly every type of cancer.

IV. Induced pluripotent stem cells—these types of cells have been extracted from a non-pluripotent cell, mostly from an adult somatic cell by treating mature cells with genes that dedifferentiate them to a pluripotent stage.

Stem cells are present in all multicellular living organisms. These stem cells are able to differentiate into specialized cell kinds depending on their potency. These include:

I. Totipotent cells—Morula, Spore, and Zygote; these types of cells have the possibility to give rise to any human cells (brain, liver, heart, or blood). They are also each capable of giving rise to a whole viable organism.
II. Pluripotent cells—Embryonic stem cells; cannot emerge to a whole organism; however, they are capable of giving rise to all types of tissues. Can create callus-like groups of similar cells.
III. Multipotent cells—Progenitor cells (hematopoietic and mesenchymal stem cells); they can emerge to a restricted variety of cells inside a tissue type.
IV. Unipotent cells—Specific originator cells, such as muscle stem cells (Hui et al., 2011).

Embryonic stem cells have two properties: to propagate themselves under defined conditions and to be able to differentiate into the ectoderm, endoderm and mesoderm germ layers. Human embryos have 50–150 cells at the blastocyst phase, post-fertilization (Ying et al., 2003). It is challenging to study cells in situ, so advancements in cell culture have been essential to understand cellular characteristics. For many years pluripotent cell lines have been isolated and maintained for in vitro cultures such as from mouse blastocysts (Evans & Kaufman, 1981). Also, Martin (1981) demonstrated observing these cultures that were derived from single cells that could differentiate into various cell types. This made it possible to study early mammalian development. Undifferentiated human embryonic stem cells and embryonic germ cells were cultured by James Thomson and his group in 1998. This study provides an understanding of differentiation and how human tissue functions as well as new approaches for drug discovery and analysis (Thomson & Odorico, 2000).

Adult stem cells are derived from mature tissue. Examples comprise stem cells obtained from placental tissues like human amnion epithelial cells and mesenchymal stem cells. These cells have been indicated to be able to reduce inflammation and augment repair of animal replica injury model studies. Even though these cells can segregate into tissues in vitro that are found from the separate parts of the germ cell layers in vitro, they have limited differentiation capacity. Adult stem cells benefit since cells sourced from self or family use may have reduced ethical and biological concerns (Kolios & Moodley, 2013).

Cancer stem cells have many different characteristics that allow them to assist in tumourigenesis. They are discrete populations of cells within tumours. (Dalerba et al., 2007). Cancer stem cells (CSCs) have been acknowledged to be significant in leukemia and dense cancers. Researchers have suggested that cancer stem cells have the capability to self-regenerate and can then separate into distinctive forms of cancer cells. They are believed to be accountable for malignancy, evolution, metastasis, reappearance, and drug resistance. Cancer stem cells remain pluripotent, so they can lead to tumour cells with different phenotypes, leading to the development of leading cancer and the appearance of novel cancers (Chen et al., 2013).

Induced pluripotent stem cells (iPSCs) are obtained by reprogramming grown mammalian cells by the enforced expression of DNA regions involved in pluripotency and cell multiplication (Stadtfeld & Hochedlinger, 2010). iPSCs can afterwards segregate into several specific somatic cell varieties and share identical features, including but not restricted to the morphology and proliferation of ESC, self-renewal, and the possible variation into different kinds of cells (Ji et al., 2016). This capability creates a revolutionary tool for a wider range of technical methods and reformative medication (Cantz & Martin, 2010; Reibetanz et al., 2016).

Differences Between Somatic Cells and Somatic Stem Cells

Somatic cells refer to all the body cells except sperm and egg cells (germline cells). Somatic cells are diploid and have two groups of each chromosome, one from each parent. In this sense, adult human stem cells can be considered to be somatic. These cells do not have the capability to create any progenies; instead, they shape all the organism's structures and tissues. Several types of these originator cells have been separated in adult tissues; hence most tissues have their own specific stem cells (Fig. 1.2). For example, the neural stem cells in the subventricular region (outside the lateral ventricle wall in vertebrates) contain epithelial stem cells. The spongy marrow tissue of the bone is where hematopoietic stem cells reside and function to refill cells into the blood that have deceased in function or accumulated pathological processes. These stem cells are normally located in proximity to their normal somatic cell's tissue position. In the case of blood, the unmineralized hollow bones serving as a site for this organ.

All somatic cells, both normal and ASCs, can divide through mitosis to replace and maintain the tissue continuity across a multicellular system's lifetime. The ASCs are unspecialized cells with self-renewal capacity. The source of stem cells starts with a totipotent egg, which can segregate to the placenta and all the kinds of tissue in the body. This blastocyst forms after seven to eight-cell divisions of an egg that has been fertilized. The blastocyst's outer wall will be altered to hold it to the uterine wall along with the inner cell mass, which contains the pluripotent cells that can segregate into any other cell type with their more specific functions. They are known as the embryonic stem cells that differentiate into different multipotent stem cells and progenitor-specific cells (Ramesh et al., 2009). Adult stem cells are found in a particular area of tissue in tiny numbers called a stem cell niche. An essential factor is that stem cells can remain inactive for a period of time until a signal is operated internally or externally, for example a tissue injury or diseased state (Dutta, 2020). This is very advantageous as uncontrolled cell division in these cells would be very dangerous.

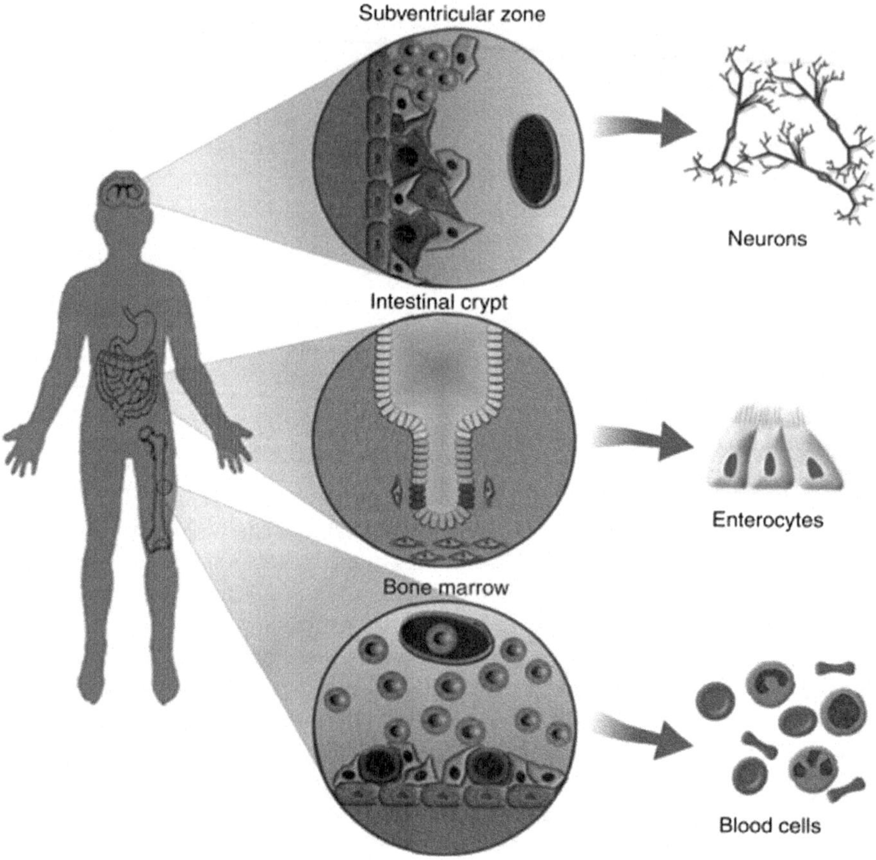

Fig. 1.2 Adult somatic stem cells (ASCs). ASCs exist in most tissues of the body. Due to most cell's dependency upon anchorage, ASCs are located close to their relevant sites. Cells of the blood being non-anchorage dependent migrate from the ASCs in bone marrow (Chagastelles & Nardi, 2011)

Normal somatic cells being differentiated and specialized in function are very often different from each other as well as from their stem cells. This is best demonstrated by studying the most extreme examples. Cells that have made the ultimate decision to terminally differentiate have become so specialized that their potency has become significantly compromised in normal circumstances. Keratinocytes in the epidermis commit to bulking up on proteins like keratin. During the latter end of the cornification process, the cell also has no need to update other structures. The cell undergoes apoptosis, and the cell morphology progresses to dead structural material. Antibody producing plasma cells upregulate endoplasmic reticulum and Golgi volume for rapid extracellular secretory protein production. Though this continued state may be undesirable if continued indefinitely. Red blood cells mature to become denucleated and so are unable to repair themselves and require replacement. Once cells make these sorts of commitments, dramatic differences occur in cellular morphologies related to function and structural requirements. Despite exceptions, in the cited cells above, once they prioritize to this extent, it seems that their lifespan is measured within a matter of months (Andraud et al., 2012; Eckhart et al., 2013; Franco, 2012).

On the other hand, stem cells do not really require anything more than the apparatus for the correct response to cell cycle signals and the faithful copying of DNA. This is to form new daughter stem cells or forcing altered gene regulation upon daughter cells for them to specialize. Stem cells and somatic cell types can be characterized according to their specialized mixtures of morphologies, organelles, proteins, etc. A reasonable, logical correlation between microscopic observations mapped biochemical pathways and expected functions often make sense. However, those valuable observational differences are underpinned by the mechanisms that regulate the products of DNA. It is the differential use of DNA in cells that drives these differences.

The basis of DNA to mRNA to protein, while central is too basic. Cells use of transcription factors, DNA methylation, histone modifications and interfering or microRNA are used to control gene products and regulate this process. Gene promotors, often upstream of the gene, contain protein binding domains that influence the ability to make mRNA. This is through proteins binding or being removed to allow repression or promotion of the production of mRNA. These transcription factors decide how much and when mRNA transcripts can access ribosomal translation. Also, how much and when regulatory RNAs can interfere with this process is an essential factor. This cascades onwards to control more and more complex feedback loops to encourage or restrict the flow of DNA to proteins. Post-translational modifications, alternative spicing, peptide trafficking via signal peptides, and the expression of receptors with signal transduction then all go back to influence the whole process of DNA again to proteins.

The study of epigenetics and imprinting has also been able to explain further why the same genomic DNA can be controlled to give different cellular outcomes. The use of methylation of CpG rich sites in promotors can downregulate the expression of a gene. While CpG is common in DNA, it seems to be higher than expected in some promotor regions. 5′-CG-3′ on one strand of DNA forms a very short palindrome of the same on the other strand. This creates methylation on each strand's C bases, offset from each other. When DNA replication occurs, both strands contain a template with methylation on the C bases. This allows the ability to maintain a DNA methylation template to each daughter cell when the other strand is replicated. There are also enzymatic abilities to remove these methylations to change the outcome of gene expression in a cell. Stem cells and somatic cells have differential methylation patterns that are noted in these islands of CpG rich areas (Kim & Costello, 2017).

Histone protein modifications alter the readability of the chromatid by altering the density of the DNA windings upon the nucleosomes. This plays a role in the accessibility of RNA polymerase II to make mRNA and their proteins. Cellular proteins have to be produced in the correct time, place and amount to allow cell differentiation and replication. Via these proteins, the non-proteinaceous parts of the cell can also be altered. Enzymatically cells are controlled, and they are supported by turnovers and pools of lipids, sugars, tRNA, rRNA, amino acids, cholesterol, steroids, NAD(P)H, ATP and many other supporting factors. Proteins can be are cleaved, glycosylated, switched on/off by phosphorylation, require redox supplies, interact with each other, be exported to control other cells. With this unexhausted list of factors above, an impressive array of variables is prominent. So are, therefore, why so many types of cells are possible. In summary, one set of cellular instructions can be read very many ways to produce variable and adaptive cellular outcomes.

Stem Cells in Different Species (Animals and Humans)

Animal development has been studied to help scientists understand animal biology for many years. There are many reasons for this, including personal curiosity, investigating animal welfare and farming, studying simple systems to help understand complex ones, understanding evolutionary trajectories and even exploring the potential for changing human health. Biologists have characterized animals capable of partial or full regeneration body parts for more than 100 years. So, the notion of tissue re/generation and its potential for exploitation long predates identifying the possible mechanisms. Observations in the nineteenth century were also very important. Karl Ernst von Baer's chordate embryo collections displayed some similarity to each other at early stages, but he noted they diverged in development to look dissimilar from other animals' adults. Ernst Haeckel's recapitulation theory of embryos moving through ancestral primitive forms to achieve a higher state of embryo did not provide a useful answer to this (Richtsmeier, 2018). Either way, the level of understanding at the time would have not allowed for a modern cell biology model to be formed. That is, animal stem cells acting in both very early life all the way through to the repair towards their end of life. Long before the discovery of DNA, the notion of stem cells differentiating towards specific cell types via altered access to—and regulation of their proteomes and regulatory RNAs against a backdrop of epigenetic factors—was too complex to put together.

With the right mindset, asking questions about the complex phenotypical changes that occur in frogs, many insects and butterflies, etc. alludes to a certain level of unequal pattern formation fluidity in animals. Moving forward, at the core of driving animal stem cell curiosity, there is juxtaposition in this work between offering hope to human disease, protecting rare species from extinction and the way society views what is ethical. This work has the potential to extend human life and so challenge the resources of Earth but, at the same time, perhaps provide an answer by enabling distant space travel and terraforming in other worlds. One of the simplest animals, the rotifer, can produce embryos with stem cells capable of dormancy (García-Roger et al., 2019). Germination of ancient plant embryos encased in their seed cases is also possible (Sallon et al., 2020). These two examples seem to be in extreme contrast to higher animal capabilities, but human embryos have been viable after decades of ultralow temperature storage. This all indicates that, once understood, animal stem cell applications may be pushed into novel situations.

Multicellular eukaryotic organisms exhibit diversity in their structural complexities, genotypes, transcriptomes, proteomes, non-coding gene products, post-translational modifications and epigenetics. They also exhibit diversity in their cell biology, reproduction, gestation, life span, propulsion, food, social interactions, and the value of individual colony members, behaviours, predation, disease states and environmental parameters. All of these factors are likely to have played some role in the evolution and divergence of stem cell abilities and how they behave in animals. It is important to understand the key times when animals may need stem cell activities. Embryogenesis, homeostatic maintenance, growth/development, metamorphosis and repair. Clearly, these processes are not equal, some essential and other non-existent in certain animals. Some human cells and tissues are never subjected to remodelling or replacement, so have reduced need for stem cells. The animals that have greater levels of repair in their structures analogous to human tissues are obviously of great interest. This offers hope for learning what may stimulate analogous mechanisms in humans as some key proteins and transcription factors are highly conserved among animals and highlight routes for enhanced repair outcomes (McGurk et al., 2015).

Clearly essential to our understanding in mammalian and animal embryogenesis is to use a model of totipotent progenitor cells moving towards specialist cell types, or even terminal differentiation. Once we understand this, the next most important consideration will be in animal cells as we ask if this process may run in reverse. In situ study or harvesting stem cells for in vitro work in humans has challenges regarding ethics, legality, religion, variable clinical outcomes, cost and time taken for long term mechanisms and effects to be fully understood. Therefore, understanding stem cells both in vitro and in animals is not only beneficial for understanding those animals. It also offers potential for simple models to be created and inspire, what may be possible? Many human diseases can be understood through very simple models. Using yeast cells for modelling pathways in human neurodegenerative diseases seems unlikely but is accepted (Khurana et al., 2017). So far, simple single-celled organisms have not been so useful for studying in place of animal stem cells, but it is noted that there are sometimes similarities. The ability of animal stem cells to perform asymmetric cell division, producing either daughter stem cells or differentiated cells. This process can be seen in some single-cell organisms and organisms can influence each other demonstrating colony/multicellular-like behaviours like quorum sensing. So currently the most flexible model systems are in animals. This is useful to understand the animal itself but sometimes offers limited specific insight into the full overlapping systems in human stem cells. There are also too many variables to make use of single experiments. However, the sum of multiple data to drive computational models will hopefully make animals more redundant in this area (Zhou et al., 2019). This is a complex area, and it is important to remember the scope of animal's stem cells. Regarding what we can learn, a basic structure repair might in an animal be adequate for function and survival. However, in human medicine the expectation for full functional replacements is more likely. Presuming the regeneration of a tail after autotomy in a lizard has been evolutionarily optimized, it is still sometimes a compromised structure and not fully identical to the original.

Some animals demonstrate unusual forms of asexual reproduction like pathrithogenesis, haploid oocytes can undergo meiosis without subsequent cytokinesis to retain diploidy. This is of interest for the potential to make autologous high potency stem cells from gametes or even differentiated somatic cells. However, many animals form an embryo from the fertilization of oocytes. Either way, a totipotent diploid zygote is the expected outcome. The zygote in animals forms the morula ball of cells within the confines of the zygote. Changing the ratio of increasing cell numbers to decreasing cytoplasm volume perhaps serves to allow a feedback to control this process (Neurohr et al., 2019).

The next major difference in animals is the use of placenta in mammals and those that form eggs. Sometimes eggs are retained and in or near the body or left to the environment for different lengths of times. Based on this and other factors, the most obvious difference is regarding the formation of lesser or substantive yolks. These factors influence the way that the cleavage of the stem cells occurs. Cleavage methods have diversified in animals in the transition towards the blastula. In different animals, there are several directional ways in which cells perform cleavage both in relation to each other and the yolk (Houston, 2016). These differences complicate the use of some simple animals from fully modelling human stem cell embryogenesis.

Another important factor in understanding how early stage animal stem cells grow is their transition from the maternal influence of the oocyte. The oocyte contains one haploid copy of maternal DNA, the proteins, regulatory RNAs and many other factors. As the blastula develops, the diploid influence changes the landscape of proteins and these other elements. Later on, in mammals, the development of the placenta again allows further maternal influence of the stem cells of the developing embryo via molecules crossing the blood-placenta barrier. By studying these processes in diverse species like marsupials that birth early in development and the fullest range of animals, it could be possible to better characterize stem cell behaviour by compare and contrast. However, to fully understand animal stems and maximize their potential it is really important to understand the whole complex environment that they are exposed to. Despite the divergent pathways in animals, a polarity in embryogenesis generally occurs in the blastula stage. The designated area of higher activity the animal pole and the lower activity vegetal pole that develops into the extraembryonic membranous structures that provide the immediate and correct environment to protect and nourish the embryo.

In the same way as the cell membrane allows compartmentalisation, the multicellular nature of animals allows the development of layers of cells for more complex structures and functions. Many animals produce three main layers of stem cells during embryogenesis, though some simple aquatic animals only produce one or two. All triploblastic animals have three germ layers. From a human health point of view, understanding all three of these germ cell types is critical, each of them gives rise to highly functional and specialized tissue types and so can build organs. Due to this commonality, even in basic animals like worms, we can appreciate that stem cells give rise to an analogous digestive tract, nervous system, a gas exchange/vascular system, an outer surface and other specialized tissues.

The next major distinction in animals is the presence of a notochord in chordates which give rise to the vertebrate animals we best recognize. This comes from the ectoderm during gastrulation. Gastrulation in animals does have variations but follows similar patterns of organogenesis and pattern formation of the animal from the germ layers. After this the stem cells are pluripotent and gradually differentiate towards loosing potency as the organism's tissues become more specialized.

Post Development Growth/Repair/Maintenance

Putting aside the complex metamorphosis that occurs in some animals, later development in animals is to increase complexity, size and repair. In mammals, the process of post developmental stem cell activity in maintenance is well documented via hematopoietic stem cells in their generation of a large number of blood cells from an individual kind of stem cell. This model in place by the 1960s from the study of mice but has been well documented also in humans since (Till et al., 1964). This led to offering stem cell transplants in humans for immunodeficiencies, cancers and autoimmune diseases.

From the realization that adult animals have activate stem cells, it has driven research into the possibilities that cells can increase their potency levels and whether any stem cells can be used to produce gametes. Somatic cells have been induced to become pluripotent stem cells in mice and human cells. Interestingly, this has been through the control of transcription factors. This discovery allows complex control of the genes that can increase cell potency with minimal need to understand and control every single gene and factor involved (Mullen & Wrana, 2017). More promising is that these transcription factors can be encoded transiently by introduction into cells as RNA species. If this process is then continued to make viable germ cells and gametes in mice, this transient alteration could mean that the mouse need not retain retroviral genetic modifications in offspring. So being as close to naturally wild type as possible.

So far there has been limited success in producing in vitro germ cells from mammalian embryonic stem cells for both male and female forms except in mice. This offers potential for in vitro gamete and even zygote production from stem cells. As well as for humans this offers great potential for rare and difficult to breed animals (Goszczynski et al., 2019).

Certain animals have been well represented in the understanding of stem cells in embryogenesis and adult repair. For example, the large size of the xenopus oocyte and its early discovery to translate exogenous mRNA; the mouse that has been one of the most adaptable as mammalian species; the zebrafish that demonstrates regenerative repair of tissues in humans that are incapable of repair; drosophila for their genetic homologies with development and diseases of humans; salamanders which can regenerate highly functional limbs; and lizards which can regenerate full or partial tails after autotomy.

These organisms and systems all highlight great overlap in animal stem cell abilities and mechanisms. They also suggest that many cells can increase their potencies and then decrease them by specializing; there is also the possibility that they could even go on to form viable gametes and offspring. This offers an exciting future for not only human and animal survival but also quality of life.

Location and Function of Stem Cells in the Human Body

Stem cells have the potential to grow into any kind of the known 220 cells in the human body of an adult. Stem cells are obtained from bone marrow; however, they are also derived from the umbilical cord blood. To become erythrocytes, leukocytes or platelets, each stem cell receives chemical signals. These mechanisms are controlled by growth factors and epigenetic processes (chromatin remodelling and DNA methylation). The bone marrow space is where this growth activity takes place whereby only the mature cells are released into the bloodstream at a distant (Tuch, 2006). Tissues and organs have adult stem cells recognized in them; each tissue has a small number of stem cells present. In order to maintain a ready pool of supply for replication to take place at a controlled rate, stem cells are present in certain regions of every tissue (Kumar et al., 2010).

There Are Many Known Sites of Stem Cell Activities

Embryonic stem cells. They are particular kind of stem cells obtained from a human embryo that are about three to seven days old from the developmental phase. This blastocyst stage is known as the embryo. A blastocyst is a thin-walled globe of 150 cells, which can divide into more stem cells of an individual. Therefore, embryonic stem cells are considered to be capable to repair or revive unhealthy organs for regenerative therapeutics. Adult stem cells. The adult stem cells are found in small numbers, which makes them difficult to be found for study and harvesting. Unlike embryonic stem cells, adult stem cells have a restricted ability to provide growth to most cells. Scientists had believed that the use of AS cells could only just provide blood cells in the bone marrow. However, the scope for adult stem cells is greater and people with heart or neural diseases are being tested for the regenerative potential adult stem cells.

Induced pluripotent stem cells. It is possible to alter the genes in adult stem cells to give rise to induced pluripotent stem cells in the laboratory. This alteration is done by reprogramming the cells to perform like embryonic stem cells. Reprogramming techniques (a procedure used to revert fully grown cells into induced pluripotent stem cells) may allow an alternative of embryonic stem cells and put an end to immune system rejection of the newly grown stem cells in allogenic transplants. Yet, it has not been confirmed if using altered adult cells will cause any risky effects in humans. In several cases, connective tissue cells have been reprogrammed to become efficient heart cells. New heart cells have been injected to animals having a heart failure, leading to an improvement to the heart function and increased survival period. Perinatal stem cells. Researchers have discovered stem cells in both umbilical cord blood and placenta. To protect a growing foetus in the uterus a liquid called amniotic fluids fills the sac. Stem cells have been seen in samples of amniotic fluid from pregnant women. Amniocentesis is carried out to examine possibilities of developmental.

Role of Stem Cells in Human Development and Embryology

Section "Stem Cells in Different Species (Animals and Humans)" describes the common roles of stem cells in animals and humans leading up to the generation of three germ layers in triploblasts. After this, human stem cells are still highly active on the formation towards a foetus. However, over time with the final body pattern in place, stem cells can reduce in potency, becoming more specific in their roles. Loss of potency obviously is not the same as loss of activity. This is complex and human tissues must always balance the pulls from cell cycle repression, growing in size, growth in complexity, housekeeping, mechanical tissue repair, infection, DNA repair, endogenous and exogenous chemical signals, tumourigenesis, epigenetic regulation, oxidative damage, ageing and apoptosis. To remain functional certain tissues must also resist the urge for both stem cell and sometimes also immunological cellular activities to occur. This is because remodelling or repair may limit functionality of the tissue. In humans, it may be too simplistic but useful to consider stem cell activities to be related to either the structural or functional natures of the tissue types. The stem cells giving rise to dermal fibroblasts or neurones, for example not requiring the same level of activities in adults and embryos. It is this lifetime compromise in the needs of embryo development differing from ageing that can lead to disease in later life. Attempting to change the nature of this balance of compromise is what offers hope for stem cells correcting disease states and, perhaps, monitoring and preventing problematic embryogenesis. This starts with understanding embryogenesis.

After gastrulation forms a cavity globe of cells, the stem cells of the developing embryo's three germ layers are highly active and potent. They have lost the ability to form any type of cell, but each layer still retains potency to form the great number of cells and tissues specific to its layer. The outer layer of cells as the external facing ectoderm, mesenchymal cells as a central connecting layer form the mesoderm and the inner cells of the developing gut from the endoderm. This early pattern of cellular organization is transferred through to structures retained throughout life. The endodermal gut development process forms a pathway from the buccal cavity to the anus. As well as the obvious structures of the digestive system, the lungs terminate off this branch, glands and organs such as the liver that interact with the digestive system and endocrine also form from this layer.

From the mesoderm, a connectivity of motor-related and spatial matrices forms bones, joints, muscles and connective tissues and so their ultimate proteinaceous depositions such as collagen and others in the dermis. The vascular system including heart and the tubes of the urinary and reproductive systems also originate from the mesoderm. The mesoderm also gives rise to the dorsal midline. This gives the body a symmetrical and central axis point based on the notochord which then goes on to form the vertebrae of the spinal column. The mesodermal stem cells' areas of activity are organized into blocks. These blocks called somite's and arrange themselves either side of the notochord in discontinuous segments along the length of the notochord. The blocks allow corresponding levels of regional muscles to form in situ, but they also have migratory abilities.

The ectoderm goes on to form the skin epidermis but is also able to form the central nervous system. The mesodermal derived notochord is externally covered in ectoderm which is several cells thick and forms a neural plate. This then pulls itself up from a flat layer of cells to meet together, forming a central hollow neural tube above the notochord and below a new layer of sealing ectoderm. This neural tube gives space to form the spinal cord and brain. Between the neural tube and the ectoderm, a grouping of ectodermal cells forms a neural crest. The cells from this area are able to move through the mesoderm and populate it with cells of the peripheral nervous system (PNS). However, in the head they are also able to form bones, joints and connective tissues. A further complexity is the three sense organs of the head are not all derived from the same germ layers (Alberts, 2017).

The key points in the ongoing development of the germ layers are that some of these processes occur in three dimensions through extensions and elongations, respecting a directional governance. Also, that the origin of the stem cells in a region may have originally migrated from another germ layer. The applications attempting to use the stem cells in medicine are complex. For example, to repair significant damage to the spinal cord itself and also re-joining it to peripheral nerves is problematic as they did not have the same origins. Compared to embryogenesis, such an attempted repair may lack respect for the natural directional abilities, migratory origin of the cells and the cell signalling that initially gave rise to neurulation. Attempts to use scaffolding may mitigate some of these problems (Katoh et al., 2019).

As well as the mesodermal somites arranged around the rear of the embryo, the lateral plate mesoderm separates the rest of the endoderm from the ectoderm around the rest of the forming body. The final external body pattern of the embryo is not in place until the limb buds of the body develop from this lateral plate. During and after this, regional stem cells produce many layers of compartmentalisation and tissue types required for development towards near the final body pattern in miniature to form a foetus. The foetus is formed towards the end of trimester one.

The high level of the potency in very early embryo development is evident as it can sustain loss of cells to create twins or repair a cell lost. This level of potency is required as the cells that distinguish the germ layers transition to different and specialized tissues and organs have yet to form. Stem cells do not retain their full potency as part differentiated but as lower potency role specific stem cells. It could be that a reversible lower potency state could be controlled only by general gene and transcription factor regulation. However, this is perhaps risky and a degree of cellular memory and commitment to not normally revert is held in place as specialized stem cells. In the developing embryo, cells must maintain their differentiated states. The reversal of state or migration of higher potency stem cells without the context of their original spatial and chemical signals could lead to a non-viable embryo or structures in compromised positions.

In post embryo humans, an increase in potency or retention of stem cells may be desirable against disease, ageing and damage repair. However, the consequences of retaining cells potentially able to form a full repertoire of tissues out of context, capable of rapid cell division and migration might be a step too close to tumourigenesis (Wuputra et al., 2020).

Accepting this model and the risks of embryo interference, recent stem cell harvesting from the placenta to retain hematopoietic progenitor and hematopoietic stem cell populations for autologous use against certain blood cancers and autoimmune diseases seems to provide potential for a low risk for high rewards. Placental tissue also contains mesenchymal stem cells that can form a very diverse range of cells in the laboratory when given specific cell signalling reagents (Wang & Zhao, 2010).

Apart from a constant general growth in size and brain complexity throughout childhood and towards adulthood, puberty leads to rapid changes in cellular activities. However, compared to embryogenesis the changes are slight and slow. Throughout the process though, there appears to be a theme that cells have a built-in memory for retaining correct potency levels and some kind of clock activity for timing change. On the route from foetus to adult, it is clear that there are reducing needs for stem cell potencies but still great need for stem cell activities. In adults, it is noted that most cells have finite lifespans requiring turnover. Some cells like the heart muscle are also seemingly mechanically and functionally expected to need turnover yet they are not capable of doing so. It is also expected to think of skin, muscles, and connective and synovial tissues of having high need for stem cell turnover for repair just from normal mechanical stress. Yet these cells often form scar tissue and joints demonstrate little ability to renew from arthritis.

The best documented example of ongoing stem cell activities in adults is in the generation of blood cells. The necessity being the expiration of the nonnucleated red cells incapable of dividing and repair. The white immunological cells also require strict control of their levels of activities. The liver is also well-known for its regenerative capacity, it can receive a lot of oxidative damage from ethanol metabolism via cytochrome P450 CYP2E1, a great deal of functional stress (Abdelmegeed et al., 2017). In adults, the regenerative capacities of tissues would appear to be sometimes illogical. It may be expected that stem cell regeneration of the hip joints to prevent osteoarthritis and allow food gathering would be of evolutionary advantage over the detoxification of exogenous compounds in the food. However, humans have had the ability of social cooperation to share the social needs of food gathering. Humans have also been able to elongate the gap between reproductive age and death, reducing the drive for adaptations to counteract ageing. This contrasts against the level of systemic sickness that can occur from liver damage and can cause premature death.

If the biological evolutionary aim for human life is to reach an age of securing the independence of the next generation, then we may consider the natural evolutionary use of stem cells beyond maintenance and repair evolutionary redundant within a generation of reproductive age.

Multipotent Stem Cells Versus Pluripotent Stem Cells

Multipotent stem cells only segregate into certain cell kinds once they perceive signals, such stem cells known as hematopoietic, neural and mesenchymal. The hematopoietic stem cells transform into various blood cells; neural stem cells into astrocytes, oligodendrocytes and neurons; mesenchymal stem cells can give rise to adipocytes, osteoblasts and chondrocytes. Hematopoietic stem cells become an oligopotent cell after differentiation and later differentiate into many cell types. An example of this is a myeloid stem cell that can split into white blood cells excluding red blood cells. Found in small numbers, multipotent stem cells are available in specialized tissues mostly in adult tissues and are able to restore cells that are damaged.

Pluripotent stem cells are categorized into perinatal, induced pluripotent and embryonic stem cells. The embryonic stem cells are harvested inside the range of the first four cell division of the embryo, they are called totipotent. Totipotent cells can differentiate into extraembryonic tissues in a constant direction with germ layer obtained tissues. Another specialty of embryonic stem cell is that they are able to split up in vivo, this is different from the adult stem cells that split separate from each other during death of the cell or tissue damage. The pluripotent stem cells play a vital role in forming endoderm, ectoderm and mesoderm (germ layers) except for the structures located outside the embryo.

Unlike pluripotent stem cells, multipotent stem cells can adapt to separate into cells of very many given cell descents. Hematopoietic stem cell is an example of this adaption, this kind of stem cell can give rise to many kinds of blood cells by hematopoiesis. Multipotent stem cells can further become an oligopotent cell after differentiation. The ability to differentiate will be restricted to cells of its derivation. Yet, these cells are able to convert into discrete cell kinds, which leads to identifying them as pluripotent cells (Fig. 1.3.) (Zakrzewski et al., 2019).

All cell types that can form an adult are *pluripotent*. From zygote to very early embryo and formation of the extraembryonic tissues of the foetus are *totipoten*t. In adults, stem cells can be limited to being multipotent. If a tissue has cells off a single differentiated lineage and stem cells are used to maintain this lineage, they are called *unipotent.* Adult stem cells can be reprogrammed to pluripotency by somatic nuclear transfer. Practical applications of inducing/reversing pluripotency in adult cells have become evident, with the formation of iPSCs. The transcription factors (TF) Oct4, Sox2, KLF4 and Myc are greatly expressed in ES cells. So, upregulation of such TF and so their target genes are thought to lead cells like fibroblasts (cells that produce collagen and fibres) to become pluripotent (Watt & Driskell, 2010).

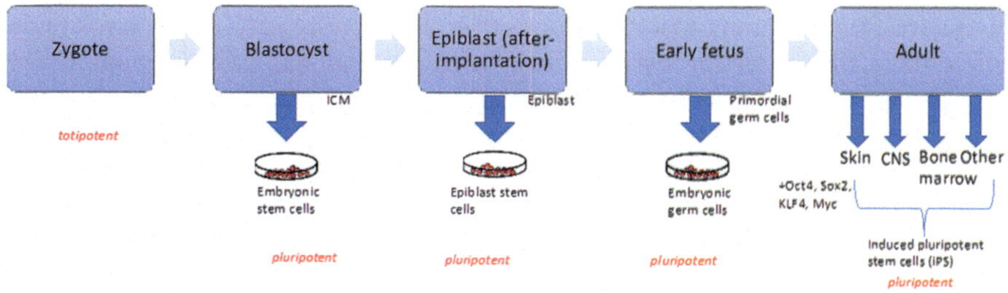

Fig. 1.3 Origin of stem cells

Both pluripotent and multipotent stem cells have pros and cons in their potential for exploration. Firstly, the ability of pluripotent cells to develop to any cell kind could be a curative benefit such as treating unhealthy or ageing tissues in which multipotent stem cells are inadequate. Secondly, proliferation of pluripotent stem cells is rapid, leading to a level of cellular production that may not be useful (Drukker et al., 2002). The benefit of this work is that the material is regarded as self. The immune system responses to particular epitopes presented on the surface of cells via the major histocompatibility complexes for immune cells to sample for signals as to if a cell is infected or has acquired non-self peptides and haptens. Self-confirming epitopes can also be displayed. Cells derived from the same (autologous) multipotent stem cells comprise of the patient's own particular proteins on the surface that permit it to considered self by the immune system of the host. Pluripotent stem cells are partly immune privileged and do not normally display the same signals for the immune system to form a rejection response. So, their use to treat other people as allogenic transplants may be valuable (Biehl & Russell, 2009). However, this is not well understood and seemingly even autologous iPSCs have been noted to be targeted by the immune system.

Nobel Prize Research on Stem Cells

In 2012, Sir John Gurdon was presented with a Nobel Prize when he was able to modify a frog egg cell so that it could grow into a typical tadpole. This was by substituting the nucleus of a Xenopus frog zygote cell with a nucleus from a tadpole enterocyte. Moreover, his work led to this sort of concept of animal cloning and modified the field of cell specialization and growth. Larry Goldstein also mentioned how Sir John Gurdon's work will be vital to teach how reprogrammed adult cells can behave like embryonic stem cells, including how to utilize the full repertoire of DNA instructions to form many kinds of adult cells. During the 1960s, it was unsure of what the future of the curative applications of this information might be.

Later in 2006, a researcher named Shinya Yamanaka helped respond what the future held for this field. He demonstrated how to reprogram a whole and transformed mouse cell into becoming a pluripotent stem cell by adding just four features to the cell culture. Yamanaka successfully introduced these as induced pluripotent stem cells. He did not waste any time in demonstrating a similar process can be useful to transform human cells. This work was important in research as well as to demonstrate how to prompt adult cells to recall how to act like embryonic stem cells. These researchers and others have generated the daybreak of dawning of a new era of how cells types can be coordinated and defined. Furthermore, how we can put this skill to into clinical options to improve new treatments for several deadly aliments (Holmes, 2012).

Mario Capecchi, Martin Evans and Oliver Smithies were awarded the 2007 Nobel prize in physiology or medicine for their introductory work in mouse embryonic stem cell technology, also named as "knockout mouse". It was believed that their achievements would show the vital genomic causes and effects in human disorders. It has now become an important tool in the laboratory research into the role of genes. Both Capecchi and Smithies revealed how adjustments can be carried out by means of chromosomal crossover. This procedure allows the exchange of genetic material by homologous recombination. This mimics the natural genomic DNA exchange and encourages the appearance of new traits. They thought this chromosomal crossover could be utilized to knock out hereditary information from a genome. This allows the ability to study chromosomal loci in knockout mice which can target specific genes for an animal's development. Martin Evans also worked on the alteration of genetic material in animals by modifying specific DNA sequences. He grew stem cells from early mouse embryos in culture. Then he injected modified stem cells from a mouse strain inside the embryo of a different mouse strain and implanted the embryo into a surrogate mother. This procedure showed that the mice that developed were comprised of both mouse strains as a chimeric embryo. Additionally, he managed to combine the genes into the stem cells by using a retrovirus to modify the mice DNA to form a stable cell line. As this was done in embryonic stem cells, this could then give the ability to produce many further mice by changing the germline to ensure every cell in their progenies conveyed the foreign DNA. As well as studying embryogenesis this created stains of disease model mice (Watts, 2007).

References

Abdelmegeed, M., Ha, S., Choi, Y., Akbar, M., & Song, B. (2017). Role of CYP2E1 in mitochondrial dysfunction and hepatic injury by alcohol and non-alcoholic substances. *Current Molecular Pharmacology, 10*(3), 207–225.

Alberts, B. (2017). *Molecular biology of the cell* (6th ed).

Alvarez, C., Garcia-Lavandeira, M., Garcia-Rendueles, M., Diaz-Rodriguez, E., Garcia-Rendueles, A., Perez-Romero, S., Vila, T., Rodrigues, J., Lear, P., & Bravo, S. (2012). Defining stem cell types: Understanding the therapeutic potential of ESCs, ASCs, and iPS cells. *Journal of Molecular Endocrinology, 49*(2), R89–R111.

Andraud, M., Lejeune, O., Musoro, J., Ogunjimi, B., Beutels, P., & Hens, N. (2012). Living on Three Time Scales: The Dynamics of Plasma Cell and Antibody Populations Illustrated for Hepatitis A

Virus. *PLoS Computational Biology, 8*(3), e1002418. Available at: https://journals.plos.org/ploscompbiol/article?id=10.1371/journal.pcbi.1002418. Accessed on November 10, 2020.

Biehl, J., & Russell, B. (2009). Introduction to stem cell therapy. *The Journal of Cardiovascular Nursing, 24*(2), 98–103.

Cantz, T., & Martin, U. (2010). Induced pluripotent stem cells: Characteristics and perspectives. *Bioreactor Systems for Tissue Engineering, II*, 107–126.

Carpenedo, R., & McDevitt, T. (2013). Stem cells. *Biomaterials Science*, 487–495.

Chagastelles, P., & Nardi, N. (2011). Biology of stem cells: An overview. *Kidney International Supplements, 1*(3), 63–67.

Chen, K., Huang, Y., & Chen, J. (2013). Understanding and targeting cancer stem cells: Therapeutic implications and challenges. *Acta Pharmacologica Sinica, 34*(6), 732–740.

Dalerba, P., Cho, R., & Clarke, M. (2007). Cancer Stem cells: Models and concepts. *Annual Review of Medicine, 58*(1), 267–284.

Drukker, M., Katz, G., Urbach, A., Schuldiner, M., Markel, G., Itskovitz-Eldor, J., Reubinoff, B., Mandelboim, O., & Benvenisty, N. (2002). Characterization of the expression of MHC proteins in human embryonic stem cells. *Proceedings of the National Academy of Sciences, 99*(15), 9864–9869.

Dutta, D. (2020). *Differences between stem cells and somatic cells.* [online] News-Medical.net. Available at: https://www.azolifesciences.com/article/Differences-Between-Stem-Cells-and-Somatic-Cells.aspx#:~:text=The%20major%20difference%20between%20embryonic,thus%2C%20into%20any%20cell%20type. Accessed on September 22, 2020.

Eckhart, L., Lippens, S., Tschachler, E., & Declercq, W. (2013). Cell death by cornification. *Biochimica et Biophysica Acta (BBA)—Molecular Cell Research, 1833*(12), 3471–3480.

Eguizabal, C., Aran, B., Chuva de Sousa Lopes, S., Geens, M., Heindryckx, B., Panula, S., Popovic, M., Vassena, R., & Veiga, A. (2019). Two decades of embryonic stem cells: a historical overview. *Human Reproduction Open*, [online] 2019(1). Available at: https://www.ncbi.nlm.nih.gov/pmc/articles/PMC6396646/. Accessed on September 29, 2020.

Evans, M., & Kaufman, M. (1981). Establishment in culture of pluripotential cells from mouse embryos. *Nature, 292*(5819), 154–156.

Franco, R. (2012). Measurement of red cell lifespan and aging. *Transfusion Medicine and Hemotherapy, 39*(5), 302–307.

García-Roger, E., Lubzens, E., Fontaneto, D., & Serra, M., (2019). Facing adversity: Dormant embryos in rotifers. *The Biological Bulletin, 237*(2), 119–144. [online]. Available at: https://www.journals.uchicago.edu/doi/full/10.1086/705701?mobileUi=0&

Gonzalez, M., & Bernad, A. (2012). Characteristics of adult stem cells. *Advances in Experimental Medicine and Biology,* 103–120.

Goszczynski, D., Denicol, A., & Ross, P. (2019). Gametes from stem cells: Status and applications in animal reproduction. *Reproduction in Domestic Animals.*

Gucciardo, L., Lories, R., Ochsenbein-Kölble, N., Done', E., Zwijsen, A., & Deprest, J. (2008). Fetal mesenchymal stem cells: isolation, properties and potential use in perinatology and regenerative medicine. *BJOG: An International Journal of Obstetrics & Gynaecology, 116*(2), 166–172.

Holmes, D. (2012). Stem cell scientists share 2012 Nobel Prize for medicine. *The Lancet, 380*(9850), 1295.

Houston, D. (2016). Vertebrate axial patterning: From egg to asymmetry. *Advances in Experimental Medicine and Biology,* 209–306.

Hui, H., Tang, Y., Hu, M., & Zhao, X. (2011). Stem cells: General features and characteristics. *Stem Cells in Clinic and Research.*

Ji, P., Manupipatpong, S., Xie, N., & Li, Y. (2016). Induced pluripotent stem cells: Generation strategy and epigenetic mystery behind reprogramming. *Stem Cells International, 2016*, 1–11.

Katoh, H., Yokota, K., & Fehlings, M. (2019). Regeneration of spinal cord connectivity through stem cell transplantation and biomaterial scaffolds. *Frontiers in Cellular Neuroscience, 13.*

Khurana, V., Peng, J., Chung, C., Auluck, P., Fanning, S., Tardiff, D., Bartels, T., Koeva, M., Eichhorn, S., Benyamini, H., Lou, Y., Nutter-Upham, A., Baru, V., Freyzon, Y., Tuncbag, N.,

Costanzo, M., San Luis, B., Schöndorf, D., Barrasa, M., ... Lindquist, S. (2017). Genome-scale networks link neurodegenerative disease genes to α-synuclein through specific molecular pathways. *Cell Systems, 4*(2), 157-170.e14.

Kim, M., & Costello, J. (2017). DNA methylation: An epigenetic mark of cellular memory. *Experimental & Molecular Medicine, 49*(4), e322–e322.

Kolios, G., & Moodley, Y. (2013). Introduction to stem cells and regenerative medicine. *Respiration, 85*(1), 3–10.

Kumar, R., Sharma, A., Pattnaik, A., & Varadwaj, P. (2010). Stem cells: An overview with respect to cardiovascular and renal disease. *Journal of Natural Science, Biology and Medicine, 1*(1), 43.

Łos, M., Skubis, A., & Ghavami, S. (2019). Stem cells. *Stem Cells and Biomaterials for Regenerative Medicine*, 5–16.

Martin, G. (1981). Isolation of a pluripotent cell line from early mouse embryos cultured in medium conditioned by teratocarcinoma stem cells. *Proceedings of the National Academy of Sciences, 78*(12), 7634–7638.

Mayo Clinic. n.d. *Stem Cells*. [online] Available at: https://www.mayoclinic.org/tests-procedures/bone-marrow-transplant/in-depth/stem-cells/art-20048117#:~:text=These%20stem%20cells%20come%20from,of%20cell%20in%20the%20body. Accessed on November 12, 2020.

McGurk, L., Berson, A., & Bonini, N. (2015). Drosophilaas an in vivo model for human neurodegenerative disease. *Genetics, 201*(2), 377–402. [online]. Available at: https://www.ncbi.nlm.nih.gov/pmc/articles/PMC4596656/#bib230. Accessed on October 28, 2020.

Morrison, S., & Kimble, J. (2006). Asymmetric and symmetric stem-cell divisions in development and cancer. *Nature, 441*(7097), 1068–1074.

Mullen, A., & Wrana, J. (2017). TGF-β Family Signaling in Embryonic and Somatic Stem-Cell Renewal and Differentiation. *Cold Spring Harbor Perspectives in Biology, 9*(7), p.a022186.

Neurohr, G., Terry, R., Lengefeld, J., Bonney, M., Brittingham, G., Moretto, F., Miettinen, T., Vaites, L., Soares, L., Paulo, J., Harper, J., Buratowski, S., Manalis, S., van Werven, F., Holt, L., & Amon, A. (2019). Excessive cell growth causes cytoplasm dilution and contributes to senescence. *Cell, 176*(5), 1083-1097.e18.

Pavlović, M. and Radotić, K. (2017). Essential characteristics of stem cells: Self-renewal, and plasticity. *Animal and Plant Stem Cells*, 17–21.

Ramesh, T., Lee, S., Lee, C., Kwon, Y., & Cho, H. (2009). Somatic cell dedifferentiation/reprogramming for regenerative medicine. *International Journal of Stem Cells, 2*(1), 18–27.

Reibetanz, U., Hübner, D., Jung, M., Liebert, U., & Claus, C. (2016). Influence of growth characteristics of induced pluripotent stem cells on their uptake efficiency for layer-by-layer microcarriers. *ACS Nano, 10*(7), 6563–6573.

Richtsmeier, J. (2018). A century of development. *American Journal of Physical Anthropology, 165*(4), 726–740.

Sallon, S., Cherif, E., Chabrillange, N., Solowey, E., Gros-Balthazard, M., Ivorra, S., Terral, J., Egli, M., & Aberlenc, F. (2020). Origins and insights into the historic Judean date palm based on genetic analysis of germinated ancient seeds and morphometric studies. *Science Advances, 6*(6), eaax0384.

Shihadeh, H. (2015). History and Recent Advances of Stem Cell Biology and the Implications for Human Health. [online]. Paper 421. Available at: https://digitalcommons.uri.edu/cgi/viewcontent.cgi?article=1432&context=srhonorsprog. Accessed on September 29, 2020.

Snoeckx, R., Bogaert, K., & Verfaillie, C. (2009). Stem cells. *Genomic and Personalized Medicine*, 599–609.

Stadtfeld, M., & Hochedlinger, K. (2010). Induced pluripotency: History, mechanisms, and applications. *Genes & Development, 24*(20), 2239–2263.

Thomson, J., & Odorico, J. (2000). Human embryonic stem cell and embryonic germ cell lines. *Trends in Biotechnology, 18*(2), 53–57.

Till, J., McCulloch, E., & Siminovitch, L. (1964). A stochastic model of stem cell proliferation based on the growth of spleen colony-forming cells. *Proceedings of the National Academy of Sciences, 51*(1), 29–36.

Tuch, B. (2006). Stem cells--a clinical update. *Australian family physician*, [online] 35, pp.719–721. Available at: https://pubmed.ncbi.nlm.nih.gov/16969445/. Accessed on September 20, 2020.

Wakui, T. (2017). Method for evaluation of human induced pluripotent stem cell quality using image analysis based on the biological morphology of cells. *Journal of Medical Imaging, 4*(04), 1.

Wang, Y., & Zhao, S. (2010). *Vascular biology of the placenta.* Morgan & Claypool Life Sciences.

Watt, F., & Driskell, R. (2010). The therapeutic potential of stem cells. *Philosophical Transactions of the Royal Society B: Biological Sciences, 365*(1537), 155–163.

Watts, G. (2007). Nobel prize is awarded for work leading to "knockout mouse". *BMJ*, 335(7623), 740.1–740.

Wuputra, K., Ku, C., Wu, D., Lin, Y., Saito, S., & Yokoyama, K. (2020). Prevention of tumor risk associated with the reprogramming of human pluripotent stem cells. *Journal of Experimental & Clinical Cancer Research*, 39(1) [online]. Available at: <https://www.ncbi.nlm.nih.gov/pmc/articles/PMC7268627/. Accessed on October 28, 2020.

Ying, Q., Nichols, J., Chambers, I., & Smith, A. (2003). BMP induction of id proteins suppresses differentiation and sustains embryonic stem cell self-renewal in collaboration with STAT3. *Cell, 115*(3), 281–292.

Zakrzewski, W., Dobrzyński, M., Szymonowicz, M., & Rybak, Z. (2019). Stem cells: past, present, and future. *Stem Cell Research & Therapy, 10*(1).

Zhou, J., Schor, I., Yao, V., Theesfeld, C., Marco-Ferreres, R., Tadych, A., Furlong, E., & Troyanskaya, O. (2019). Accurate genome-wide predictions of spatio-temporal gene expression during embryonic development. *PLOS Genetics, 15*(9), e1008382.

Chapter 2
Types and Classification of Stem Cells

Aayush A. Shah and Firdos Alam Khan

Abstract The stem cells can be classified based on the location in which they reside; for example, the stem cells which are present in an adult person are called as adult stem cells, stem cells which are located in an embryo are called as embryonic stem cells, and also, the stem cells which are present in the umbilical cord are called as cord blood stem cells, respectively. Besides, stem cells are also classified based on their differentiation characteristics; for example, stem cells are differentiated into five distinct types like totipotent, pluripotent, multipotent, oligopotent and unipotent. A stem cell line is defined as a clonal, self-renewing cell population that is multipotent and thus can generate several differentiated cell types and is primarily distinguished according to their potency, origin and lineage progression. In this chapter, we have discussed different types of stem cells which are present in human body and also discussed about induced pluripotent stem cells and their significance.

Keywords Stem cells · Classification of stem cells · Adult stem cells · Embryonic stem cells · Induced pluripotent stem cells

Introduction

The classification and identification of stem cells are some of the crucial methodological questions in stem cell biology. A stem cell line is defined as a clonal, self-renewing cell population, that is multipotent and thus can generate several differentiated cell types (Melton, 2014) and is primarily distinguished according to their potency, origin and lineage progression.

A. A. Shah (✉)
Department of Biochemistry and Molecular & Cellular Biology, Georgetown University, Washington, DC, USA

F. A. Khan
Department of Stem Cell Biology, Institute for Research and Medical Consultations, Imam Abdulrahman Bin Faisal University, Post Box No. 1982, Dammam 31441, Saudi Arabia
e-mail: fakhan@iau.edu.sa

F. A. Khan (ed.), *Advances in Application of Stem Cells: From Bench to Clinics*, Stem Cell Biology and Regenerative Medicine 69,
https://doi.org/10.1007/978-3-030-78101-9_2

Classification of Stem Cells Based on Their Potency

Corresponding to their potency, stem cells are differentiated into five distinct types, viz. totipotent, pluripotent, multipotent, oligopotent and unipotent (Fig. 2.1).

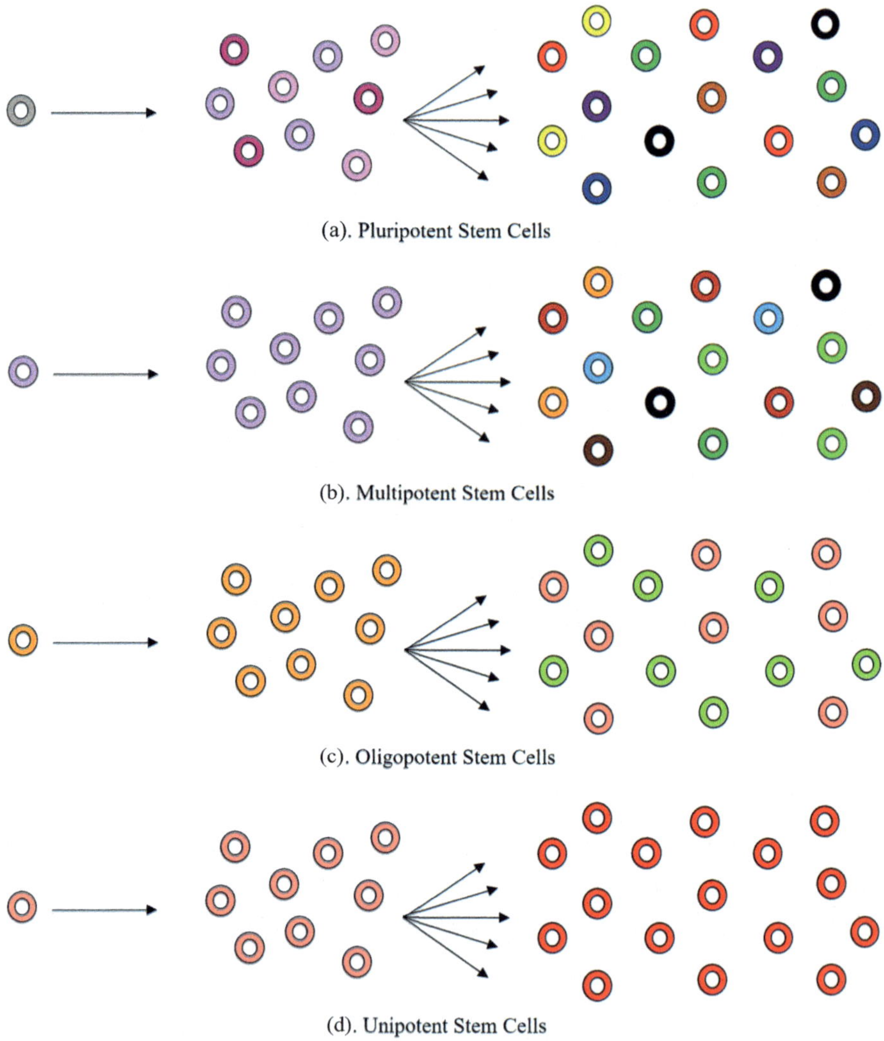

Fig. 2.1 Various stem cell potencies. a Pluripotent stem cells can distinguish into any of the three embryonic germ layers, which eventually give rise to any type of fetal or adult stem cells. **b** Multipotent stem cells are derived from a single germ layer and are found in various tissues of the body. **c** Oligopotent stem cells have limited differential potential and can differentiate into two or more lineages. **d** Unipotent stem cells have low differentiation potential and can differentiate into mature cells of the same lineage

Totipotent

These cells have the ability to divide and generate various differentiated cells in an organism, as well as the extra-embryonic tissues. For example, zygote, the single-celled embryo formed by the fertilization of the oocyte by the sperm, presents the ability to mature into an entire embryo including all the cells required for the development of a human as well as to the support structure of the placenta which is crucial for the fetal development (Mitalipov & Wolf, 2009).

Pluripotent

These cells have the capability to differentiate into one of the embryonic germ layers: ectoderm, mesoderm and endoderm—which can then give rise to any type of fetal or adult stem cells for tissue development (de Miguel et al., 2010). However, a single pluripotent cell or a group of cells lack the ability to develop into an entire fetal or adult animal as they cannot organize into an embryo. But some of the inner cell mass (ICM) cells in the embryo are pluripotent and can essentially form any type of somatic or germ cell type in the body. The pluripotency of ICM cells is retained by the expression of endogenous factors. These cells which occur transiently in vivo can be isolated in an undifferentiated state, then altered and proliferated *in vitro* as embryonic stem cells (ESCs) (Evans & Kaufman, 1981; Martin, 1981). In 2006, Shinya Yamanaka and Kazutoshi Takahashi (Takahashi, 2007; Takahashi & Yamanaka, 2006) demonstrated the creation of a new type of pluripotent stem cells termed as induced pluripotent stem cells (iPSCs) by reprogramming somatic cells. These cells were seen to have comparable attributes to ESCs.

Multipotent

The cells which have been differentiated from a single germ later are multipotent in nature and can be found in most of the tissues in the body (Ratajczak et al., 2012). The most widely known multipotent cell type is the mesenchymal stem cells (MSCs), which originate from numerous tissues such as bone marrow, adipose tissue, bone, Wharton's Jelly, umbilical cord blood and peripheral blood (Augello et al., 2010; Wuchter et al., 2011). Various researchers have established that these types of cells can generate a variety of cell forms and produce different mesoderm-derived tissues such as bone, cartilage, muscles, adipose tissues and their related tissues (Augello et al., 2010; Bibber, 2013; Bruder et al., 1997; Campagnoli, 2001; Kolios & Moodley, 2012; Prockop, 1997; Sobhani et al., 2017). One of the more important forms of MSCs which have got the researchers interested is the neuronal tissue. Recently, researchers have successfully differentiated MSCs into neural cells that are ectoderm derived, via transdifferentiation (Barzilay et al., 2009). These neural cells further differentiate into nerve cells which may suggest for possible treatments of brain and spinal cord injuries (Butti et al., 2014; Kennea & Mehmet, 2002).

Oligopotent

These stem cells have the capability to self-regenerate and differentiate in two or more lineages within the particular tissue. For example, hematopoietic stem cells are considered to be oligopotent stem cells as they differentiate into both the lymphoid and the myeloid lineages (Bender Kim, 2005).

Unipotent

These cells have the unique characteristics of having the narrowest differentiation potential and high self-renewal. For instance, satellite cells develop into mature muscle cells, aiding in muscle regeneration (Beck & Blanpain, 2012; Bentzinger, 2013; Overturf et al., 1997). The latter attribute renders them as a promising tool for therapeutic usage in regenerative medicine.

Classification of Stem Cells Based on Their Origin

The easiest way to classify stem cells based on their origin is by dividing them into four major kinds, viz. embryonic stem cells, fetal stem cells, umbilical cord stem cells and the adult stem cells. The classification of stem cells by origin is illustrated in Fig. 2.2.

Human Embryonic Stem Cells (hESCs)

The fusion of the sperm and the ovum during the mammalian fertilization process produces the zygote that ultimately gives rise to a whole embryo. The zygote undergoes several cleavage divisions, producing equipotent blastomeres. These divisions further result in the formation of the blastocyst, a microscopic hollow sphere of cells. These blastocysts are comprised of two distinctive cell types: the trophectoderm (TE) and the inner cell mass (ICM) which forms the hypoblast and the epiblast, further stimulating the fetal development. A 5–6 day old human blastocyst's ICM is the primary source of the pluripotent hESCs (Evans & Kaufman, 1981).

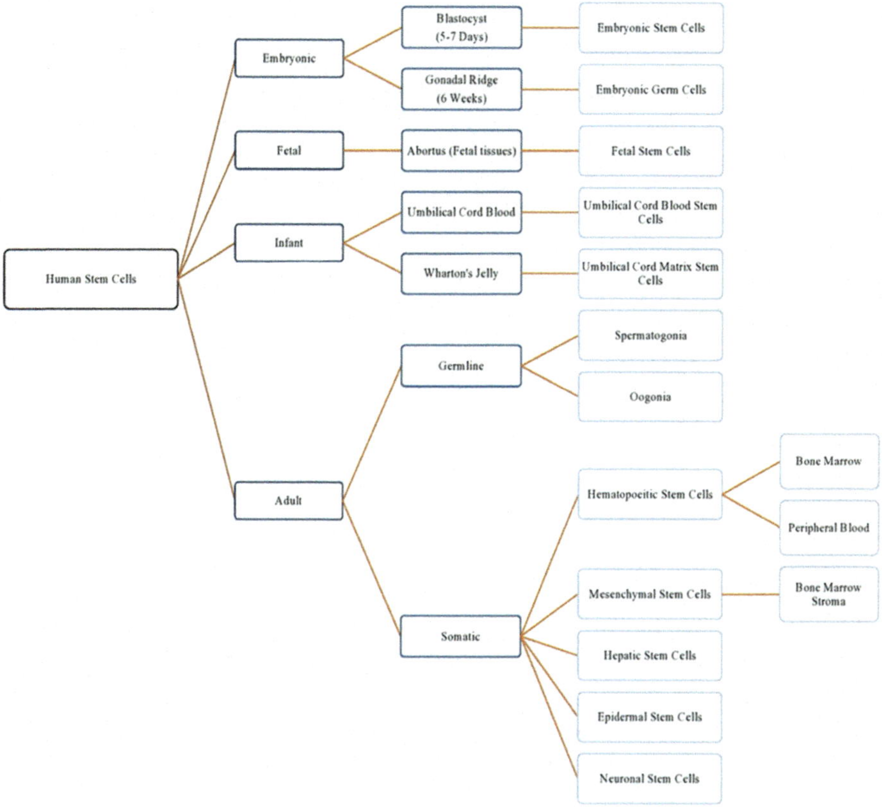

Fig. 2.2 Classification of stem cells by their origin

The embryonic stem cells are self-replicating and pluripotent in nature, and through the process of embryogenesis, they can differentiate into one of the three germ layers—endoderm, mesoderm and ectoderm (Fonseca, 2014; Yao, 2006). Once the hESCs differentiate into a particular germ layer, they turn into multipotent stem cells. ESC lines can be established upon transferring the hESCs in an undifferentiated state from the ICM into a culture dish under certain conditions (Bongso, 2006). The presence of these cells can be detected by the presence of NANOG and Oct4 transcription factors which confirm the "stemness" and self-renewal potential of these cells (Hambiliki, 2012; Wang, 2012). The perseverance and maintenance of the ESC cell lines are regulated by the certain culture conditions such as the usage of anti-differentiation cytokine leukemia inhibitory factor (LIF) or mouse embryonic fibroblast cell feeder layer (Doetschman et al., 1985; Hamazaki, 2001; Heydarkhan-Hagvall, 2012; Shiroi, 2005; Thoma, 2012).

Human Fetal Stem Cells (hFSCs)

The fetal stem cells are a kind of primal embryonic stem cells which can be obtained from two separate sources—the fetal tissues like blood, bone marrow (BM), pancreas, kidney, liver, spleen and the extra-embryonic tissues for instance the cord blood, placenta and the amniotic fluid (Cananzi et al., 2009; Marcus & Woodbury, 2008; Saha & Jaenisch, 2009; Spinelli et al., 2013). hFSCs such as the stem cells of the neural crest, hematopoietic stem cells and fetal progenitor islet cells are extracted from abortuses (Beattie et al., 1997) or the fetal blood during the first trimester (Abdulrazzak et al., 2010; Spinelli et al., 2013). The phenotypic features and properties of the hFSCs as well as their cell expression markers are heterogeneous and differ according to their tissue of origin and gestational age.

Adult Stem Cells

These undifferentiated multipotent stem cells are found scattered in different organs and tissues of the body, post embryonic development and have the capability to develop no less than one type of the differentiated progeny (Foster et al., 2002; Fu et al., 2011; Li & Xie, 2005). Their principal function is to maintain the organs or the tissues of an organism by replenishing the apoptotic cells and regenerate injured tissues. Although they have a narrow differential potential, i*n vitro*, these cells have shown differentiation into tissues from all the three different germ layers (Ilancheran et al., 2009; Moodley, 2010). There are numerous advantages of adult stem cells which make them an attractive alternative to embryonic stem cells for research studies and cell therapies. Being autologous cells, these stem cells do not involve any ethical or political concerns, or rejections by the host immune system (Körbling & Estrov, 2003; McCormick & Huso, 2010). Recently, there has been a growing interest in these stem cells for their use in regenerative medicine, with several research groups exhibiting the potential of adult stem cells to restore damaged tissues when transplanted in vivo, as seen in the re-engineering of the bone tissue by osteogenesis or application of bone marrow cells to improve cardiac function post-myocardial infarction (Chimutengwende-Gordon & Khan, 2012; Erbs, 2005; Orlic, 2001; Perin, 2003; Strauer, 2002).

Hematopoietic Stem Cells (HSCs)

The preservation of the adult human hemopoiesis necessitates the body to replace around 10^{10} leucocytes and 2×10^{11} erythrocytes per day. These differentiated cells are incompetent of growth and regeneration and thus are replenished by more immature cells called hematopoietic stem cells (HSCs) inhabiting the bone marrow, which proliferate and develop to form various functional cell types of the blood (Goldman,

1982). The growth and proliferation of these cells, under certain circumstances such as injuries and diseases, are arbitrated by inducer colony-stimulating factor (CSF) and interleukins counteracted by inhibitory cytokines (Zipori, 1992).

During birth, the HSCs are located in the bone marrow (BM) which is the primary site for hemopoiesis in humans, while in the occurrence of disease or stress, hemopoiesis will also occur in the liver, spleen or other sites during the span of life. The BM microenvironment known as the "niche" helps in providing the essential factors required for the self-renewal and differentiation of the HSCs (Szade et al., 2018). Several studies have shown the niche influencing quiescence and stem cell cycle entry, regulating the stem cells' fate and limiting the rate of mutation in these cells (Scadden, 2014; Schofield, 1978a, 1978b).

The pluripotent HSCs have the capability to develop various functional cell types (Fig. 2.3) like erythrocytes, monocytes, eosinophils, neutrophils, basophils, platelets, lymphocytes and dendritic cells. In the hematopoietic model of development, the HSCs are found to be of two types, the long-term repopulating cells (LTRC) and the short-term repopulating cells (STRC). LTRCs have extensive self-renewal potential and are competent to develop all types of blood cells for the complete duration of the individual's life span (Harrison & Zhong, 1992; Morrison & Weissman, 1994). STRCs can differentiate into the multipotent progenitors (MPP), the direct progeny of the HSCs, which can further differentiate into the common lymphoid progenitors (CLP) that engender the adult lymphoid effector cells, including T-cells, B-cells, dendritic and natural killer cells and the common myeloid progenitors (CMP) which engender the adult erythrocytes, monocytes, granulocytes and platelets (Morrison, 1997; Morrison & Weissman, 1994).

Generally, a set of cell surface markers are used to identify HSCs in the absence of lineage markers (lin^-). One of them is CD34 which is expressed on about 5% of the BM cells in humans, mainly observed on the premature progenitor cells (Civin et al., 1990). CD34 along with other surface markers have been used for identification of premature cell population like the $CD34^+CD38^-$ cells population. Upon comparison to the $CD34^+CD38^{low}$ cell population, $CD34^+Cd38^-$ cell population has been found to have more LTRCs. The human lymphoid committed progenitors (CLP) are defined by the co-expression of CD10, $CD34^+$ or CD07 (Bellantuono, 2004; Hao, 2001; Ziegler, 1999), while the (lin^-)$CD34^+$ $CD38^+$ cell population's expression of $CD123^{low}$ $CD45RA^-$ defines the human myeloid progenitors (CMP) (Bellantuono, 2004; Manz et al., 2002). There have been other cell markers such as CD133 and KDR that have also been employed along with or in place of CD34 (Bellantuono, 2004; Bhatia, 2015; Cimato et al., 2019; Yin, 1997). But HSCs can only be conclusively assessed by their capability to differentiate into the CMP and CLP lineage post transplantation in an irradiated host (Bellantuono, 2004).

Mesenchymal Stem Cells (MSCs)

These cells were at first discovered in mouse bone marrow and were described as colony-forming unit fibroblasts (CFU-F) (Friedenstein et al., 1966, 1976, 1987).

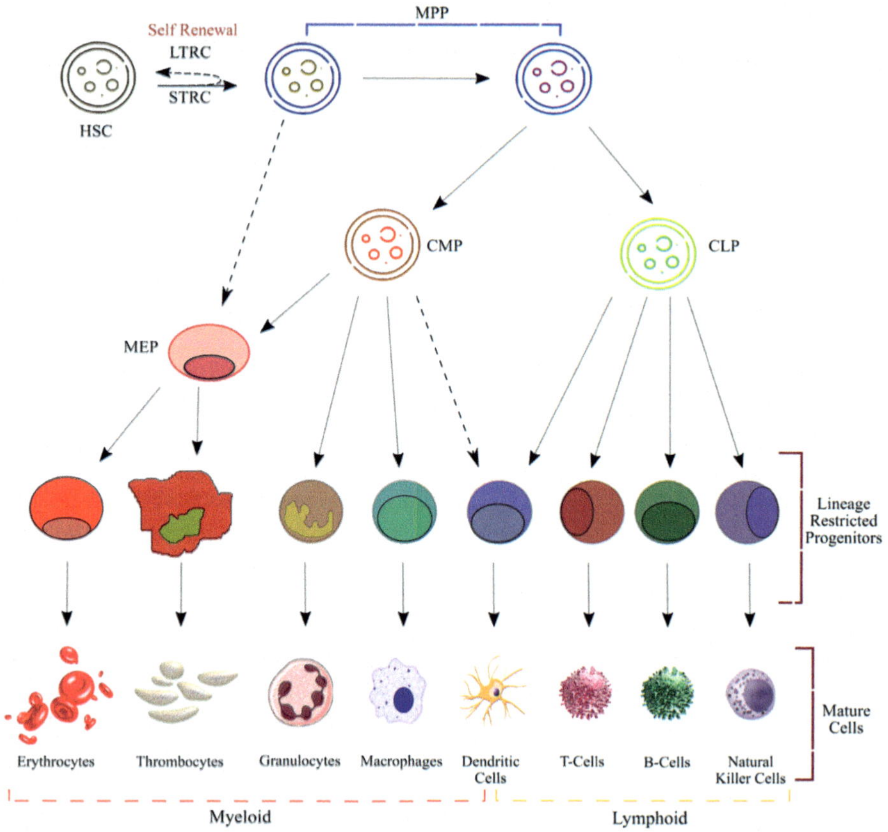

Fig. 2.3 Hematopoietic developmental hierarchy model. Hematopoietic stem cells (HSCs) seem to be either long-term repopulating cells (LTRC) which are self-renewable and can repopulate a host via transplantation, or they are short-term repopulating cells (STRC) which transform into multipotent progenitors (MPP). These progenitors produce the oligopotent progenitors like the common lymphoid progenitors (CLP) or the common myeloid progenitors (CMP) that further metamorphose into lineage-restricted progenitors. The CMPs further give rise to erythrocytes, thrombocytes, granulocytes and macrophages, while the CLPs differentiate into mature T-cells, B-cells and natural killer (NK) cells. CMP and CLP both have proposed to develop dendritic cells

Friedenstein et al. studies showed that the bone marrow stromal cells derived from these CFU-Fs were found to serve as feeder layers to culture HSCs and these spindle-shaped and clonogenic cells could differentiate both in vivo and ex *vivo* into adipocytes, osteocytes and chondrocytes (Friedenstein, 1974). According to Caplan et al., these stromal cells from the bone marrow were deemed to be stem cells and coined the term mesenchymal stem cells (MSCs), owing to their ability to self-renew and differentiate (Caplan, 1991; Uccelli et al., 2008a). These MSCs can be isolated from several tissue sources like the umbilical cord, bone marrow, adipose tissues, menses blood, endometrial polyps, etc. These tissues sources seem to be most suitable for research and medical application due to the cell numbers obtained and the ease of harvest (Ding, 2006; Ding et al., 2007, 2011).

The MSCs can differentiate under precise in vitro conditions into several lineages of three germ layers—mesoderm, ectoderm and endoderm like the muscle, fat, bone, neuron, chondrocyte (Fig. 2.4), liver cells and islet cells (Kuo et al., 2009; Oishi, 2009). As a result of the lineage differentiation, the MSCs form several vital functional elements of the HSC niche for supporting hematopoiesis such as the bone marrow stromal cells, osteocytes, osteoblasts, endothelial cells, pericytes, adipocytes and myofibroblasts (Muguruma, 2006; Uccelli et al., 2008a). In the bone marrow, the immature hematopoietic cells are surrounded and maintained in a state of quiescence by the niche stromal cells which provide a sheltering microenvironment along with controlling their differentiation and the release of mature progeny (Schofield, 1978a, 1978b; Uccelli et al., 2008b). In addition to the preservation of the HSC pool, the

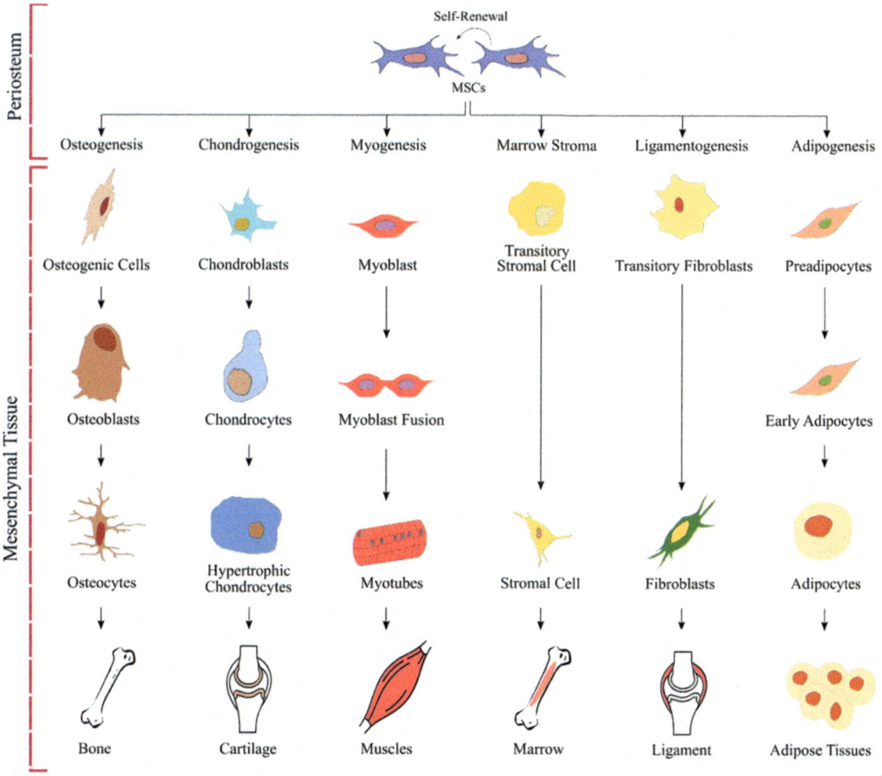

Fig. 2.4 Mesengenic development hierarchy model. In vivo, the MSCs have the capability to self-renew within the periosteum and can differentiate down the mesodermal lineage to form mesenchymal tissues such as the bone, bone marrow, muscle, ligament, fat and cartilage cells (Kemp et al., 2005). These cells can also transdifferentiated in ectodermal and mesodermal lineages

stromal cells are hypothesized to be having anti-proliferative properties at all phases of maturation, which are similar to the stromal cell niches like the above-mentioned HSC niche. Partially differentiated stromal cells like the chondrocytes and fibroblasts have shown to possess the same anti-proliferative properties as their precursors, as exhibited by their capability to hinder T-cell proliferation (Bocelli-Tyndall, 2006; Jones, 2007).

The characterization of in vitro cultures of MSCs is restricted due to the deficiency of some of the key surface CD markers. They generally do not express the hematopoietic markers CD14, CD34, HLA-DR and CD45 (Fox, 2007) or the costimulatory proteins like CD40, CD80 and CD86. However, these cultures are characterized by their variable expression of the surface antigen endoglin (CD105) and other antigens like THY1 (CD44 and CD90), transferrin receptor (CD71), ecto-5'-nucleotidase (CD73), ALCAM (CD166) and β1-integrin (CD29) (Chamberlain, 2007; Roubelakis et al., 2012), while Stro-1 is definite for clonogenic MSCs. The studies showing in vivo characterization of the functionality of MSCs validated that Stro-1^+ stromal cells are present in the dental pulp and the bone marrow confirming the presence of MSCs in the perivascular sites.

Adipose-Derived Stem Cells (ADSCs)

The adipose tissue (AT) was always believed to be just an energy reservoir and used to be discarded post-liposuction. In recent years, various studies have recognized this tissue as an endocrine organ that could control metabolic activity and immunity (Karastergiou & Mohamed-Ali, 2010; Kershaw & Flier, 2004; Tateishi-Yuyama, 2002; Zhang, 1994). In 2001, Zuk et.al. pioneered the study which considered AT as a new source of adult stem cells (Zuk, 2001). Although the bone marrow is the most common source of MSCs for cell therapies, harvesting them from the iliac crest is quite excruciating and increases the chances of infection (Macrin et al., 2017; Zuk et al., 2001). On the contrary, the isolation of the MSCs from the adipose tissue yields up to 500 times more cells with zero complications (Fraser, 2006). These adipose tissues are derived from the mesoderm and are situated in the subcutaneous tissues, and intraperitoneal compartments surround the vital organs and structures of the body (Cinti, 2005).

AT is home to various cell types such as preadipocytes, adipocytes, vascular smooth muscle cells and the multipotent adipose-derived stem cells (ADSCs) which are the source for adipogenesis and expansion of AT. They reside in the perivascular niches within the white adipose tissue and can be isolated from the stromal vascular fraction (SVF) (Barba et al., 2017). The SVF cells are collected via liposuction, and the lipoaspirate is processed by washing, digestion and centrifugation before being plated and cultured to acquire ADSCs.

Aside from adipogenesis, these ADSCs are competent to differentiate in vitro into various mesodermal lineages (Fig. 2.5) such as osteocytes (Catalano, 2017; Kargozar et al., 2018a; Logovskaya, 2013; Tirkkonen, 2013), cardiomyocytes (Safaeijavan, 2014), vascular and visceral smooth muscle cells (Marra et al., 2011; Park, 2012;

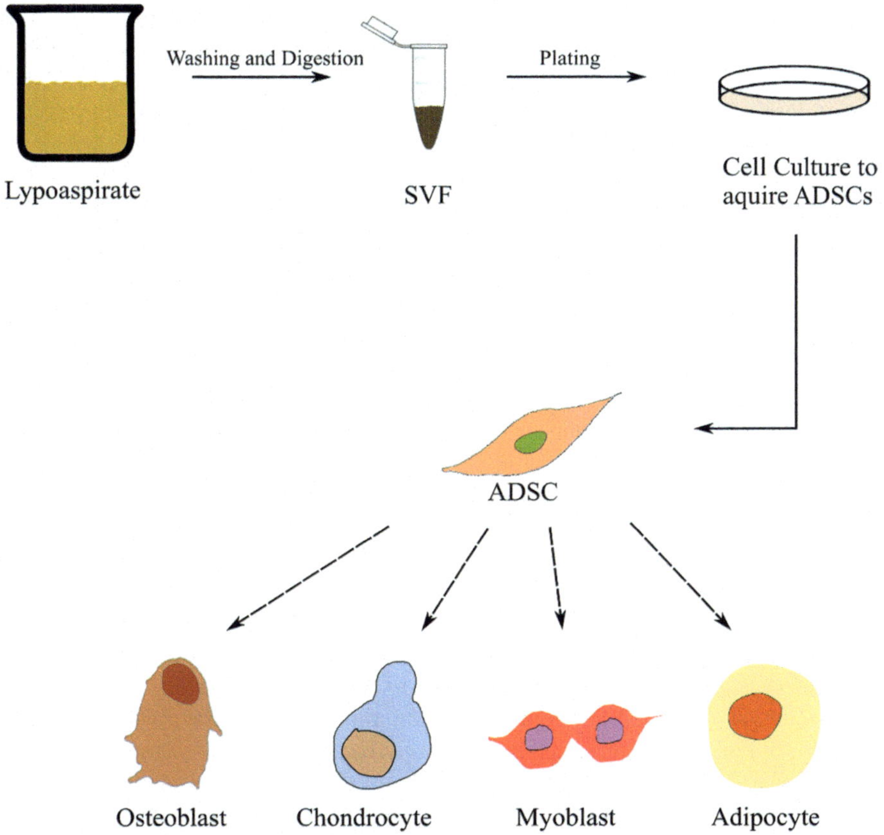

Fig. 2.5 ADSC Isolation Process & their Multilineage Potential. The adipose-derived stem cells (ADSCs) are isolated from the lipoaspirate obtained from the liposuction. Under suitable environments, these cells can differentiate into different mesodermal lineage cells

Wang et al., 2010), skeletal myocytes (Forcales, 2015), articular and tracheal chondrocytes (Batioglu-Karaaltin et al., 2015; Forget, 2016; Mardani, 2016), dermal fibroblasts (Sivan et al., 2016) and endothelial cells (Bekhite et al., 2014; Fischer, 2009; Marino, 2012).

In contrast to other MSCs, ADSCs have been shown to have delayed senescence (Ding, 2013; Kokai et al., 2014) and a higher proliferation rate (Barba et al., 2017). They also seem to be extremely active in producing various immunomodulators and growth factors like insulin-like growth factor-1 (IGF-1), vascular endothelial growth factor-D (VEGF-D) and interleukin-8 (IL-8) in comparison with other stem cells (Hsiao et al., 2012). When compared to the bone marrow MSCs (equal cell number), the production of different immunomodulators like interleukin-6 (IL-6) and TGF-B1 by ADSCs is also greater (Melief, 2013). However, these cells have lower osteogenic differentiation potential than the bone marrow-derived MSCs (Kargozar et al., 2018b; Przekora, 2017). In cell culture, ADSCs express the cell surface markers which are akin to those expressed by other MSCs like CD29 (fibronectin receptor) CD44 (THY-1), CD73 (ecto-5'-nucleotidase), CD90 (THY-1), CD105 (endoglin), CD146 (melanoma cell adhesion molecule) Stro-1 and SH3. However, they do not express the endothelial marker CD31 (PECAM-1), CD104 (integrin β4), SMA or the hematopoietic markers CD45 (De Ugarte, 2003; Gronthos, 2001; Marappagounder, 2013; Rodriguez, 2005; Wickham, 2003), while other cells within the SVF of whole AT, such as pericytes express CD31, CD34 and CD146 markers.

Induced Pluripotent Stem Cells (iPSCs)

Over the last seventy years, the development and advancement of scientific principles and technologies in the field of cellular reprogramming have led to the discovery of induced pluripotency. Developments for instance—[a] SCNT or somatic cell nuclear transfer which shows that the differentiated cells preserve the genetic material identical to the premature embryonic cells. [b] the technical progress which enabled scientists to culture and study pluripotent cell lines and [c] the research studies which confirm the role of transcription factors in influencing the cell fate through enforced expression that can switch one type of mature cell to another.

In 2006, Shinya Yamanaka and Kazutoshi Takahashi originally demonstrated the production of ESC-like stem cells from mouse fibroblasts by concurrently inserting four genes, and termed these cells as induced pluripotent stem cells (iPSCs) (Takahashi & Yamanaka, 2006). In 2007, they validated that an identical method can be relevant for human fibroblasts using various factors which would generate human-induced pluripotent stem cells (hiPSCs) (Takahashi et al., 2007; Yu, 2007).

The induction of pluripotency in the fibroblasts was attained by a "cocktail" of four transcription factors, namely *Oct3/4, c-Myc, Sox2* and *Klf4* or *Nanog, Sox2, Oct4* and *Lin28* (Takahashi & Yamanaka, 2006; Yu et al., 2007). In the following years, various research groups succeeded in getting the same outcome by selectively altering either the cell source, inducing agents or the delivery vectors (Table 2.1). Nevertheless, using these factors and inserting them into the genome has restricted the efficiency of these cells due to a higher probability of insertional mutation. Additionally, using oncogene copies such as *c-Myc* has also shown to produce tumors in mice and can increase the hazard when the cells are used for regenerative medicine (Okita et al., 2007). In the past, the method of transmission of the transcription factors was done by nucleic acid-based delivery. While certain studies demonstrated the use of retroviruses as a mode of delivery, major technological advancements in the field led to the use of recombinant proteins, small molecules and plasmids, thereby minimizing the risk of mutagenesis (Lyssiotis et al., n.d.; Okita et al., 2007; Woltjen, 2009; Yusa, 2009; Zhou et al., 2009).

The capability of iPSCs to differentiate into other lineages is similar to ESCs, along with the prospect of them being patient-specific has revealed an array possible implementations, especially in regenerative medicine (Jaenisch & Young, 2008). A primary use of iPS cell reprogramming is the development of in vitro disease models (Fig. 2.6). The hiPSCs could be differentiated from patients with a particular disease to create superior disease models for various conditions (Park, 2008). Normally, these disease models are challenging to develop due to the inability of the human primary cells to be sustained in culture for a prolonged interval (Maury, 2012). The ex vivo "disease in a dish" models developed using iPSCs could hold key information for

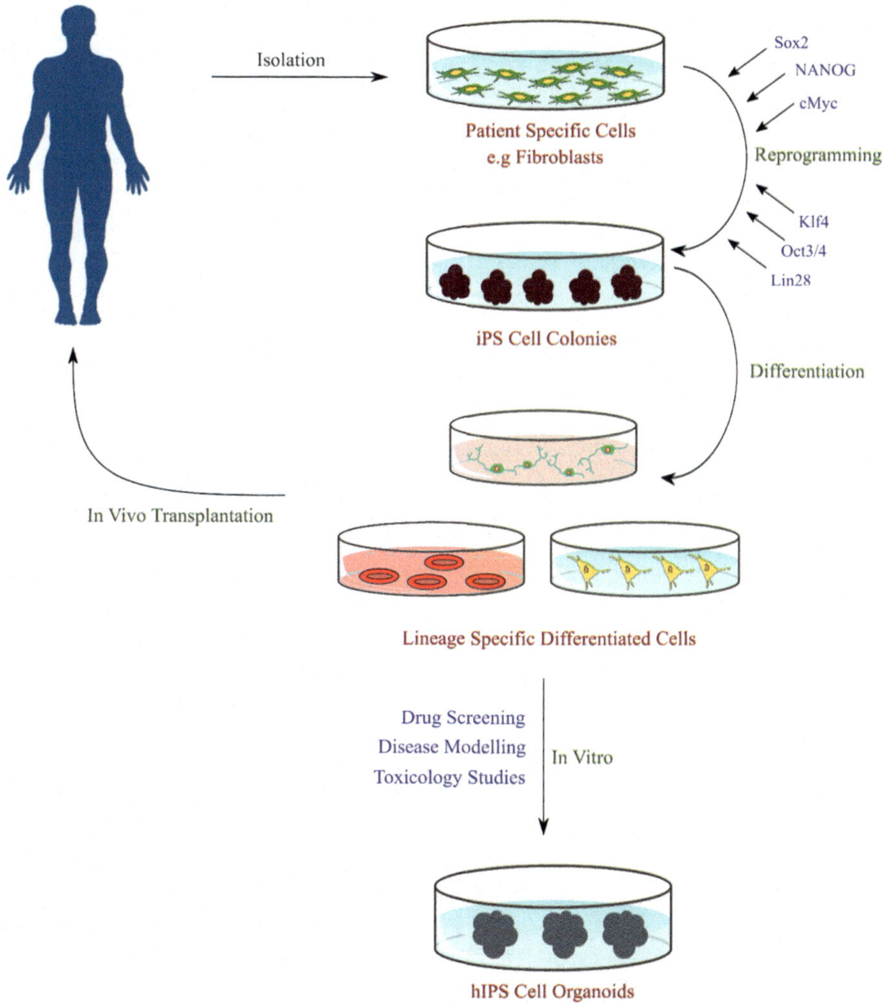

Fig. 2.6 Development and applications of hiPSCs. The IPS cells are developed using patient-specific cells (e.g., fibroblasts) to create various cell lines across all germ layers for their potential application in disease modeling, toxicological studies, pharmaceutical drug testing and in vivo transplantation for regenerative cell therapies

understanding various disease mechanisms and can accurately mimic the phenotype of the diseased cells (Merkle & Eggan, 2013).

The use of iPSCs for the screening of novel compounds for drug toxicity studies could overcome the issues caused by the unpredictability of the human metabolism which is the prime reason for withdrawal of new drug projects. For instance, until a few years back, the human drug models used were developed using cancer cell lines or from liver biopsy cells. The problem with these hepatocytes is that they lose functionality when cultured in vitro and are difficult to predict the hepatotoxicity properties. Hence, iPSC-derived hepatocytes have been used to develop various drug models to address these issues and study the drug-induced injuries to the liver (Maury et al., 2012; Ware et al., 2015). In the past decade, a major revolution in the area of stem cells was the development of the "organ in a dish" model, i.e., the organoids. These organoids are 3D stem cell structures which mimic the microenvironment and lineage progression of a particular tissue. This technology in tandem with the development of patient-derived iPSCs has aided in precise disease modeling, and it emphasizes the capability to transform translational medicine, biomedical applications and cell therapies (Aihara et al., n.d.; Bredenkamp, 2014; Dye et al., 2015; Eiraku, 2011; Eiraku & Sasai, 2012; Jo, 2016; Lancaster, 2013; Longworth-Mills, Koehler & Hashino, 2015; Song, 2013; Takasato, 2014, 2015; Takebe, 2013; Völkner, 2016) (Table 2.1).

Apart from the in vitro applications, researchers have also studied the prospect of in vivo application of iPSCs in regenerative medicine. There have been various studies experimenting the use of iPSCs in vivo such as the production of cardiovascular precursor cells that differentiate down to cardiomyocytes, endothelial cells and smooth muscle cells in SCID mice models, which could potentially be used to explain the pathogenesis of various cardiac ailments (Moretti, 2010). Another study has reported the production of iPSC-derived Factor-VIII producing endothelial progenitor cells, which upon transplantation into the liver of hemophilic mice have shown an increased rate of survival of up to 3 months, suggesting the use iPSCs in treating hemophilic patients (Xu et al., 2009).

Parkinson's disease has been greatly studied using ESCs and dopaminergic neurons in cell therapies. Recently, a study has reported the production of neural progenitor cells from iPSC differentiation. These cells upon transplantation into a fetal rat brain differentiated into DA neurons and glial cells incorporating into the brain and re-establishing neuronal activity and improved behavior (Doi, 2014; Kikuchi, 2011; Wernig, 2008). These studies have given a positive view of the viability of iPSCs in the treatment of Parkinson's disease. Another approach of iPSC technology was implemented to treat sickle cell anemia. Here, the researchers reprogrammed the patient-specific iPSCs and genetically modified them in vitro to correct the gene defect using homologous recombination (Hanna, 2007; Ye et al., 2009a, b). Upon differentiating the iPSCs into hematopoietic progenitor cells, they were transplanted into the rat model and were observed to treat the anemic mice (Hanna et al., 2007).

In the past decade, iPS cell technology has transformed the field of regenerative medicine and personalized medicine. It has allowed for a unique possibility to

Table 2.1 List of all the factors involved in the creation of various induced pluripotent stem cell lines

Cell sources	Inducing agents	Target cell types	Mode of delivery
Fibroblasts (Takahashi et al., 2007)	NANOG + Oct4 + Sox2 + Lin28 (Yu et al., 2007)	Hepatocytes (Li, 2010)	Retrovirus (Takahashi & Yamanaka, 2006; Takahashi et al., 2007)
Keratinocytes (Li, 2009)	Oct3/4 + Sox2 + c-Myc + Klf4 (Takahashi & Yamanaka, 2006; Takahashi et al., 2007)	Osteoblasts (Tashiro, 2009)	Lentivirus (Carey et al., 2009)
ADSCs (Sun et al., 2009)	Oct4 & Sox2 (Huangfu, 2008)	Cardiomyocytes (Kuzmenkin, 2009)	Adenovirus (Stadtfeld et al., 2008a, 2008b)
Endothelial cells (Lagarkova et al., 2010)	Oct4 only for neural cells (Kim, Greber, et al., 2009)	Muscle cells (Xie, 2009)	Plasmids (Okita, 2008, 2011)
Neural cells (Kim, Greber, et al., 2009; Kim, Sebastiano, et al., 2009)	Small molecules like Kenpaullone (Lyssiotis et al., n.d.)	Hematopoietic cells (Choi, 2009)	Recombinant proteins (Kim et al., 2009)
Blood cells (Choi et al., 2009; Ye et al., 2009a, b)		Renal cells (Morizane et al., 2009)	Sendai virus (Fusaki et al., 2009)
Liver cells (Aoi, 2008)		Neural cells (Dimos, 2008)	Small molecules (Bertolotti, 2009)
Stomach cells (Aoi et al., 2008)		Pancreatic β-cells (Tashiro et al., 2009; Tateishi, 2008)	
Pancreatic β-cells (Stadtfeld et al.,)		Adipocytes (Tashiro et al., 2009)	
B Lymphocytes (Hanna, 2008)		Retinal cells (Buchholz, 2009; Hirami, 2009)	
Progenitor cells			
Umbilical cord blood (Okita, 2013)			

interpret various pathophysiological processes and causes of various diseases along with helping in improving our understanding of human biology at a cellular stage. While still in its infancy, the use of this technology in various clinical studies and animal disease models to treat rare conditions, neural diseases and tissue injuries has given encouraging results, thereby cementing its future in the field of regenerative medicine. The advent of newer technologies such as gene-editing tool CRISPR-Cas9, next-generation sequencing, organ-on-a-chip model, drug discovery platforms will greatly drive the application of iPSC-based cell therapies in the future.

References

Abdulrazzak, H., et al. (2010) Biological characteristics of stem cells from foetal, cord blood and extraembryonic tissues. *Journal of the Royal Society Interface. Royal Society*. https://doi.org/10.1098/rsif.2010.0347.focus

Aihara, E., et al. (n.d.). Characterization of stem/progenitor cell cycle using murine circumvallate papilla taste bud organoid. *nature.com*. Available at: https://www.nature.com/articles/srep17185. Accessed on October 14, 2020.

Aoi, T., et al. (2008). Generation of pluripotent stem cells from adult mouse liver and stomach cells. *Science, 321*(5889), 699–702. https://doi.org/10.1126/science.1154884

Augello, A., Kurth, T. B., & de Bari, C. (2010). Mesenchymal stem cells: A perspective from in vitro cultures to in vivo migration and niches. *European Cells and Materials*, 121–133. AO Research Institute Davos. https://doi.org/10.22203/eCM.v020a11

Barba, M., Di Taranto, G., & Lattanzi, W. (2017). Adipose-derived stem cell therapies for bone regeneration. *Expert Opinion on Biological Therapy, 17*(6), 677–689. https://doi.org/10.1080/14712598.2017.1315403

Barzilay, R., Melamed, E., & Offen, D. (2009). Introducing transcription factors to multipotent mesenchymal stem cells: Making transdifferentiation possible. *Stem Cells*, 2509–2515. https://doi.org/10.1002/stem.172

Batioglu-Karaaltin, A., et al. (2015). In Vivo tissue-engineered allogenic trachea transplantation in rabbits: A preliminary report. *Stem Cell Reviews and Reports, 11*(2), 347–356. https://doi.org/10.1007/s12015-014-9570-8

Beattie, G. M., et al. (1997) Functional β-cell mass after transplantation of human fetal pancreatic cells: Differentiation or proliferation? *Diabetes, 46*(2), 244–248. https://doi.org/10.2337/diab.46.2.244

Beck, B., & Blanpain, C. (2012). Mechanisms regulating epidermal stem cells. *EMBO Journal*, 2067–2075. https://doi.org/10.1038/emboj.2012.67

Bekhite, M. M., et al. (2014). Hypoxia, leptin, and vascular endothelial growth factor stimulate vascular endothelial cell differentiation of human adipose tissue-derived stem cells. *Stem Cells and Development, 23*(4), 333–351. https://doi.org/10.1089/scd.2013.0268

Bellantuono, I. (2004). Haemopoietic stem cells. *International Journal of Biochemistry and Cell Biology, 36*(4), 607–620. https://doi.org/10.1016/j.biocel.2003.10.008

Bender Kim, C. F., et al. (2005). Identification of bronchioalveolar stem cells in normal lung and lung cancer. *Cell. Cell, 121*(6), 823–835. https://doi.org/10.1016/j.cell.2005.03.032

Bentzinger, C. F., et al. (2013). The emerging biology of muscle stem cells: Implications for cell-based therapies. *BioEssays, 35*(3), 231–241. https://doi.org/10.1002/bies.201200063

Bertolotti, R. (2009). Translational perspectives transient epigenetic gene therapy: Hazard-free cell reprogramming approach and rising arm of a universal stem cell gene therapy platform. *Gene Therapy and Regulation*, 11–39. https://doi.org/10.1142/S1568558609000102

Bhatia, S., et al. (2015). Control of AC133/CD133 and impact on human hematopoietic progenitor cells through nucleolin. *Leukemia, 29*(11), 2208–2220. https://doi.org/10.1038/leu.2015.146

Bibber, B., et al. (2013). A review of stem cell translation and potential confounds by cancer stem cells. *Stem Cells International*, 2013.https://doi.org/10.1155/2013/241048

Bocelli-Tyndall, C., et al. (2006). Human articular chondrocytes suppress in vitro proliferation of anti-CD3 activated peripheral blood mononuclear cells. *Journal of Cellular Physiology, 209*(3), 732–734. https://doi.org/10.1002/jcp.20789

Bongso, A. (2006). Blastocyst culture for deriving human embryonic stem cells. Methods in molecular biology (Clifton, N.J.). *Methods in Molecular Biology, 331*, 13–22. https://doi.org/10.1385/1-59745-046-4:13

Bredenkamp, N., et al. (2014). An organized and functional thymus generated from FOXN1-reprogrammed fibroblasts. *Nature Cell Biology, 16*(9), 902–908. https://doi.org/10.1038/ncb3023

Bruder, S. P., Jaiswal, N., & Haynesworth, S. E. (1997). Growth kinetics, self-renewal, and the osteogenic potential of purified human mesenchymal stem cells during extensive subcultivation and following cryopreservation. *Journal of Cellular Biochemistry, 64*(2). https://doi.org/10.1002/(SICI)1097-4644(199702)64:2<278::AID-JCB11>3.0.CO;2-F

Buchholz, D. E., et al. (2009). Derivation of functional retinal pigmented epithelium from induced pluripotent stem cells. *Stem Cells, 27*(10), 2427–2434. https://doi.org/10.1002/stem.189

Butti, E., et al. (2014). Neurogenic and non-neurogenic functions of endogenous neural stem cells. Frontiers in Neuroscience, 92. https://doi.org/10.3389/fnins.2014.00092

Campagnoli, C., et al. (2001). Identification of mesenchymal stem/progenitor cells in human first-trimester fetal blood, liver, and bone marrow. *Blood, 98*(8), 2396–2402. https://doi.org/10.1182/blood.V98.8.2396

Cananzi, M., Atala, A., & De Coppi, P. (2009). Stem cells derived from amniotic fluid: New potentials in regenerative medicine. *Reproductive BioMedicine Online*, 17–27. https://doi.org/10.1016/S1472-6483(10)60111-3

Caplan, A. I. (1991). Mesenchymal stem cells. *Journal of Orthopaedic Research, 9*(5), 641–650. https://doi.org/10.1002/jor.1100090504

Carey, B. W., et al. (2009). Reprogramming of murine and human somatic cells using a single polycistronic vector. *Proceedings of the National Academy of Sciences of the United States of America, 106*(1), 157–162. https://doi.org/10.1073/pnas.0811426106

Catalano, M. G., et al. (2017). Extracorporeal shockwaves (ESWs) enhance the osteogenic medium-induced differentiation of adipose-derived stem cells into osteoblast-like cells. *Journal of Tissue Engineering and Regenerative Medicine, 11*(2), 390–399. https://doi.org/10.1002/term.1922

Chamberlain, G., et al. (2007). Concise review: Mesenchymal stem cells: Their phenotype, differentiation capacity, immunological features, and potential for homing. *Stem Cells, 25*(11), 2739–2749. https://doi.org/10.1634/stemcells.2007-0197

Chimutengwende-Gordon, M., & Khan, S. W. (2012). Advances in the use of stem cells and tissue engineering applications in bone repair. *Current Stem Cell Research & Therapy, 7*(2), 122–126.https://doi.org/10.2174/157488812799219036

Choi, K.-D., et al. (2009). Hematopoietic and endothelial differentiation of human induced pluripotent stem cells. *Stem Cells, 27*(3), 559–567. https://doi.org/10.1634/stemcells.2008-0922

Cimato, T. R., et al. (2019). CD133 expression in circulating hematopoietic progenitor cells. *Cytometry Part B—Clinical Cytometry, 96*(1), 39–45. https://doi.org/10.1002/cyto.b.21732

Cinti, S. (2005). The adipose organ. *Prostaglandins Leukotrienes and Essential Fatty Acids*, 9–15. https://doi.org/10.1016/j.plefa.2005.04.010

Civin, C. I., et al. (1990). Positive stem cell selection--basic science. *Progress in Clinical And Biological Research, 333*, 387–401; discussion 402. Available at: https://jhu.pure.elsevier.com/en/publications/positive-stem-cell-selection-basic-science-4. Accessed on August 10, 2020.

Dimos, J. T., et al. (2008). Induced pluripotent stem cells generated from patients with ALS can be differentiated into motor neurons. *Science, 321*(5893), 1218–1221. https://doi.org/10.1126/science.1158799

Ding, D.-C., et al. (2006). Current concepts in adult stem cell therapy for stroke. *Current Medicinal Chemistry, 13*(29), 3565–3574. https://doi.org/10.2174/092986706779026237

Ding, D. C., et al. (2007). The role of endothelial progenitor cells in ischemic cerebral and heart diseases. *Cell Transplantation*, 273–284. https://doi.org/10.3727/000000007783464777

Ding, D. C., et al. (2013). Human adipose-derived stem cells cultured in keratinocyte serum free medium: Donor's age does not affect the proliferation and differentiation capacities. *Journal of Biomedical Science, 20*(1), 59. https://doi.org/10.1186/1423-0127-20-59

Ding, D. C., Shyu, W. C., & Lin, S. Z. (2011). Mesenchymal stem cells. *Cell Transplantation*, 5–14. https://doi.org/10.3727/096368910X

Doetschman, T. C., et al. (1985). The in vitro development of blastocyst-derived embryonic stem cell lines: formation of visceral yolk sac, blood islands and myocardium. *Development, 87*(1).

Doi, D., et al. (2014). Isolation of human induced pluripotent stem cell-derived dopaminergic progenitors by cell sorting for successful transplantation. *Stem Cell Reports, 2*(3), 337–350. https://doi.org/10.1016/j.stemcr.2014.01.013

Dye, B. R., et al. (2015). In vitro generation of human pluripotent stem cell derived lung organoids. *eLife, 2015*(4), 1–25. https://doi.org/10.7554/eLife.05098

Eiraku, M., et al. (2011). Self-organizing optic-cup morphogenesis in three-dimensional culture. *Nature, 472*(7341), 51–58. https://doi.org/10.1038/nature09941

Eiraku, M., & Sasai, Y. (2012). Self-formation of layered neural structures in three-dimensional culture of ES cells. *Current Opinion in Neurobiology*, 768–777. https://doi.org/10.1016/j.conb.2012.02.005

Erbs, S., et al. (2005). Transplantation of blood-derived progenitor cells after recanalization of chronic coronary artery occlusion: First randomized and placebo-controlled study. *Circulation Research, 97*(8), 756–762. https://doi.org/10.1161/01.RES.0000185811.71306.8b

Evans, M. J., & Kaufman, M. H. (1981). Establishment in culture of pluripotential cells from mouse embryos. *Nature, 292*(5819), 154–156. https://doi.org/10.1038/292154a0

Fischer, L. J., et al. (2009). Endothelial differentiation of adipose-derived stem cells: Effects of endothelial cell growth supplement and shear force. *Journal of Surgical Research, 152*(1), 157–166. https://doi.org/10.1016/j.jss.2008.06.029

Fonseca, N. A. C. da. (2014). Targeted intracellular delivery of synergistic drug combinations: Tackling drug resistance in human breast cancer. Available at: https://estudogeral.sib.uc.pt/handle/10316/27016. Accessed on October 28, 2020.

Forcales, S. V. (2015). Potential of adipose-derived stem cells in muscular regenerative therapies. *Frontiers in Aging Neuroscience*, *7*(JUN), 123. https://doi.org/10.3389/fnagi.2015.00123

Forget, J., et al. (2016). Differentiation of human mesenchymal stem cells toward quality cartilage using fibrinogen-based nanofibers. *Macromolecular Bioscience, 16*(9), 1348–1359. https://doi.org/10.1002/mabi.201600080

Foster, C. S., et al. (2002). An introduction to stem cells. *Journal of Pathology*, 419–423. https://doi.org/10.1002/path.1187

Fox, J. M., et al. (2007). Recent advances into the understanding of mesenchymal stem cell trafficking. *British Journal of Haematology*. https://doi.org/10.1111/j.1365-2141.2007.06610.x

Fraser, J. K., et al. (2006). Fat tissue: An underappreciated source of stem cells for biotechnology. *Trends in Biotechnology, 24*(4), 150–154. https://doi.org/10.1016/j.tibtech.2006.01.010

Friedenstein, A. J., et al. (1974). Stromal cells responsible for transferring the microenvironment of the hemopoietic tissues. Cloning in vitro and retransplantation in vivo. *Transplantation, 17*(4), 331–340. https://doi.org/10.1097/00007890-197404000-00001

Friedenstein, A. J., Chailakhyan, R. K., & Gerasimov, U. V. (1987). Bone marrow osteogenic stem cells: In vitro cultivation and transplantation in diffusion chambers. *Cell Proliferation, 20*(3), 263–272. https://doi.org/10.1111/j.1365-2184.1987.tb01309.x

Friedenstein, A. J., Gorskaja, U. F., & Kulagina, N. N. (1976). Fibroblast precursors in normal and irradiated mouse hematopoietic organs. *Experimental Hematology, 4*(5), 267–274.

Friedenstein, A. J., Piatetzky-Shapiro, I. I., & Petrakova, K. V. (1966). Osteogenesis in transplants of bone marrow cells. *Journal of Embryology and Experimental Morphology*, *16*(3), 381–390.

Fu, R. H., et al. (2011). Differentiation of stem cells: Strategies for modifying surface biomaterials. *Cell Transplantation*, 37–47. https://doi.org/10.3727/096368910X532756

Fusaki, N., et al. (2009). Efficient induction of transgene-free human pluripotent stem cells using a vector based on Sendai virus, an RNA virus that does not integrate into the host genome. *Proceedings of the Japan Academy Series B: Physical and Biological Sciences, 85*(8), 348–362. https://doi.org/10.2183/pjab.85.348

Goldman, J. (1982). Blood and its disorders. In R. M. Hardisty & D. J. Weatherall (Eds.), (pp. 601–627). Blackwell Scientific Publications.

Gronthos, S., et al. (2001). Surface protein characterization of human adipose tissue-derived stromal cells. *Journal of Cellular Physiology, 189*(1), 54–63. https://doi.org/10.1002/jcp.1138

Hamazaki, T., et al. (2001). Hepatic maturation in differentiating embryonic stem cells in vitro. *FEBS Letters, 497*(1), 15–19. https://doi.org/10.1016/S0014-5793(01)02423-1

Hambiliki, F., et al. (2012). Co-localization of NANOG and OCT4 in human pre-implantation embryos and in human embryonic stem cells. *Journal of Assisted Reproduction and Genetics, 29*(10), 1021–1028. https://doi.org/10.1007/s10815-012-9824-9

Hanna, J., et al. (2007). Treatment of sickle cell anemia mouse model with iPS cells generated from autologous skin. *Science, 318*(5858), 1920–1923. https://doi.org/10.1126/science.1152092

Hanna, J., et al. (2008). Direct reprogramming of terminally differentiated mature B lymphocytes to pluripotency. *Cell Press, 133*(2), 250–264. https://doi.org/10.1016/j.cell.2008.03.028

Hao, Q. L., et al. (2001). Identification of a novel, human multilymphoid progenitor in cord blood. *Blood, 97*(12), 3683–3690. https://doi.org/10.1182/blood.V97.12.3683

Harrison, D. E., & Zhong, R. K. (1992). The same exhaustible multilineage precursor produces both myeloid and lymphoid cells as early as 3–4 weeks after marrow transplantation. *Proceedings of the National Academy of Sciences of the United States of America, 89*(21), 10134–10138. https://doi.org/10.1073/pnas.89.21.10134

Heydarkhan-Hagvall, S., et al. (2012). The effect of vitronectin on the differentiation of embryonic stem cells in a 3D culture system. *Biomaterials, 33*(7), 2032–2040. https://doi.org/10.1016/j.biomaterials.2011.11.065

Hirami, Y., et al. (2009). Generation of retinal cells from mouse and human induced pluripotent stem cells. *Neuroscience Letters, 458*(3), 126–131. https://doi.org/10.1016/j.neulet.2009.04.035

Hsiao, S. T. F., et al. (2012). Comparative analysis of paracrine factor expression in human adult mesenchymal stem cells derived from bone marrow, adipose, and dermal tissue. *Stem Cells and Development, 21*(12), 2189–2203. https://doi.org/10.1089/scd.2011.0674

Huangfu, D., et al. (2008). Induction of pluripotent stem cells from primary human fibroblasts with only Oct4 and Sox2. *Nature Biotechnology, 26*(11), 1269–1275. https://doi.org/10.1038/nbt.1502

Ilancheran, S., Moodley, Y., & Manuelpillai, U. (2009). Human fetal membranes: A source of stem cells for tissue regeneration and repair? *Placenta, 30*(1), 2–10. https://doi.org/10.1016/j.placenta.2008.09.009

Jaenisch, R. & Young, R. (2008). Stem cells, the molecular circuitry of pluripotency and nuclear reprogramming. *Cell*, 567–582. https://doi.org/10.1016/j.cell.2008.01.015

Jo, J., et al. (2016). Midbrain-like organoids from human pluripotent stem cells contain functional dopaminergic and neuromelanin-producing neurons. *Cell Stem Cell, 19*(2), 248–257. https://doi.org/10.1016/j.stem.2016.07.005

Jones, S., et al. (2007). 'The antiproliferative effect of mesenchymal stem cells is a fundamental property shared by all stromal cells. *The Journal of Immunology, 179*(5), 2824–2831. https://doi.org/10.4049/jimmunol.179.5.2824

Karastergiou, K., & Mohamed-Ali, V. (2010). The autocrine and paracrine roles of adipokines. *Molecular and Cellular Endocrinology*, 69–78. https://doi.org/10.1016/j.mce.2009.11.011

Kargozar, S., et al. (2018a). Osteogenic potential of stem cells-seeded bioactive nanocomposite scaffolds: A comparative study between human mesenchymal stem cells derived from bone, umbilical cord Wharton's jelly, and adipose tissue. *Journal of Biomedical Materials Research Part B: Applied Biomaterials, 106*(1), 61–72. https://doi.org/10.1002/jbm.b.33814

Kargozar, S., et al. (2018b). Osteogenic potential of stem cells-seeded bioactive nanocomposite scaffolds: A comparative study between human mesenchymal stem cells derived from bone, umbilical cord Wharton's jelly, and adipose tissue. *Journal of Biomedical Materials Research Part B: Applied Biomaterials, 106*(1), 61–72. https://doi.org/10.1002/jbm.b.33814

Kemp, K. C., Hows, J., & Donaldson, C. (2005). Bone marrow-derived mesenchymal stem cells. *Leukemia and Lymphoma*, 1531–1544. https://doi.org/10.1080/10428190500215076

Kennea, N. L., & Mehmet, H. (2002). Neural stem cells. *The Journal of Pathology, 197*(4), 536–550. https://doi.org/10.1002/path.1189

Kershaw, E. E., & Flier, J. S. (2004). Adipose tissue as an endocrine organ. *Journal of Clinical Endocrinology and Metabolism*, 2548–2556. https://doi.org/10.1210/jc.2004-0395

Kikuchi, T., et al. (2011). 'Survival of human induced pluripotent stem cell-derived midbrain dopaminergic neurons in the brain of a primate model of Parkinson's disease. *Journal of Parkinson's Disease, 1*(4), 395–412. https://doi.org/10.3233/JPD-2011-11070

Kim, D., et al. (2009). Generation of human induced pluripotent stem cells by direct delivery of reprogramming proteins. *Cell Stem Cell*, 472–476. https://doi.org/10.1016/j.stem.2009.05.005

Kim, J. B., Greber, B., et al. (2009) Direct reprogramming of human neural stem cells by OCT4. *Nature, 461*(7264), 649–653.https://doi.org/10.1038/nature08436

Kim, J. B., & Sebastiano, V., et al. (2009) Oct4-induced pluripotency in adult neural stem cells. *Cell, 136*(3), 411–419.https://doi.org/10.1016/j.cell.2009.01.023

Kokai, L. E., Marra, K., & Rubin, J. P. (2014). Adipose stem cells: Biology and clinical applications for tissue repair and regeneration. *Translational Research*, 399–408. https://doi.org/10.1016/j.trsl.2013.11.009

Kolios, G., & Moodley, Y. (2012). Introduction to stem cells and regenerative medicine. *Respiration*, 3–10. https://doi.org/10.1159/000345615

Körbling, M., & Estrov, Z. (2003). Adult stem cells for tissue repair—A new therapeutic concept? *New England Journal of Medicine*, 570–582. https://doi.org/10.1056/NEJMra022361

Kuo, T. K., Ho, J. H., & Lee, O. K. (2009). Mesenchymal stem cell therapy for nonmusculoskeletal diseases: Emerging applications. *Cell Transplantation*, 1013–1028. https://doi.org/10.3727/096368909X471206

Kuzmenkin, A., et al. (2009). Functional characterization of cardiomyocytes derived from murine induced pluripotent stem cells in vitro. *The FASEB Journal, 23*(12), 4168–4180. https://doi.org/10.1096/fj.08-128546

Lagarkova, M. A., et al. (2010). Induction of pluripotency in human endothelial cells resets epigenetic profile on genome scale. *Cell Cycle, 9*(5), 937–946. https://doi.org/10.4161/cc.9.5.10869

Lancaster, M. A., et al. (2013). Cerebral organoids model human brain development and microcephaly. *Nature, 501*(7467), 373–379. https://doi.org/10.1038/nature12517

Li, L., & Xie, T. (2005). Stem cell niche: Structure and function. *Annual Review of Cell and Developmental Biology, 21*(1), 605–631. https://doi.org/10.1146/annurev.cellbio.21.012704.131525

Li, W., et al. (2009). Generation of human-induced pluripotent stem cells in the absence of exogenous Sox2. *Stem Cells, 27*(12), 2992–3000. https://doi.org/10.1002/stem.240

Li, W., et al. (2010). Generation of functional hepatocytes from mouse induced pluripotent stem cells. *Journal of Cellular Physiology, 222*(3), 492–501. https://doi.org/10.1002/jcp.22000

Logovskaya, L. V., et al. (2013). Induction of osteogenic differentiation of multipotent mesenchymal stromal cells from human adipose tissue. *Bulletin of Experimental Biology and Medicine, 155*(1), 145–150. https://doi.org/10.1007/s10517-013-2100-x

Longworth-Mills, E., Koehler, K. R., & Hashino, E. (2015). Generating inner ear organoids from mouse embryonic stem cells. *Methods in Molecular Biology*, 391–406. https://doi.org/10.1007/7651_2015_215

Lyssiotis, C. A., et al. (n.d.). *Reprogramming of murine fibroblasts to induced pluripotent stem cells with chemical complementation of Klf4, National Acad Sciences*. Available at: www.pnas.org/cgi/content/full/. Accessed on October 6, 2020.

Macrin, D., et al. (2017). Eminent sources of adult mesenchymal stem cells and their therapeutic imminence. *Stem Cell Reviews and Reports*, 741–756. https://doi.org/10.1007/s12015-017-9759-8

Manz, M. G., et al. (2002). Prospective isolation of human clonogenic common myeloid progenitors. *Proceedings of the National Academy of Sciences of the United States of America, 99*(18), 11872–11877. https://doi.org/10.1073/pnas.172384399

Marappagounder, D., et al. (2013). 'Differentiation of mesenchymal stem cells derived from human bone marrow and subcutaneous adipose tissue into pancreatic islet-like clusters in vitro. *Cellular and Molecular Biology Letters, 18*(1), 75–88. https://doi.org/10.2478/s11658-012-0040-5

Marcus, A. J., & Woodbury, D. (2008). Fetal stem cells from extra-embryonic tissues: Do not discard: Stem cells review series. *Journal of Cellular and Molecular Medicine, 12*(3), 730–742. https://doi.org/10.1111/j.1582-4934.2008.00221.x

Mardani, M., et al. (2016). Induction of chondrogenic differentiation of human adipose-derived stem cells by low frequency electric field. *Advanced Biomedical Research, 5*(1), 97. https://doi.org/10.4103/2277-9175.183146

Marino, G., et al. (2012). 'Growth and endothelial differentiation of adipose stem cells on polycaprolactone. *Journal of Biomedical Materials Research Part A, 100A*(3), 543–548. https://doi.org/10.1002/jbm.a.33296

Marra, K. G., Brayfield, C. A., & Rubin, J. P. (2011). Adipose stem cell differentiation into smooth muscle cells. *Methods in Molecular Biology (clifton, N.J.), 702*, 261–268. https://doi.org/10.1007/978-1-61737-960-4_19

Martin, G. R. (1981). Isolation of a pluripotent cell line from early mouse embryos cultured in medium conditioned by teratocarcinoma stem cells. *Proceedings of the National Academy of Sciences of the United States of America, 78*(12 II), 7634–7638. https://doi.org/10.1073/pnas.78.12.7634

Maury, Y., et al. (2012). Human pluripotent stem cells for disease modelling and drug screening. *BioEssays, 34*(1), 61–71. https://doi.org/10.1002/bies.201100071

McCormick, J. B., & Huso, H. A. (2010). Stem cells and ethics: Current issues. *Journal of Cardiovascular Translational Research*, 122–127. https://doi.org/10.1007/s12265-009-9155-0

Melief, S. M., et al. (2013). Adipose tissue-derived multipotent stromal cells have a higher immunomodulatory capacity than their bone marrow-derived counterparts. *Stem Cells Translational Medicine, 2*(6), 455–463. https://doi.org/10.5966/sctm.2012-0184

Melton, D. (2014). "Stemness": Definitions, criteria, and standards. In *Essentials of stem cell biology* (3rd ed, pp. 7–17). Elsevier Inc. https://doi.org/10.1016/B978-0-12-409503-8.00002-0

Merkle, F. T., & Eggan, K. (2013). Modeling human disease with pluripotent stem cells: From genome association to function. *Cell Stem Cell*, 656–668. https://doi.org/10.1016/j.stem.2013.05.016

de Miguel, M. P., Fuentes-Julián, S., & Alcaina, Y. (2010). Pluripotent stem cells: Origin, maintenance and induction. *Stem Cell Reviews and Reports*, 633–649. https://doi.org/10.1007/s12015-010-9170-1

Mitalipov, S., & Wolf, D. (2009). 'Totipotency, pluripotency and nuclear reprogramming. *Advances in Biochemical Engineering/biotechnology, 114*, 185–199. https://doi.org/10.1007/10_2008_45

Moodley, Y., et al. (2010). Human amnion epithelial cell transplantation abrogates lung fibrosis and augments repair. *American Journal of Respiratory and Critical Care Medicine, 182*(5), 643–651. https://doi.org/10.1164/rccm.201001-0014OC

Moretti, A., et al. (2010). Mouse and human induced pluripotent stem cells as a source for multipotent Isl1 + cardiovascular progenitors. *The FASEB Journal, 24*(3), 700–711. https://doi.org/10.1096/fj.09-139477

Morizane, R., Monkawa, T., & Itoh, H. (2009). Differentiation of murine embryonic stem and induced pluripotent stem cells to renal lineage in vitro. *Biochemical and Biophysical Research Communications, 390*(4), 1334–1339. https://doi.org/10.1016/j.bbrc.2009.10.148

Morrison, S. J., et al. (1997). Identification of a lineage of multipotent hematopoietic progenitors. *Development, 124*(10), 1929–1939.

Morrison, S. J., & Weissman, I. L. (1994). The long-term repopulating subset of hematopoietic stem cells is deterministic and isolatable by phenotype. *Immunity, 1*(8), 661–673. https://doi.org/10.1016/1074-7613(94)90037-X

Muguruma, Y., et al. (2006). Reconstitution of the functional human hematopoietic microenvironment derived from human mesenchymal stem cells in the murine bone marrow compartment. *Blood, 107*(5), 1878–1887. https://doi.org/10.1182/blood-2005-06-2211

Oishi, K., et al. (2009). Differential ability of somatic stem cells. *Cell Transplantation, 18*(5–6), 581–589. https://doi.org/10.1177/096368970901805-614

Okita, K., et al. (2008). Generation of mouse induced pluripotent stem cells without viral vectors. *Science, 322*(5903), 949–953. https://doi.org/10.1126/science.1164270
Okita, K., et al. (2011). A more efficient method to generate integration-free human iPS cells. *Nature Methods, 8*(5), 409–412. https://doi.org/10.1038/nmeth.1591
Okita, K., et al. (2013). An efficient nonviral method to generate integration-free human-induced pluripotent stem cells from cord blood and peripheral blood cells. *Stem Cells, 31*(3), 458–466. https://doi.org/10.1002/stem.1293
Okita, K., Ichisaka, T., & Yamanaka, S. (2007). Generation of germline-competent induced pluripotent stem cells. *Nature, 448*(7151), 313–317. https://doi.org/10.1038/nature05934
Orlic, D., et al. (2001). Bone marrow cells regenerate infarcted myocardium. *Nature, 410*(6829), 701–705. https://doi.org/10.1038/35070587
Overturf, K., et al. (1997). Serial transplantation reveals the stem-cell-like regenerative potential of adult mouse hepatocytes', *American Journal of Pathology, 151*(5), 1273–1280. Available at: /pmc/articles/PMC1858091/?report=abstract. Accessed on June 25, 2020.
Park, I. H., et al. (2008). Disease-specific induced pluripotent stem cells. *Cell, 134*(5), 877–886. https://doi.org/10.1016/j.cell.2008.07.041
Park, I. S., et al. (2012). Synergistic effect of biochemical factors and strain on the smooth muscle cell differentiation of adipose-derived stem cells on an elastic nanofibrous scaffold. *Journal of Biomaterials Science, Polymer Edition, 23*(12), 1579–1593. https://doi.org/10.1163/092050611X587538
Perin, E. C., et al. (2003). Transendocardial, autologous bone marrow cell transplantation for severe, chronic ischemic heart failure. *Circulation, 107*(18), 2294–2302. https://doi.org/10.1161/01.CIR.0000070596.30552.8B
Prockop, D. J. (1997). Marrow stromal cells as stem cells for nonhematopoietic tissues. *Science*, 71–74. https://doi.org/10.1126/science.276.5309.71
Przekora, A., et al. (2017). Evaluation of the potential of chitosan/β-1,3-glucan/hydroxyapatite material as a scaffold for living bone graft production in vitro by comparison of ADSC and BMDSC behaviour on its surface. *Biomedical Materials (Bristol), 12*(1), 015030. https://doi.org/10.1088/1748-605X/aa56f9
Ratajczak, M. Z., et al. (2012). Pluripotent and multipotent stem cells in adult tissues. *Advances in Medical Sciences*, 1–17. https://doi.org/10.2478/v10039-012-0020-z
Rodriguez, A. M., et al. (2005). Transplantation of a multipotent cell population from human adipose tissue induces dystrophin expression in the immunocompetent mdx mouse. *Journal of Experimental Medicine, 201*(9), 1397–1405. https://doi.org/10.1084/jem.20042224
Roubelakis, M. G., Trohatou, O., & Anagnou, N. P. (2012). Amniotic fluid and amniotic membrane stem cells: Marker discovery. *Stem Cells International*. https://doi.org/10.1155/2012/107836
Safaeijavan, R., et al. (2014). Comparison of random and aligned pcl nanofibrous electrospun scaffolds on cardiomyocyte differentiation of human adipose-derived stem cells. *Iranian Journal of Basic Medical Sciences, 17*(11), 903–911. https://doi.org/10.22038/ijbms.2014.3743
Saha, K., & Jaenisch, R. (2009). Technical challenges in using human induced pluripotent stem cells to model disease. Cell Stem Cells, 584–595. https://doi.org/10.1016/j.stem.2009.11.009
Scadden, D. T. (2014). Nice neighborhood: Emerging concepts of the stem cell niche. *Cell*, 41–50. https://doi.org/10.1016/j.cell.2014.02.013
Schofield, R. (1978a). The relationship between the spleen colony-forming cell and the haemopoietic stem cell. A hypothesis. *Blood Cells*, *4*(1–2), 7–25.
Schofield, R. (1978b). The relationship between the spleen colony-forming cell and the haemopoietic stem cell. A hypothesis. *Blood Cells*, *4*(1–2), 7–25. Available at: https://www.ncbi.nlm.nih.gov/pubmed/747780. Accessed on July 30, 2020.
Shiroi, A., et al. (2005). Differentiation of embryonic stem cells into insulin-producing cells promoted by Nkx2.2 gene transfer. *World Journal of Gastroenterology, 11*(27), 4161–4166. https://doi.org/10.3748/wjg.v11.i27.4161
Sivan, U., Jayakumar, K., & Krishnan, L. K. (2016). 'Matrix-directed differentiation of human adipose-derived mesenchymal stem cells to dermal-like fibroblasts that produce extracellular

matrix. *Journal of Tissue Engineering and Regenerative Medicine, 10*(10), E546–E558. https://doi.org/10.1002/term.1865

Sobhani, A, et al. (2017). Multipotent stem cell and current application. *Acta Med Iran, 55*(1), 6–23. Available at: http://acta.tums.ac.ir/index.php/acta/article/view/4962

Song, J. J., et al. (2013). Regeneration and experimental orthotopic transplantation of a bioengineered kidney. *Nature Medicine, 19*(5), 646–651. https://doi.org/10.1038/nm.3154

Spinelli, V., Guillot, P. V., & De Coppi, P. (2013). Induced pluripotent stem (iPS) cells from human fetal stem cells (hFSCs). *Organogenesis*, 101–110. https://doi.org/10.4161/org.25197

Stadtfeld, M., et al. (2008a). Induced pluripotent stem cells generated without viral integration. *Science, 322*(5903), 945–949.https://doi.org/10.1126/science.1162494

Stadtfeld, M., Brennand, K., & Hochedlinger, K. (2008b). Reprogramming of pancreatic β cells into induced pluripotent stem cells. *Current Biology, 18*(12), 890–894.https://doi.org/10.1016/j.cub.2008.05.010

Strauer, B. E., et al. (2002). Repair of infarcted myocardium by autologous intracoronary mononuclear bone marrow cell transplantation in humans. *Circulation, 106*(15), 1913–1918. https://doi.org/10.1161/01.CIR.0000034046.87607.1C

Sun, N., et al. (2009). Feeder-free derivation of induced pluripotent stem cells from adult human adipose stem cells. *Proceedings of the National Academy of Sciences of the United States of America, 106*(37), 15720–15725. https://doi.org/10.1073/pnas.0908450106

Szade, K., et al. (2018). Where hematopoietic stem cells live: the bone marrow niche. *Antioxidants and Redox Signaling*, 191–204. https://doi.org/10.1089/ars.2017.7419

Takahashi, K., et al. (2007). Induction of pluripotent stem cells from adult human fibroblasts by defined factors. *Cell, 131*(5), 861–872. https://doi.org/10.1016/j.cell.2007.11.019

Takahashi, K., & Yamanaka, S. (2006). Induction of pluripotent stem cells from mouse embryonic and adult fibroblast cultures by defined factors. *Cell, 126*(4), 663–676. https://doi.org/10.1016/j.cell.2006.07.024

Takasato, M., et al. (2014). Directing human embryonic stem cell differentiation towards a renal lineage generates a self-organizing kidney. *Nature Cell Biology, 16*(1), 118–126. https://doi.org/10.1038/ncb2894

Takasato, M., et al. (2015). Kidney organoids from human iPS cells contain multiple lineages and model human nephrogenesis. *Nature*. https://doi.org/10.1038/nature15695

Takebe, T., et al. (2013). Vascularized and functional human liver from an iPSC-derived organ bud transplant. *Nature, 499*(7459), 481–484. https://doi.org/10.1038/nature12271

Tashiro, K., et al. (2009). Efficient adipocyte and osteoblast differentiation from mouse induced pluripotent stem cells by adenoviral transduction. *Stem Cells, 27*(8), 1802–1811. https://doi.org/10.1002/stem.108

Tateishi-Yuyama, E., et al. (2002). Therapeutic angiogenesis for patients with limb ischaemia by autologous transplantation of bone-marrow cells: A pilot study and a randomised controlled trial. *Lancet, 360*(9331), 427–435. https://doi.org/10.1016/S0140-6736(02)09670-8

Tateishi, K., et al. (2008). Generation of insulin-secreting islet-like clusters from human skin fibroblasts. *Journal of Biological Chemistry, 283*(46), 31601–31607. https://doi.org/10.1074/jbc.M806597200

Thoma, E. C., et al. (2012). Parallel differentiation of embryonic stem cells into different cell types by a single gene-based differentiation system. *Cellular Reprogramming, 14*(2), 106–111. https://doi.org/10.1089/cell.2011.0067

Tirkkonen, L., et al. (2013). Osteogenic medium is superior to growth factors in differentiation of human adipose stem cells towards bone-forming cells in 3D culture. *European Cells and Materials, 25*, 144–158. https://doi.org/10.22203/ecm.v025a10

Uccelli, A., Moretta, L., & Pistoia, V. (2008a). Mesenchymal stem cells in health and disease. *Nature Reviews Immunology*, 726–736. https://doi.org/10.1038/nri2395

Uccelli, A., Moretta, L., & Pistoia, V. (2008b). Mesenchymal stem cells in health and disease. *Nature Reviews Immunology*, 726–736. https://doi.org/10.1038/nri2395

De Ugarte, D. A., et al. (2003). Comparison of multi-lineage cells from human adipose tissue and bone marrow. *Cells, Tissues, Organs, 174*(3), 101–109. https://doi.org/10.1159/000071150
Völkner, M., et al. (2016). Retinal organoids from pluripotent stem cells efficiently recapitulate retinogenesis. *Stem Cell Reports, 6*(4), 525–538. https://doi.org/10.1016/j.stemcr.2016.03.001
Wang, C., et al. (2010). Differentiation of adipose-derived stem cells into contractile smooth muscle cells induced by transforming growth factor-β1 and bone morphogenetic protein-4. *Tissue Engineering—Part A, 16*(4), 1201–1213. https://doi.org/10.1089/ten.tea.2009.0303
Wang, Z., et al. (2012). Distinct lineage specification roles for NANOG, OCT4, and SOX2 in human embryonic stem cells. *Cell Stem Cell, 10*(4), 440–454. https://doi.org/10.1016/j.stem.2012.02.016
Ware, B. R., Berger, D. R., & Khetani, S. R. (2015). Prediction of drug-induced liver injury in micropatterned co-cultures containing iPSC-derived human hepatocytes. *Toxicological Sciences, 145*(2), 252–262. https://doi.org/10.1093/toxsci/kfv048
Wernig, M., et al. (2008). Neurons derived from reprogrammed fibroblasts functionally integrate into the fetal brain and improve symptoms of rats with Parkinson's disease. *Proceedings of the National Academy of Sciences, 105*(15), 5856–5861. https://doi.org/10.1073/pnas.0801677105
Wickham, M. Q., et al. (2003). Multipotent stromal cells derived from the infrapatellar fat pad of the knee. *Clinical Orthopaedics and Related Research, 412*, 196–212. https://doi.org/10.1097/01.blo.0000072467.53786.ca
Woltjen, K., et al. (2009). PiggyBac transposition reprograms fibroblasts to induced pluripotent stem cells. *Nature, 458*(7239), 766–770. https://doi.org/10.1038/nature07863
Wuchter, P., Wagner, W., & Ho, A. D. (2011). Mesenchymal stem cells: An oversimplified nomenclature for extremely heterogeneous progenitors. *Regenerative Medicine: From Protocol to Patient*, 377–395. https://doi.org/10.1007/978-90-481-9075-1_16
Xie, C. Q., et al. (2009). A comparison of murine smooth muscle cells generated from embryonic versus induced pluripotent stem cells. *Stem Cells and Development, 18*(5), 741–748. https://doi.org/10.1089/scd.2008.0179
Xu, D., et al. (2009). Phenotypic correction of murine hemophilia a using an iPS cell-based therapy. *Proceedings of the National Academy of Sciences of the United States of America, 106*(3), 808–813. https://doi.org/10.1073/pnas.0812090106
Yao, S., et al. (2006). Long-term self-renewal and directed differentiation of human embryonic stem cells in chemically defined conditions. *Proceedings of the National Academy of Sciences of the United States of America, 103*(18), 6907–6912. https://doi.org/10.1073/pnas.0602280103
Ye, L., et al. (2009a). Induced pluripotent stem cells offer new approach to therapy in thalassemia and sickle cell anemia and option in prenatal diagnosis in genetic diseases. *Proceedings of the National Academy of Sciences of the United States of America, 106*(24), 9826–9830. https://doi.org/10.1073/pnas.0904689106
Ye, Z., et al. (2009b). Human-induced pluripotent stem cells from blood cells of healthy donors and patients with acquired blood disorders. *Blood, 114*(27), 5473–5480.https://doi.org/10.1182/blood-2009-04-217406
Yin, A. H., et al. (1997). AC133, a novel marker for human hematopoietic stem and progenitor cells. *Blood. American Society of Hematology, 90*(12), 5002–5012. https://doi.org/10.1182/blood.v90.12.5002
Yu, J., et al. (2007). Induced pluripotent stem cell lines derived from human somatic cells. *Science, 318*(5858), 1917–1920. https://doi.org/10.1126/science.1151526
Yusa, K., et al. (2009). Generation of transgene-free induced pluripotent mouse stem cells by the piggyBac transposon. *Nature Methods, 6*(5), 363–369. https://doi.org/10.1038/nmeth.1323
Zhang, Y., et al. (1994). Positional cloning of the mouse obese gene and its human homologue. *Nature, 372*(6505), 425–432. https://doi.org/10.1038/372425a0
Zhou, H., et al. (2009). Generation of induced pluripotent stem cells using recombinant proteins. *Cell Stem Cell*, 381–384. https://doi.org/10.1016/j.stem.2009.04.005
Ziegler, B. L., et al. (1999). KDR receptor: A key marker defining hematopoietic stem cells. *Science, 285*(5433), 1553–1558. https://doi.org/10.1126/science.285.5433.1553

Zipori, D. (1992). The renewal and differentiation of hemopoietic stem cells. *The FASEB Journal, 6*(9), 2691–2697. https://doi.org/10.1096/fasebj.6.9.1612293

Zuk, P. A., et al. (2001). Multilineage cells from human adipose tissue: Implications for cell-based therapies. *Tissue Engineering, 7*(2), 211–228. https://doi.org/10.1089/107632701300062859

Chapter 3
Isolation and Characterization of Stem Cells

Hassan Ahmed Khan

Abstract Embryonic stem cells can be differentiated into all body cells. It was revealed that human embryonic stem cells can only be isolated from inner cell mass of human embryo. Several methods were practiced for isolation of inner cell mass from human embryo including mechanical dissection, laser-assisted dissection and immunosurgical procedures which are preferred on the basis of advantages. Characterization of stem cells is based on the creation of embryoid bodies is a routine process used to classify human-induced pluripotent stem cells into separate lineages of cells. Embryoid bodies can be formed by suspension culture, hanging drop culture and semisolid media. Growth factors play regulatory roles in cellular functions, comprises adhesion, proliferation, migration, matrix synthesis and cell differentiation are potential agents to target specific tissue reactions. Stem cells can also be characterized at cellular level by immunofluorescent techniques and at molecular level by PCR and RT-PCR and by formation of teratoma and genetic analysis (karyotyping).

Keywords Mechanical isolation · Laser isolation · Microdissection · Immunosurgery · Trophoblast · Embryoid body · Suspension culture · Immunofluorescence · Genetic analysis

Discovery of Isolation and Characterization Methods

First isolation of embryonic stem cells was made in 1981 from mouse embryo; Gail R. Martin coined the term "Embryonic stem cells" for the first time. In 1998, Thomson and his team used a technique to isolate human embryonic stem cells from human embryo. Thereafter, it was revealed that these embryonic stem cells can differentiate into all body cells including neural cells, hepatic cells, cardiomyocyte and beta cells of the islets of Langerhans. It was challenging for the researchers to find out technique of isolation, later it was revealed that hESCs can be exclusively isolated from inner cell mass (ICM) of human embryo. It was reported that both fresh and frozen embryo

H. A. Khan (✉)
Amson Vaccines and Pharma PVT Ltd., Islamabad, Pakistan
e-mail: hassanahmed@amson.org.pk

F. A. Khan (ed.), *Advances in Application of Stem Cells: From Bench to Clinics*, Stem Cell Biology and Regenerative Medicine 69,
https://doi.org/10.1007/978-3-030-78101-9_3

can yield inner cell mass. Thereafter, several methods were practiced for isolation of ICM from human embryo including mechanical dissection, laser-assisted dissection and immunosurgical procedures (Khan et al., 2018).

Isolation Techniques

Tissue reconstruction can be obtained naturally by isolation and characterizing stem cells to mature cells. Different isolation techniques are deployed for stem cells.

Mechanical Isolation/Dissection

Human embryonic stem cells (hESCs) is subcultured by mechanical dissection to maintain cell-to-cell interaction and an optimal cluster size. This is a preferred method for the progression of hESC lines, at least in the initial stages of the culture of cell lines.

Methodology

As a process for hESC derivation, mechanical dissection and mechanical isolation of the ICMusing flexible metal needles with sharpened tips have been used.

Mechanical Dissection of Colony with hESCs

The size of colony differs. Every colony usually comprises of about 30,000–50,000 cells. Prior to dissection, differentiated cells located at the margin of the colony will be separated. Using a cutting instrument, the colony is dissected by horizontal and vertical cuts into microscopic clumps (usually about 500–1000 cells). The clumps or clusters are then moved to a capillary before passage or cryopreservation and deposited in medium (Hunt, 2007).

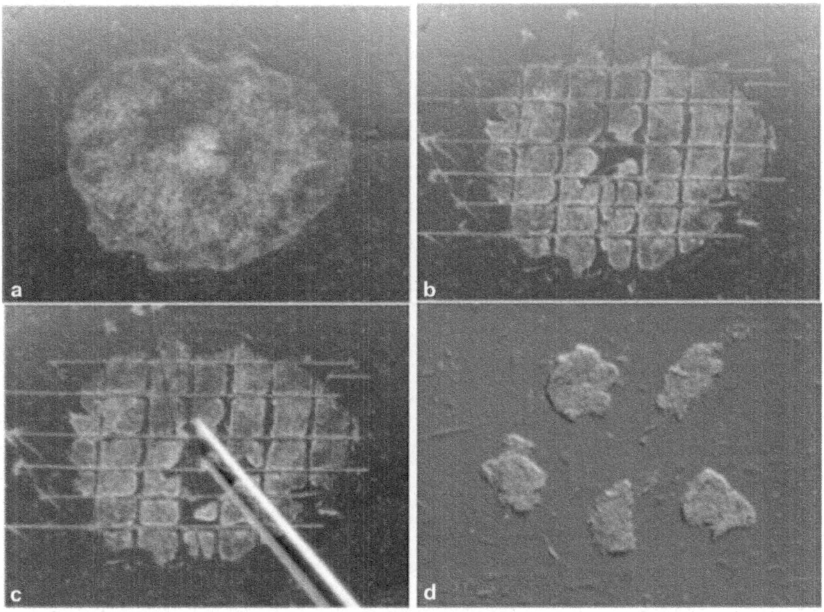

(Photographs courtesy of P. Timmons, copyright NIBSC)

Advantages: This mechanical transfer method is useful during the early phases of hESCs because the hESCs also differ very quickly. When producing consistent sizes of embryoid bodies or obtaining similar sized of hESC colonies, this is especially advantageous (Oh et al., 2005). The mechanical dissection of the ICM has proven to be an efficient method of deriving new hESC lines. This technique is swift and needs no xeno components (Khan et al., 2018).

Disadvantages: The process of mechanical transfer is laborious and time-consuming. Since some experiments require larger volumes of hESCs, this technique is less than optimal (Oh et al., 2005).

Laser Isolation

Methodology

Laser-assisted ICM (LCM) isolation was carried out as described: Two holding pipettes were held by blastocycts with the ICM placed at 9 o'clock. Approximately, ten infrared laser pulses were thrown to separate the blastocyst into two unequal pieces once adequate tension was applied, the smaller comprising of ICM, the larger comprising exclusively of trophoblast (Tanaka et al., 2006).

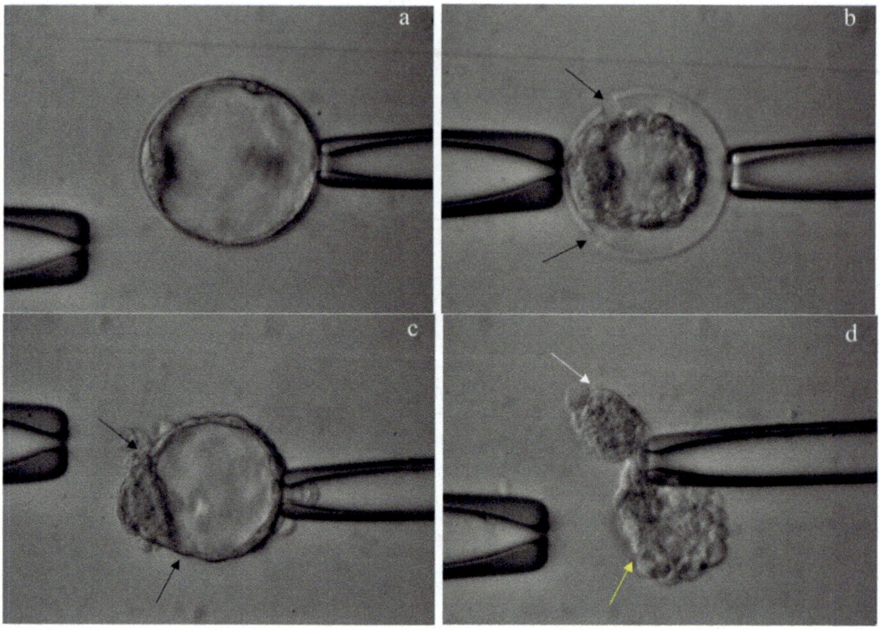

Advantages: There are several benefits of LCM. It allows an extremely limited number of cells to be precisely segregated (i.e., a homogenous cell population from a heterogeneous population) or isolated from a single cell. Cell organelles can be isolated these days due to the recent improvement of optic resolution in LCM. LCM also allows live cells or single cell in a culture dish to be isolated and recultured. It may also retain the morphology of the tissue during dissection. LCM is therefore a faster method of cell separation than other methods of microdissection, which is important for genomic molecules to be preserved (Chung & Shen, 2015).

Disadvantages: While LCM has brought a number of advantages to biomedical research, there are still drawbacks to it. Lastly, it is one of the most costly instruments. LCM will cost more than a million dollars with a microscope, and the rest of the consumables are more than 5 times more costly than usual consumables, such as nuclease-free membrane slides and tube collection. Also, due to its exposure to fixatives and staining reagents, there is a significant risk that the consistency of microdissected tissues cannot meet the standard quality for further study. Therefore, the durability of cellular materials as well as the morphology of the section can also be affected by the dehydration of sections caused by the absence of cover slips and the mounting medium on sections. To ensure quality outcomes, researchers should also consider investing a considerable amount of time on training and troubleshooting (Chung & Shen, 2015).

Immunosurgery

In most laboratories in which Embryonic stem cells derivation takes place, the isolation of ICM by immunosurgery (a technique that was introduced by Solter and Knowles in 1975) has become a standard practice (Kim et al., 2005).

Methodology

In a solution containing antibodies against humans from whom the embryo is derived, the blastocyst is incubated as a whole. The antibodies will bind to the embryo's trophectoderm cells but not to the inner cell mass as the antibodies are unable to reach the layer of epithelial trophectoderm cells covering the inner cell mass. In this way, the cells of inner cell mass are covered. If the embryo is subsequently exposed to complementary blood serum factors, they accept the attached antibodies and cause an immune response in which the complement continues to act as a protease. Enzymes that break down proteins are proteases. Thus, the trophectoderm is directly damaged by the action of protease but the inner cell mass cells stay unchanged because they do not have bonded antibodies and thus do not activate proteases. In a pipette, the internal cell mass cells can be harvested and moved for culture in vitro to a plastic dish. However, there is an inherent tendency for the inner cell mass to divide into the specialized forms of cells that make up the fetus (Melton, 2007).

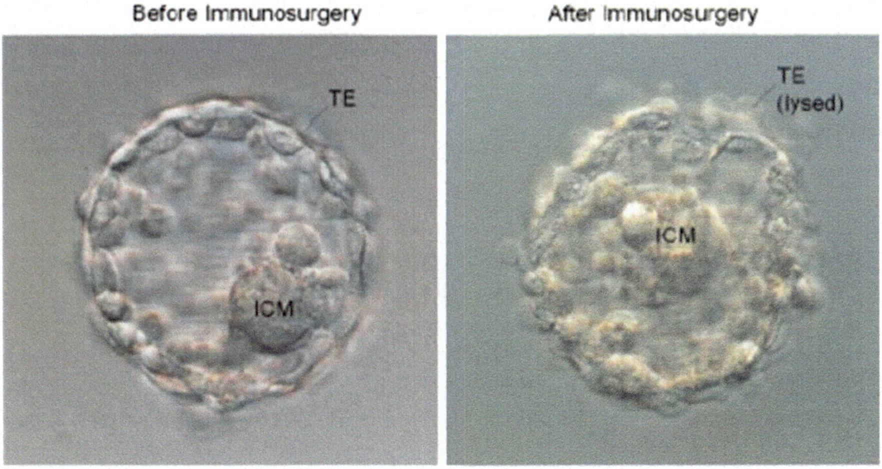

Advantages: It gives high rate of ICM isolation (Khan et al., 2018).

Disadvantages: However, in all situations, immunosurgery may not be the option of choice according to our experience. Not all embryos develop into blastocysts that are distinguished by large distinct ICMs that are well extended. In certain cases, smaller ICMs are found to possess the embryos. Blastocysts also exhibit poorly described ICMs in many situations (Kim et al., 2005). Animal-derived antibodies have been deployed, making them less appropriate for medicinal application (Findikli et al., 2006). Popular ICM isolation immunosurgical procedures require the use of allogenic antibodies and supplements that can also add unnecessary epitopes that make some ESC derivatives unfit for more applications (Tanaka et al., 2006).

Microdissection

Laser capture microdissection (LCM) is a method used to split target cells or regions often from solid tissue samples using a narrow laser beam. LCM can separate cell populations from heterogeneous mixtures by microscopic visualization using cell morphology or particular histological and immunological staining. When dealing with limited sample sizes, LCM is especially helpful. First mentioned in the early twentieth century, laser dissection technology essentially entails splitting areas or cells of our interest using a narrow beam laser from tissue parts. Liotta and Emmert-Buck later established a more advanced, microscope-based method of laser dissection technique called laser capture microdissection (LCM) in the NIH to separate pure cell populations from heterogeneous tissues based on their natural morphology or unique histological/immunological staining. LCM has allowed researchers to isolate tumors with high precision from healthy tissues, stroma stem cells and parenchyma epithelial cells. LCM may also be used to separate single cells in combination with high precision surgical techniques. For applications such as proteomics, genomics and molecular characterization of cancer cells, diagnostic pathology, tissue regeneration and basic cell biology, sophisticated LCM devices have been developed.

Methodology

LCM instrument comprises inverted microscope, a laser diode, a laser control unit, a joystick-controlled stage, a CCD camera and a computer for additional controls and image visualization. LCM devices can be operated robotically as well as manually. LCM systems have two types, infrared (IR) and ultraviolet (UV) depending on the type of laser. The minimum LCM microscope laser beam diameter is 7.5 μm and the mean diameter of about 30 μm. The tissues are possibly heated to a temperature of 90 °C at most only a few milliseconds as well maintaining the molecular structure and macromolecules intact. Based on the type of laser used, there are two major categories of LCM platforms—UV and IR. The Arcturus Veritas TM instrument is also available as a hybrid automated IR/UV system (Choudhury, 2017).

Advantages: It allows an exceedingly limited number of cells to be accurately segregated (i.e., a homogeneous cell population from a heterogeneous population) or isolated from a single cell, and cell organelles can be separated these days due to recent increase in optical resolution of laser capture microdissection (LCM). It allows live cells or a single cell in a culture dish to be isolated and recultured. It may also retain the shape of the tissue during dissection. It is therefore a faster method of cell isolated than other types of microdissection, which is important for genomic molecules to be retained (Shen, 2015).

Disadvantages: While LCM has brought a range of advantages to biomedical research, there are also drawbacks to it. Lastly, it is one of the most expensive instruments. LCM with a microscope will cost more than a million dollars, and the majority of the consumables, such as nuclease-free membrane slides and collecting tubes are more than 5 times more expensive than regular consumables. Often due to its sensitivity to fixatives and staining reagents, there is a significant danger that the consistency of microdissected tissues cannot meet the standard quality for further study. Therefore, the durability of cellular materials as well as the morphology of the segment can also be influenced by the dehydration of sections caused by the absence of cover slips and the mounting medium on sections. Therefore, researchers should consider investing considerable time on testing and troubleshooting in order to get quality results (Shen, 2015).

Minimized Trophoblast Cell Differentiation

This technique is used to derive hESCs from both normal and abnormal embryos can be used successfully under feeder-free conditions with higher production.

Methodology

On plates coated with either CELLstart or Matrigel in Nutristem medium under hypoxic conditions, the ICMs isolated by MID were cultured. The ICM clump was manually removed from the adjacent TE cells after 3 days of cultivation and moved to a new culture plate containing human foreskin fibroblasts (HFFs). At this phase, every day during the entire cultural period, half of the cultural media was restored by fresh medium. The culture was analyzed regularly, and mechanically the cells were subcultured on reaching the colonies on optimal size (Laowtammathron et al., 2018).

Advantages: In contrast to other methodologies, this approach can be successfully used to extract hESCs from both normal and abnormal embryos under feeder-free conditions with better quality. With this MTP system, under feeder-free conditions in a well-defined, xeno-free medium, hESC lines can be successfully derived (Laowtammathron et al., 2018).

Disadvantages: The only disadvantage of this technique is its success rate which is 50% (Khan et al., 2018).

Characterization Techniques

Formation of Embryoid Body (EB)

EBs are three-dimensional cell aggregates that are combination of the all three germ layers. The undifferentiated HIPSCs are put in suspension in this method, which facilitates stochastic differentiation into cells on all three germ layers. The creation of EBs is a usual process used to classify human-induced pluripotent stem cells (hiPSCs) into separate lineages of cells. One of the key benefits of this technique is that it is carried out in vitro using normal procedures and resources for tissue culture, thus eliminating the regulatory concerns and substantial costs involved with the maintenance of immune-deficient mice. In the following pages, the various schemes that have been evolved to produce EBs along with the existing analyses deployed for their pluripotent evaluation are outlined (Sheridan et al., 2012). Method of producing EBs varies in their ability to shape appropriate size aggregates and their long-term viability maintenance.

Suspension Culture

The suspension culture is an ingenious way to either grow or differentiate cells, and these are often incorporated into a single bioprocess. The different characteristics of cells, static suspension culture and rotary suspension culture are shown in Table 3.1.

Methodology

Large number of EBs can be produced by using static suspension culture through inoculating suspension of embryonic stem (ES) cells on petri plates that can be bacteriological grade, ultra-low adherence or plate coated using cell adhesion inhibitor like

Table 3.1 Characteristics of static suspension culture and rotary suspension culture

Characteristic		Static suspension culture	Rotary suspension culture
Number of cells		Initiated with 4×10^5 cells per milliliter	Initiated with 4×10^5 cells per milliliter
EBs formation	02 days	Smaller and spherical	Regular, spherical after 2 days
	04 days	Fewer, smaller spherical EBs	Regular, spherical and larger EBs
	07 days	Spherical, uniform and larger in size than at day 4	Regular, spherical, uniform and larger in size
Cell aggregates		Smaller and less defined	Larger, irregular shaped

poly 2-hydroxyethyl methacrylate that allowed the cells to aggregate into spheroids. This method is simple enough but could not give enough control over size and shape of EBs. This ranked into periodic agglomeration of EBs in the form of large, irregular masses because ES cells have maximum probability of encountering. A drawback of this technique is that EBs may attach prematurely onto the plate due to surface chemistry of vessel resulted to loss of EBs from suspension culture. This method has gained its importance in some applications like differentiation of ES cells into neuronal lineage (Rungarunlert et al., 2009).

Advantages: Suspension culture can be employed to generate EBs in large volumes (hundreds to thousands) by dispensing ES cells on a non-adherent surface in a differentiation medium with a density of 10^4–10^6 cells per milliliter, whereupon the cells impulsively assemble into spheroids. Although this technique can be used to generate more EBs and separated cells than hanging drop technique, it has no control over the size and form of EBs and sometimes ends in agglomeration into large, irregular masses (Carpenedo et al., 2007).

Disadvantages: EBs in suspension culture tend to agglomerate, dissociate and bind to the culture dish resulting in low efficiency of EB formation and poor viability (Pettinato et al., 2015). The suspension culture in neoplastic stem cells differs from the initial neoplastic stem cells and not all cells that share the characteristics of the neoplastic stem cells can live in the suspension culture (Zhao & Zheng, 2012).

Hanging Drop Culture

A commonly used EB formation introduction process is the hanging drop culture. The rounded bottom of the handing drop causes ES cells to be aggregated, which can provide a good environment for mesenchymal embryonic stem cells to form EBs. By altering the number of cells in the early cell suspension to be suspended as a drop from the lid of the petri dish, the number of embryonicstem cells aggregated in a hanging drop can be restrained. We can repeatedly form homogenous EBs from a predetermined number of embryonic stem cells using this technique (Wang & Yang, 2008).

Methodology

By distributing equal numbers of embryonic stem cells onto physically isolated droplets of media suspended from the lid of a petri dish, the hanging drop approach offers standardized sizes of EBs. This approach provides a similar environment by gravity-induced aggregation of the cells for the creation of individual EBs within each decline. For this cause, abundant cell types such as neuronal cells, hematopoietic cells, vascular cells, cardiomyocyte and chondrocytes have been produced by this technique. For the evaluation of molecular processes arising in early embryogenesis of every cell type, the hanging drop procedure is extremely useful. However because of its laborious nature, this technique is primarily employed for testing purposes and is not ideal for large-scale EB production; a standard 100 mm petri dish may comprise no more than 100 drops, and each drop typically produces only one EB. Other drawbacks of this approach include significant problems with the limited amount of medium (less than 50 μl that can quickly evaporate) being shared or exploited without disrupting the EBs. The hanging drop process typically comprises two steps: the aggregation of ES cells in drops and the maturation of EB aggregates using low adherence bacterial petri dishes in suspension culture. Various elements of the process may be problematic, like loss of EBs during pipette pickup of the formed EBs and premature EBs attachment to petri dishes (Rungarunlert et al., 2009).

Advantages: The greatest attraction of hanging drop cultures is the need for just a limited volume of tissue. This is especially functional in the field of clinical testing in which biopsy tissue supply is minimal. In addition, the volume of culture medium necessary is small in such a way that only a small fraction of chemical or biological reagents are needed to alter the tissue culture environment. Another major value of the hanging drop culture is that all forms of cells are retained in the similar microenvironment in which cell–cell associations are unchanged. Special equipment or reagents are also not needed for hanging drop culture. The ease of stimulating the results of advanced 3D cellular culture is especially enticing (Wang & Yang, 2008).

Disadvantages: The drawback of this method is that the number of cells would not regularize in each microwell, which will allow the spheroid's diameter to be uneven. In the other side, the spheroids on the same device could not be interrogated by most current instruments. Such instruments also require the manual handling of samples, which may be vulnerable to variability and error (Huang et al., 2020). The limitations of the hanging drop system are as follows: Due to holding hanging drop on the lid by surface friction, the liquid volume of a drop is reduced to less than 50 μl, and it is difficult to adjust the medium for hanging drops. It is also very difficult during cultivation to detect the development of EBs in drops directly using microscopy. In addition, the hanging drop technique consists of two stages, so a sequence of steps can be problematic for the hanging drop technique (Yang, 2008).

Semisolid Media

In order to physically separate individual cells and provide a niche-like environment for EB development from individual hESCs, single cell suspension was encapsulated/cultured in a 3D semisolid matrix, such as a hydrogel (Pettinato et al., 2015).

Methodology

In double deionized water (ddH20) sterilized by autoclaving, a type A gelatin stock solution was prepared. This gelatin was pipette onto 48-well tissue culture plates in each well and incubated overnight at 4 °C. In phosphate-buffered saline (PBS), cross-linked glyceraldehyde was prepared. The gelatin-polymerized solution of cold glyceraldehyde was applied for around 24 h. In tissue culture plates, the cross-linking was done at 4 °C. A yellow color shift suggested gelatin cross-linking (Zur Nieden et al., 2015).

Advantages: Embryoid body structure in the agarose-based 3D environment showed greater efficiency with increased durability of individual EBs and lack of EB agglomeration into large clusters or fragmentation into minute non-proliferative clusters or single cells during sustained in vitro culture compared to normal suspension culture in petri dishes. Likewise, the addition of 1% methylcellulose to the hESC suspension culture media produces semisolid media, which is major to isolation of cells and developing suspended single-cell cloned EBs (Pettinato et al., 2015).

Disadvantages: Low EB yield due to the intrinsic instability of single hiPSCs during extended culture, the obstruction of mass transport requisite for efficient soluble factor treatments, as well as the complex process for differentiated cell retrieval are common problems of the semisolid culture system for EB formation (Pettinato et al., 2015).

Role of Growth Factors

The fact that various environmental factors play to the overall control of stem cell pursuit is widely accepted. Growth factors (GFs) because of their regulatory part in cellular functions, including adhesion, migration, proliferation and matrix synthesis and cell differentiation are potential agents to target specific tissue reactions (Gonçalves et al., 2013).

Non-Specific Growth Factors

Retinoid Acid

Retinoid acid (RA) is a vitamin A metabolite necessary for premature embryonic development and facilitates the specification of the neural lineage of stem cells, however limited is known about the influence of RA on mRNA transcription and microRNA levels on differentiation of embryonic stem cells (Zhang et al., 2015).

Nerve Growth Factor (NGF)

In the production and growth of nerve cells, NGF plays a significant role. Nerve growth factor (NGF) causes neuronal cell proliferation, division, survival and death.

NTs also mediate higher brain processes like learning and memory in addition to their proven roles for cell survival (Heese et al., 2007).

Lineage-Specific Growth Factors

EGF

EGF is a significant growth factor and plays an important role in the maintenance of stem cell proliferation and differentiation (Huang et al., 2020). EGF is a dynamic mitogen involved in MSCs and the proliferation of fibroblasts. In the MSC population, EGF treatment also preserves early progenitors and increases the paracrine activity of stem cells (Gonçalves et al., 2013).

FGF

FGF signaling is needed in a variety of lineages for the early stages of differentiation and is also an integral mediator of self-renewal in human stem cells (Gonçalves et al., 2013). FGF-2 is primarily expressed in mesoderm and neuroectoderm tissues and is believed to play an important role in mesoderm induction. A number of studies in recent years have identified FGF-2 fibroblast growth factors as a central regulator of a range of types of stem cells. A family of growth factors that play their role wound healing and angiogenesis are fibroblast growth factors (FGFs). In MSC-related studies showing elevated rabbit, canine and human MSC proliferation in vitro, used among the different members of this family, with the mitogenic effect being more expressed when MSCs are seeded at lower densities. Not only does b-FGF sustain the capacity for MSC proliferation, it also maintains the capacity for adipogenic, osteogenic and chondrogenic differentiation via the early mitogenic cycles; finally, however, all the MSCs separate into the chondrogenic band. In different cell types, FGF signals proliferation via the MAPK cascade (Rodrigues, 2010).

TGFβ

A growth factor that helps facilitate cell proliferation and the development of extracellular matrix is TGF-β. In the presence of a chondrogenic medium containing TGF-β1 in the 3D culture, autologous chondrocytes obtained from MSCs may undergo extension and proliferation (Kanitkar et al., 2011). TGF-beta signaling also regulates the expression of a plethora of homeostatic genes in most cell types, whose activity influences cell proliferation, development of extracellular matrix, secretion of paracrine factor, cell–cell interactions, immune response and repair of tissue. Three isoforms occur in TGFβ (TGFβ1, TGFβ2 and TGFβ3). Although all three isoforms induce MSC proliferation and chondrocyte formation, it have been observed that TGFβ3 has the most expressed effect on chondrogenesis and continually elevate MSC proliferation, making it a primary factor in the implanted MSC induction of chondrogenesis (Rodrigues, 2010).

HGF

In mouse MSCs, hepatocyte growth factor (HGF) and its c-Met receptor are exhibited at low levels. Although the low levels of HGF present in culture media are inadequate to trigger the receptor, receptor activation, affecting proliferation, migration and differentiation are triggered by the exogenous addition of HGF to MSCs. Interestingly, Ras-ERK and PI3K-Akt are activated by short-term exposure of HGF in MSCs; these are the major HGF-activated pathways in other cell types. Long-term exposure to the growth factor prevents mutagenesis, despite the activation of these pathways. In addition, penetration contributes to cytoskeletal rearrangement, migration of cells and cardiac marker expression (Rodrigues, 2010).

BMP4

A part of the transforming growth factor-βsuper family of secreted signaling molecules, bone morphogentic protein 4 (BMP4), controls the differentiation of cell proliferation, apoptosis and determination of cell fate in mammalian development. Among many tissues that are dependent on BMP4 in the embryo, hematopoietic cells are BMP4 precisely controls mesodermal cell devotion to the hematopoietic family in such a way that basic hematopoiesis does not exist in embryo without BMP4 (Goldman et al., 2009) Bone morphogenetic proteins (BMPs), especially BMP4, are key players in controlling the formation of neuronal and glial cells in the embryonic, postnatal and injured CNS from neural precursor cells (Cole et al., 2016). Bone morphogenetic protein 4 promotes the neuronal differentiation of neuronal stem cells through the ERK pathway.

Activin-A

Activin-A facilitates the division of human embryonic stem cells into endoderm and pancreatic b-cells. Activin-A (5 ng/ml) was proved to sustain self-renewal and to facilitate the long-term feeder-free culture of human ES cells (Tomizawa et al., 2011).

Cellular Characterizations of Stem Cells Using Cell Surface Markers by Immunofluorescence Technique

Using antigen-specific antibodies, immunocytochemistry is a short and simple way to assess whether a cell population is homogenous or heterogeneous with respect to a specific molecular marker. Immunocytochemistry allows for the visualization within a colony or culture of individual cells and thereby provides an overall measure of the expression of a particular marker under specific culture conditions (Nethercott et al., 2011). Detailed analysis of CD cell surface markers has drawn attention to the broad potential therapeutic applications of MSCs. Indeed, two familiar unique markers of mesenchymal cell lineages are CD90 and CD105 and CD34, and CD45 are referred to as hematopoietic markers. The expression of CD44 and CD166 in human MSCs has also been shown.

Methodology

Seed and culture cells in a 24-well plate until ready for ICC analysis. Wash each well 3 times with PBS at room temperature. Fixation of wells is made with the addition of 4& paraformaldehyde in phosphate buffer saline (PBS), and incubation is done at room temperature. Wells are washed with PBS. PBS is aspirated by the addition of Triton-X-PBS and incubating the wells at room temperature. Reaction is blocked by adding blocking buffer following incubation. Primary and secondary antibodies are added following incubation. PBS is employed for washing. Few drops of mounting medium are added to each well to stain the nuclei and preserve the samples for fluorescence microscopy imaging.

Advantages: This technique reveals localization of marker proteins. It can evaluate various markers simultaneously. This method is more effective than Western blot analysis. It can use live or fixed cells.

Limitations: It requires specialized equipment. There is potential for cross-reactivity while using multiple antibodies. It is potentially auto-fluorescence and photo-bleaching. This technique is more time taking than flow cytometer.

Molecular Characterization of Stem Cells by Transcription Factors Using PCR and RT-PCR Technique

In all human embryonic stem cells (hESCs) and cancer stem cells, transcription factors are expressed and play a key role in preserving stem features such as self-renewal and pluripotency (Sneha et al., 2019). Molecular assays depend on comparative benchmarking as a practical surrogate of in vitro and in vivo-derived cells; this is where transcription profiling is most often adopted. The goal of these studies is to compare and contrast in vitro cells with their in vivo counterparts, classify main cell-type transcriptional drivers and often make cell fate predictions. Quantitative reverse transcription PCR (RT-qPCR) and digital reverse transcription PCR produce rapid, responsive and quantitative methods for tracking any cell population's gene expression profile. The levels of transcription factors such as Oct4, NANOG, SOX2 and other synergistic factors decide if pluripotency is determined (Li, 2010). Multiplexed RT-qPCR experiments and panels offer an important means for transcription factors, and kinases and other molecules contribute in both pluripotency conservation and differentiation to be screened and quantified. The ability to do multiplex transcriptome analysis improves the sensitivity of cell status determination.

With lower concentrations of input nucleic acid, digital PCR will improve the precision of measurements.

Methodology

Human Embryonic Stem Cell Cultures and Sample Preparation

In knockout DMEM supplemented with KO serum substitution, antibiotics L-glutamine, simple fibroblast growth factor, non-essential amino acids, culture medium and beta-mercaptoethanol are incubated at 37 °C with 5% CO_2 in 6-well dishes or flasks on an approaching confluent sheet of mitomycin C treated mouse embryonic fibroblasts. By adding retinoic acid to the culture medium and removing b-FGF, differentiation of hES cells was induced. Trypsin-EDTA and glass beads were used to collect the cells. The heS cells were obtained at intervals for the experiment. There were approximately 2×105 cells separated at each time point. The hES cells using collagenase and glass beads were broken two days before the start of differentiation trial.

Microscopy: With a light microscope and a camera, phase contrast and bright field shot of the hES cell culture were procured.

RNA Isolation and RNA Quality Assessment

The cells were promptly re-suspended in TRIzol after isolation and preserved at −80 °C. Chloroform was added to the thawed samples for RNA isolation with successive step separation and purification using an RNeasy mini pack. RNA was eluted and condensed using spin columns following DNase treatment and a washing stage. For a representative collection of samples, RNA consistency was evaluated by means of microfluidic capillary electrophoresis. After calculating RNA concentrations using an RNA pack, an RNA high sense chip was used to calculate the rRNA ratio and the RNA consistency index (RQI).

cDNA Synthesis

A kit of oligo (dT) primers was used to synthesize complementary DNA (cDNA). Using a spectrophotometer, the cDNA concentration was measured with assDNA Assay package. Samples were processed at −20 °C afterward.

Reverse Transcription Quantitative PCR

For RT-qPCR, there were two related devices used. In a 384-well plate, cDNA was combined with mastermix for any reaction. Each reaction consisted of a 96-well plate, cDNA and mastermix using the sequence detection system. Primers, iTaqsupermix with ROX and water are part of the mastermix. Probes were used depending on the locus of interest or iTaqsupermix containing SYBR green was employed. They developed and tested the primers. For the study of POU5F1 and NANOG, usable TaqMan assays were used. All reactions in duple were conducted, and no template controls for all genes were used.

Data Analysis

Using the application in the program, stability analysis of the multiple sources was carried out. Using the program, relative quantification of the pluripotency markers data (Oct4 and Nanog) was measured. Doing a variance analysis (ANOVA) in R, the relative quantification data for both normalization strategies were statistically evaluated (Vossaert et al., 2013).

Advantages: Qualitative polymerase chain reaction (RT-qPCR) reverse transcription is very well adapted for pluripotency and differentiation control as it facilitates detained messenger RNA quantification of multiple samples at the same time. It identifies early changes in marker expression induced by differentiation in which various markers can be analyzed simultaneously. This method has greater sensitivity and specificity.

Limitations: This technique produces the average marker expression of a population. It does not unveil heterogeneity.

Teratoma Formation Technology

One of the important methods employed to assess the developmental capacity of pluripotent stem cells (SC) such as ESC is teratoma formation in vivo and to show the capacity of newly derived cell lines to discriminate as fraction of their process of characterization. Detailed and careful investigation of teratomas shaped from pluripotent SC is believed to impart a deeper understanding of the formation and development of tissues and may lead to advanced approaches in areas such as oncology and bioengineering. In comparison, experimentally induced teratomas including those from engrafted human embryonic stem cells (hESC) result from normal pluripotent cells transplanted into development permissive, ectopic sites are abnormal neoplastic pathologies carrying genetic defects (Aleckovic, 2008).

Methodology

We used the following components in our protocol to design a teratoma assay that is adaptive, quantitative and simple to conduct and to assess.

(1) *Quantification of the number of transplanted cells.*

To initiate transplantation of specified numbers of cells, hESC colonies were distinguished into single cell suspensions prior to inoculation.

(2) *Characterizing the phenotype and genotype of transplanted cells.*

The percentage of cells expressing pluripotency-associated cell surface markers was quantified by FACS prior to transplantation (Tra-1-60, Tra-1-81 and SSEA-4). We also authenticate that the transplanted cell karyotype was normal.

(3) *The mode and site of transplantation.*

The hESCs were co-transplanted with mitotically inactivated human fore skin-fibroblast-feeders to increase the sensitivity of the assay. Hentze et al. also noted the positive impact of co-transplantation of hESCs with their own feeders. The transplanted cells were also combined with Matrigel to further improve the sensitivity. The cells were subcutaneously inserted (s.c.) as transplantation to this site is simple and does not require an invasive surgical operation and allow easy monitoring of the growth of the teratoma. Subcutaneous hESC transplantation with Matrigel was reported to induce high-quality teratomas.

(4) *Recipient animals.*

The cells were transplanted into NOD/SCID mice to minimize the chance of immune rejection. In contrast to immune-competent mice, these mice have previously shown to withstand hESC grafts well. Since NOD/SCID mice appear to produce random tumors and to preclude unlikely tumor development by co-transplanted feeders, we included a control group of mice transplanted with mitotically inactivated feeders in each experiment.

(5) *Duration of monitoring of tumor formation.*

For a span of 30 weeks, we tracked the growth of tumors in transplanted animals. The extended follow-up improved the sensitivity of the assay by permitting a limited number of cells to identify late-appearing tumors that have formed.

(6) *Defined histological criteria for teratoma.*

If a tumor containing tissues has all three germ layers, it is defined as teratoma. A pathologist conducted the histological study.

Throughout the analysis, we adhered to these directions. With this teratoma assay procedure, we defined the identification sensitivity and tumor development time course by transplanting different numbers of hESCs ranging from 5 × 105 to 1 × 101. By nourishing the transplanted cell preparation with the p-160-Rho-associated coiled-coil kinase (ROCK) inhibitor, Y-27632, we also investigated whether the efficiency and sensitivity of the assay could be increased. Under unfavorable conditions, this inhibitor has been shown to increase hESC survival. Therefore, we hypothesized that its addition before transplantation to the dissociated hESCs may have a fruitful impact on their survival and increase the efficiency of teratoma formation from a limited number of transplanted hESCs (Gropp et al., 2012).

Advantages: Teratomas represent a special paradigm for exploring all of these components in a new way. They may provide detail into the stepwise development process and molecular pathways and mechanisms that are entailing in deciding the fate of cells as a model of early embryonic development, particularly when discussing issues of lineage commitment.

Teratomas may contribute to cell biological research, similar to EB by being a source of particular types of cells. It is possible to purify desired cell types using antibodies against lineage-specific markers. In particular, tissues derived from teratomas could be much more fitting because in vivo separation appears to be more complete than in vitro.

Limitations: The differentiation capacity measured by in vitro differentiation will be adequate for industrial and clinical applications, since only one particular cell type is always required. Therefore if evidence of pluripotency is not necessary, models of teratoma should not be employed. Indeed, stem cell differentiation in monolayers as EBs offers even more comprehensive knowledge about the production and action of stem cells. The main drawback of the teratoma assay is that the use of laboratory animals is necessary. Animal studies must be ethically justifiable, according to new law (Buta et al., 2013).

Genetic Analysis Using Chromosomes Analysis of Stem Cells

Karyotyping is an essential component of the cell line quality evaluation produced from human embryoid stem cells (hESC).

Methodology

To inquire the integrity of the chromosome complement of a cell line, many methods are currently available. In terms of sensitivity, resolution and costs, each approach has benefits and drawbacks.

A snapshot of the whole chromosome complement and the usual gross karyotype regulation of a cell line are possible with traditional banding techniques (G-, Q- or DAPI-banding). Providing 300–400 stained bands, these methods enable the identification of erroneous chromosome numbers (aneuploidies), mosaicism and significant structural chromosome defects with a resolution of 5–10 Mb, such as translocations, deletions or insertions. The progress in traditional cytogenetic banding research is reflected by spectral karyotyping (SKY) and multicolor FISH. These techniques allow the identification of submicroscopic deletions, insertions or amplifications of DNA, with a resolution of 1–2 Mb, using chromosome-specific fluorescently labeled probes. The array-based comparative genomic hybridization (array-CGH) and the single nucleotide polymorphism (SNP) array are useful tools to identify smaller genetic imbalances. They allow minor aberrations to be detected, including homo- or hemizygous deletions, copy neutral heterozygosity loss, duplications and amplifications. Their resolution varies between 1 Mb and <100 kb. The frequency of a particular abnormality within a cell population cannot be measured by both CGH and SNP-array (Rebuzzini et al., 2015).

Advantages: This technique has many advantages including its use in most clinical laboratories. Furthermore, it a well-developed approach that has been used for multiple genetic diseases and hematopoietic malignancies in routine diagnostics (Sampson, 2014).

Limitations: Any of the drawbacks of karyotype analysis includes its need for a sample containing new viable cells and its poor sensitivity for abnormality detection requiring at least 5–10% of cells analyzed for optimal detection to produce the abnormality (Sampson, 2014).

References

Aleckovic, M., Simón, C. (2008). Is teratoma formation in stem cell research a characterization tool or a window to developmental biology? *Reproductive Biomedicine, 17*(2).

Buta, C., David, R., Dressel, R., Emgård, M., Fuchs, C., Gross, U., Healy, L., Hescheler, J., Kolar, R., Martin, U., Mikkers, H., Müller, F.-J., Schneider, R. K., Seiler, A. E. M., Spielmann, H., & Weitzer, G. (2013). Reconsidering pluripotency tests: Do we still need teratoma assays? *Stem Cell Research, 11*(1).

Carpenedo, R. L., Sargent, C. Y., & Mcdevitt, T. C. (2007). Rotary suspension culture enhances the efficiency, yield, and homogeneity of embryoid body differentiation. *Stem Cells, 25*, 2224–2234.

Choudhury, A. R. (2017). Cell isolation. *Materials and Methods, 7*, 2260.

Chung, S. H., & Shen, W. (2015). Laser capture microdissection: From its principle to applications in research on neurodegeneration. *Neural Regeneration Research, 10*(6), 897–898.

Cole, A. E., Murray, S. S., & Xiao, J. (2016). Bone morphogenetic protein 4 signalling in neural stem and progenitor cells during development and after injury. *Cells International, 2016*, 16.

Findikli, N., Candan, N., & Kahraman, S. (2006). Human embryonic stem cell culture: Current limitations and novel strategies. *Reproductive Biomedicine, 13*(4), 581–590.

Goldman, D. C., Bailey, A. S., Pfaffle, D. L., Masri, A. A., Christian, J. L., & Fleming, W. H. (2009). BMP4 regulates the hematopoietic stem cell niche. *Blood, 114*(20), 4393–4401.

Gonçalves, A. I., Rodrigues, M., Lee, S.-J., Atala, A., Yoo, J. J., Reis, R. L., et al. (2013). Understanding the role of growth factors in modulating stem cell tenogenesis. *Plos ONE, 8*(12).

Gropp, M., Shilo, V., Vainer G., Gov M., Gil, Y., et al. (2012). Standardization of the teratoma assay for analysis of pluripotency of human ES cells and biosafety of their differentiated progeny. *Plos ONE, 7*(9).

Heese, K., Low, J. W., & Inoue, N. (2007). Nerve growth factor, neural stem cells and alzheimer's disease. *Neurosignals 15*(1).

Huang, S.-W., Tzeng, S.-C., Chen, J.-K., Sun, J.-S., & Lin, F.-H. (2020). A dynamic hanging-drop system for mesenchymal stem cell culture. *International Journal of Molecular Sciences, 21*, 4298.

Hunt, C. J. (2007). The banking and cryopreservation of human embryonic stem cells. *Transfusion Medicine and Hemotherapy, 34*, 293–304.

Kanitkar, M., Tailor, H. D., & Khan, W. S. (2011). The use of growth factors and mesenchymal stem cells in orthopaedics. *Open Orthopaedics Journal, 5*, 271–275.

Khan, F. A., Almohazey, D., Alomari, M., & Almofty, S. A. (2018). Isolation, culture, and functional characterization of human embryonic stem cells: Current trends and challenges. *Stem Cells International.*

Kim, H. S., Oh, S. K., Park, Y. B., Ahn, H. J., Sung, K. C., Kang, M. J., Lee, L. A., Suh, C. S., Kim, S. H., Kim, D.-W., & Moona, S. Y. (2005). Methods for derivation of human embryonic stem cells. *Stem Cells, 23*, 1228–1233.

Laowtammathron, C., Chingsuwanrote, P., Choavaratana, R., Phornwilardsiri, S., Sitthirit, K., Kaewjunun, C., Makemaharn, O., Terbto, P., Waeteekul, S., Lorthongpanich, C., U-Pratya, Y., Srisook, P., Kheolamai, P., & Issaragrisil, S. (2018). High-efficiency derivation of human embryonic stem cell lines using a culture system with minimized trophoblast cell proliferation. *Stem Cell Research & Therapy, 9*, 138.

Li, Y. Q. (2010). Master stem cell transcription factors and signaling regulation. *Cell Reprogram 12.*

Melton, A. E. C. A. D. A. (2007). Derivation of human embryonic stem cells by immunosurgery. *Journal of Visualized Experiments, 10*, 574.

Nethercott, H. E., Brick, D. J., & Schwartz, P. H. (2011). Immunocytochemical analysis of human pluripotent stem cells. *Methods in Molecular Biology, 767*, 201–220.

Oh, S. K., Kim, H. S., Park, Y. B., Seol, H. W., Kim, Y. Y., Cho, M. S., Ku, S. Y., Choi, Y. M., Kim, D.-W., & Moon, S. Y. (2005). Methods for expansion of human embryonic stem cells. *Stem Cells, 23*(5), 605–609.

Pettinato, G., Wen, X., & Zhang, N. (2015). Engineering strategies for the formation of embryoid bodies from human pluripotent stem cells. *Stem Cells and Development, 24*(14), 1595–1609.

Rebuzzini, P., Zuccotti, M., Redi, C. A., & Garagna, S. (2015). Chromosomal abnormalities in embryonic and somatic stem cells. *Cytogenetic and Genome Research, 147.*

Rodrigues, M., Griffith, L. G., & Wells, A. (2010). Growth factor regulation of proliferation and survival of multipotential stromal cells. *Stem Cell Research & Therapy, 1*(32).

Rungarunlert, S., Techakumphu, M., Pirity, M. K., & Dinnyes, A. (2009). Embryoid body formation from embryonic and induced pluripotent stem cells: Benefits of bioreactors. *World Journal of Stem Cells 1*(1).

Sampson, B., & Mcguire, A. (2014). Genetics and the molecular autopsy. Corpus ID: 68584328. https://doi.org/10.1016/B978-0-12-386456-7.06707-1.

Shen, S. H. C. A. W. (2015). Laser capture microdissection: From its principle to applications in research on neurodegeneration. *Neural Regeneration Research, 10*(6).

Sheridan, S. D., Surampudi, V., & Rao R. R. (2012). Analysis of embryoid bodies derived from human induced pluripotent stem cells as a means to assess pluripotency. *Stem Cells International.*
Sneha, S., Nagare, R. P., Manasa, P., Vasudevan, S., Shabna, A., & Ganesan A. T. S. (2019). Analysis of human stem cell transcription factors. *Cellular Reprogramming, 21*(4).
Tanaka, N., Takeuchi, T., Neri, Q. V., Sills, E. S., & Palermo, G. D. (2006). Laser-assisted blastocyst dissection and subsequent cultivation of embryonic stem cells in a serum/cell free culture system: Applications and preliminary results in a murine mode. *Journal of Translational Medicine, 4*(20).
Tomizawa, M., Shinozaki, F., Sugiyama, T., Yamamoto, S., Sueishi, M., & Yoshida, T. (2011). Activin A maintains pluripotency markers and proliferative potential of human induced pluripotent stem cells. *Experimental and Therapeutic Medicine, 2*(3).
Vossaert, L., Leary, T. O., Van Neste C., Heindryckx, B., Vandesompele, J., De Sutter, P., & Deforce, D. (2013). Reference loci for RT-qPCR analysis of differentiating human embryonic stem cells. *BMC Molecular Biology, 14.*
Wang, X., & Yang, P. (2008). In vitro differentiation of mouse embryonic stem (mes) cells using the hanging drop method. *Journal of Visualized Experiments, 17*, E825.
Yang, X. W. A. P. (2008). In vitro differentiation of mouse embryonic stem (mes) cells using the hanging drop method *17*: 825.
Zhang, J., Gao, Y., Yu, M., Wu, H., Ai, Z., Wu, Y., Liu, H., Du, J., Guo, Z., & Zhang, Y. (2015). Retinoic acid induces embryonic stem cell differentiation by altering both encoding RNA and microrna expression. *Plos ONE, 10*(7).
Zhao, D., & Zheng, Q. (2012). The application and limitation of suspension culture in the study of neoplastic stem cells. *Tumor, 32*(7), 559–563.
Zur Nieden, N. I., Turgman, C. C., Lang, X., Larsen, J. M., Granelli, J., Hwang, Y.-J., & Lyubovitsky, J. G. (2015). Fluorescent hydrogels for embryoid body formation and osteogenic differentiation of embryonic stem cells. *ACS Applied Materials & Interfaces, 7*(19).

Chapter 4
Differentiation of Stem Cells into Neuronal Lineage: In Vitro Cell Culture and In Vivo Transplantation in Animal Models

Shahid S. Siddiqui, Khaled Aboshamat, Sivakumar Loganathan, and Zeba K. Siddiqui

Abstract Neurological diseases are the major cause of disability and the second major reason of morbidity globally. In the last quarter century, the number of deaths associated with neurological diseases has risen significantly. It will not be surprising that COVID-19 may result in a significant increase of mental burden on mankind, worldwide. Therefore, there is an urgent need to combat this burden, but accessibility to the brain tissue is difficult and brain architecture of complex network of neurons and support cells is daunting. However, with the advent of stem cells, especially the ability to induce somatic cells into induced pluripotent stem cells (iPSCs) it may be possible to investigate brain structure and function in 2D and 3D model, in vivo preparations, called organoids, and such preparations have been used to study blood-brain barrier and other neurological diseases such as Parkinson's disease. Furthermore, with the techniques of molecular genetics and cellular neurobiology it is now possible to reverse neurological disease(s), such as restoration of vision in an aging animal model by reprogramming the methylation pattern of the genome (epigenome), using selected transcription factors. These developments bode well for a paradigm shift in neurological studies and have great potential for diagnosis and therapeutic approaches that were hardly imagined.

Keywords 2D and 3D cultures · Alzheimer's disease · Brain-on-a-chip · CRISPR · Blood-brain barrier · Embryonic stem cells (ESCs) · Endothelial cells ·

S. S. Siddiqui (✉) · Z. K. Siddiqui
McGenome LLC, Glenview, IL 60025-3720, USA

S. S. Siddiqui · K. Aboshamat
College of Dentistry, Kingdom of Saudi Arabia, Umm Al Qura University, Makkah, Saudi Arabia
e-mail: ktabolshamat@uqu.edu.sa

S. S. Siddiqui · S. Loganathan
Department of Medicine, University of Chicago, Chicago, IL 60637, USA

S. Loganathan
Department of Environmental Science, Periyar University, Tamil Nadu, Salem 636011, India

F. A. Khan (ed.), *Advances in Application of Stem Cells: From Bench to Clinics*,
Stem Cell Biology and Regenerative Medicine 69,
https://doi.org/10.1007/978-3-030-78101-9_4

Epigenetics · iPSCs · Neural progenitor cells (NPCs) · Organoids · Hydrogels · Parkinson's disease · Pericytes · Spinal cord injury · Stem cells

Introduction

Nervous system is the most complex organ in the universe, as it is the center of all human activities by mediating signal processing, executing commands and providing an output that ranges from mechanical movements to desire, emotions and, perhaps, the definition of a personality. To study nervous system and its function, it is important to focus that brain function is derived from a neural cellular process that is encoded in the genome of every cellular type that constitutes the nervous system and its supporting cells; thus, genetic components and their expression hold key to decipher brain. Furthermore, there is a huge diversity in neuron types and neural support cells, such as glial cells and astrocytes, and the location of these different cell types within the nervous system adds further to the diversity and complexity to the neural architecture. There is also an important issue of obtaining human brain tissue for performing experiments, which is both practically and ethically very challenging. This has led to the use of animal models, mostly rodents as a source for studying brain developmental studies and as a disease model.

One key technology that may allow a better grasp of neuronal structure and function is the use of stem cells that may be programmed to acquire different neuron types and study the molecular signaling, axonal outgrowth and neural lineage. With the contribution of the Nobel Laureate, Shinya Yamanaka from Japan in developing techniques to reprogram somatic cells by activation of introduced specific transcription factors, it has become possible to generate iPSCs (Takashi & Yamanaka, 2006; Takahashi et al., 2007, González et al., 2011), the induced pluripotent stem cells, with pluripotency that matches the ESCs, the embryonic stem cells. The iPSCs technology allows generating different neural types and tissues that were not easily accessible to obtain from living organisms. Several chapters in this book have described the generation of stem cells and their uses in a wide variety of cell types; herein, we will focus on differentiation of stem cells into neurons, in vitro cell models and their possible in vivo transplantation in live animal models.

The limitations of animal model systems are that the brain structure and development vary greatly between rodents and humans, and rodent models may not display the human neural disease's pathological and functional features. Producing differentiated human neural cells by inducing human pluripotent stem cells (hPSCs), including PSCs, mesenchymal stem cells (MSCs) and embryonic stem cells (ESCs), has now become routine and a simple and cost-effective manner. It has boosted the neuroscience research field, which was purely a fiction only a decade ago (Takashi & Yamanaka, 2006; Takahashi et al., 2007). In addition, as the patient-derived differentiated cell types can now be obtained, induced pluripotent cells can be harnessed to benefit from animal studies and preclinical investigations.

Regenerative therapy has become achievable largely due to the generation of iPSCs and the ability to coaxing them into desired cell lineages using specific transcription and growth factors, allowing pluripotency of human embryonic stem cells (hESCs), to undergo "directed differentiation," by producing large quantities of transplantable somatic cells in vitro when grown in specifically defined culture media supplemented with growth factors. With the capacity of pluripotency, self-renewal and ability to multiply hESCs can be harnessed to generate cells that can be used in transplantation experiments and a variety of cell types that can be coaxed to study human disorders, including behavioral diseases such as Alzheimer's disease and Parkinson's disease.

Differentiation of Stem Cells into Neural Lineage

Neurons have diverse variety by virtue of their position and functional activity; therefore, the neural subtypes have to be specified when developing methods to make stem cells differentiate into neurons. Embryonic tissue was the source for early experiments in attempts to differentiate stem cells into different neurons, but this approach was difficult largely from ethical grounds.

Reprogramming of somatic cells into iPSCs has been achieved by transforming cells with a combination of four transcription factors, namely Oct-3/4, Sox2, Klf4 and c-Myc, also known as the OSKM factors, giving rise to a standard protocol to generate iPSCs, following the seminal discovery in 2006 Shinya Yamanaka and his colleagues (Takahashi & Yamanaka, 2006; Takahashi et al., 2007). These methods have been routinely followed and have been used to obtain iPSCs for various neuron subtypes (González et al., 2011; Zhang, 2013, 2016; Madhavan, 2018; Marton, 2019; Cakir, 2019; Nakatake et al., 2020). However, many of these methods have different problems in the use of small molecules to direct iPSCs in neuronal differentiation, such as variable cell type heterogeneity, poor efficiency, and costly and time-consuming induction protocols for neural differentiation. To overcome these issues, Wang et al. (2018) reported the use of doxycycline inducible transcription factors (TFs) at safe-harbor loci; adapting a two-step method, these cell lines can be induced to differentiate into either lower motor neurons or cortical neurons, in a simple, quicker, scalable and highly efficient manner (Wang et al., 2018; Fernandopulle et al., 2018).

Recently, in a search for transcription factors that may allow induction of diverse cell types though differentiation from human pluripotent stem cells (hPSCs), Ng et al. (2020) reported 290 transcription factors (TFs); of these, 241 TFs were not identified previously, in only four days without changing the external growth factors and biochemical stimuli. Using four of these newly discovered TFs, they were able to reprogram hPSCs into oligodendrocytes, vascular endothelial-like cells, fibroblasts and also neurons that can mimic molecular and functional characteristics of primary

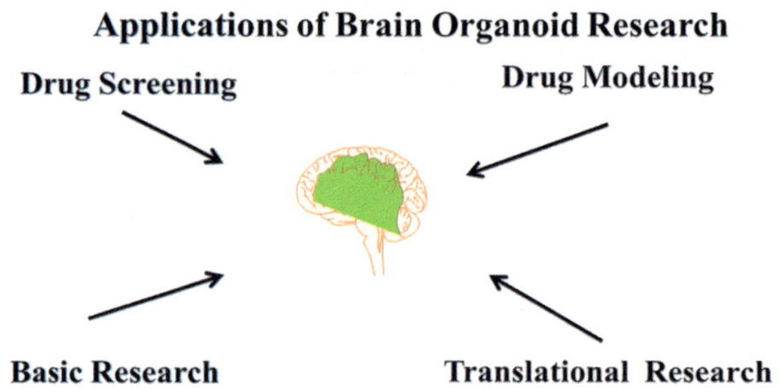

Fig. 4.1 Stem cells in organoid research. From high-throughput drug screening and drug modeling investigations; basic research questions on biochemical and physiological aspects of neural development and axon and support cell growth, etc., and the preclinical translational research; 3D brain organoids have proven their usefulness in multiple ways and elucidated molecular signaling mechanism that were not accessible in cell grown in vitro in two-dimensional format in dishes

cells, enabling programming of hPSCs into different cell types in parallel and simultaneously. The authors further showed generation of cerebral organoids with unmodified hPSCs and oligodendrocyte inducible hPSCs, which enhanced myelination in the 3D brain organoids (Ng et al., 2020) (Fig. 4.1).

In Vitro Tissue Cultures, Transition from 2D to 3D

In most mammals, new neurons are supplied from either the dentate gyrus of the hippocampus, from the cells that are present in this location and are stem cells, or the second source of neurons that is the olfactory bulb, where the neuronal stem cells reside in the lateral ventricle wall, allowing plasticity to the neural architecture and contributing to neurogenesis in adult brain. Haycock (2011) reviewed the 3D cell culture about the current approaches and techniques, arguing that cells traditionally grown in 2D in almost all tissue culture laboratories of the world fail to reproduce the physiological aspects or anatomy of a tissue for a comprehensive study, necessitating the need to develop 3D culture systems, including the consideration of scaffold to support the architecture and organization of the cell assembly or taking into account bioreactors for managing the supply of nutrients and cellular waste disposal. These efforts demand a multidisciplinary approach and expertise to make the 3D culture more appropriate and relevant to physiological recapitulation of the human tissue. Many attributes of neurons make these unique cells, such as the transport of metabolites and factors along the long axons, from the cell bodies to the synaptic region, accomplished by kinesin family of motor proteins on the network of microtubules in each neuron (Siddiqui, 2002). With an increasing sophistication of 3D culture assemblies, it is now possible to co-culture different cell types, including the integration of stem cells and iPSCs.

With the advent of 2D and 3D, organoids and brain-on-a-chip models, investigators have focused on a physiologically compatible model for developmental biology, high-throughput drug screens and preclinical studies, including modeling for neural disorders. Traditionally, neurons have been grown in 2D monolayer tissue cultures similar to other tissue culture protocols, which have been used by researchers to study molecular pathways associated with relatively simple phenomena and allowed basic understanding of neural cell biology, but these 2D monolayers do not represent the complexity of human brain, such as the development and axonal-process outgrowth to form neural networks. This necessity has required developing the 3D brain organoids that could mimic the developmental features and brain architecture in a better way than the 2D monolayer tissue culture.

Among the early research of 3D neural culture system, known as the "neurospheres" assay, this culture system has been used to characterize neural stem cells (NSCs) from the central nervous system (Reynolds & Weiss, 1992). The neural stem cells and neural progenitor cells are co-cultured without an adherent matrix, allowing single cells to multiply to produce small clumps of cells that grow in suspensions and are called "neurospheres," The multi-potent cells comprising these cell clusters can be differentiated into most of the neural subtypes of the CNS, with the exception of neurons and astrocytes (Reynolds & Weiss, 1992). Similar approaches have given rise to the production of "neural aggregates" and "neural rosettes." Neural aggregates are generated from pluripotent stem cells (PSCs), first developing an embryoid body (EB) that is cluster of PSCs recapitulating early embryonic developmental stage. The generated neural progenitor cells are used in monolayer 2D culture of neurons and glial cells by further differentiation of NPCs (Chambers, 2009). Similarly, neural rosettes are composed of neural progenitor cells that may represent the neural tube and show early neural development, and these "rosettes" can be proliferated or differentiated into a variety of mature cells, depicting regional attributes of different regions of the brain (Elkabetz, 2008). Thus, differentiation of pluripotent stem cells (PSCs), such as human embryonic stem cells (hESCs) (Thomson et al., 1998), and induced pluripotent stem cells (iPSCs) (Takahashi et al., 2007) into neural cells in 2D cultures (Zhang et al., 2001, and Chambers et al., 2009) and 3D brain organoids (Fuchs et al., 2004; Eiraku & Sasai, 2012, Lancaster et al., 2013, 2017, Park et al., 2014; Bouyer et al., 2016; Killic et al., 2016, Bordoni et al., 2018) has described experimental models to study central nervous system disorders.

Structure of Brain Organoids and the Gene Expression

Demonstrating the unique self-organizational capacity of human neocorticogenesis, a "cortical sphere" culture system was developed (Kadoshima, 2013). Similarly, the pioneering work of Sergiu Pasca, who developed human cortical spheroids from induced pluripotent cells in a medium lacking the adherent substrate, or the ECM (extra cellular matrix), and with minimal patterning factors to induce cortical spheres, containing both superficial and deep cortical neural cells (Pasca et al., 2015), ushered a new development in brain organoid cultures. Remarkably, the generated neurons in "cortical spheres" are interspersed with specific astrocyte that is hard to obtain phenotype outside the living brain tissue. These quiescent astrocytes are critical in synaptogenesis and necessary for normal neurodevelopment (Pfrieger & Barres, 1997; Ullian, 2001), and analysis of the transcription showed that 10 weeks of culturing of cortical spheres mimics the transcription pattern of developing human prenatal brain, in vivo. A number of studies have established the 3D brain organoids an important approach to study neural development and disease modeling (Sasai, 2013, Lancaster & Knoblich, 2014; Moreno et al., 2015; Bouyer et al., 2016; Killic et al., 2016; Quadrato et al., 2017; Sartore et al., 2017; Zuhang et al., 2018) (Table 4.1).

Table 4.1 Comparison of in vitro cell culture models; summary of advantages and disadvantages of 2D, 3D and organ-on-a-chip models

Types of in vitro cell culture models	Pros	Cons
2D models Culture plate Transwell membrane	Easy to set up and manipulate High reproducibility Standard and well-established technique Can be set up quickly Low cost	Uniform concentration of nutrients and drugs Not a dynamic state, quite static growth Large quantity of medium, growth factors and drug reagents
3D models Spheroid Organoid Scaffold-type	Recapitulate the 3D Architecture of the cell culture Drug response mimics in vivo concentrations Copies in vivo cell–cell interaction and cell–extracellular matrix interactions Vascularization is possible Blood perfusion is possible	Cellular waste is not removed as in an in vivo model Low reproducibility Poor nutrient transport and access Requires more time to set up Misses the dynamics of in vivo cellular environment, due to the lack of medium fluid flow
Organ-on-a-chip model Microfluidic chip	Diffusion of the medium and drugs is much better, and the microenvironment can be easily manipulated Excellent model for high-throughput screens Actuators and sensors can be easily incorporated and integrated Electronic data acquisition may be possible	Requires external gadgets, such as pumps, valves and circuits to run the experiment Ramping up the culture is difficult Standardization requires extra effort Polydimethylsiloxane, called PDMS used for the fabrication microfluidic chips, and PDMS may adsorb nutrients The microchip setup is costly and requires multidisciplinary expertise to set up

A Comparison of the 2D and 3D Brain Cultures

An important issue in 3D brain organoids is the role of cell–cell interaction that revealed a more complex cell maturation profile in the constructed organoids (Pasca et al., 2015; Kilic et al., 2016; Quadrato et al., 2017; Madhavan et al., 2018; Sloan et al., 2017; Zuhang et al., 2018; Ormel et al., 2018; Qiao et al., 2018). Transcription analysis using RNA-sequencing and whole organoid transcription data has shown similarities in cell composition and transcriptional profiles between human brain cortical organoids and human fetal neocortex (Camp et al., 2015; Kilic et al., 2016; Bouyer et al., 2016; Xiang et al., 2017; Bershteyn et al., 2017). Oftentimes, gene expression levels (mRNA) poorly correlate to expression of specific cellular markers, due to differences in translational profile (Carlyle et al., 2017), that indicated higher amounts of protein expression differences between brain regions compared to the RNA transcription. These differences in the RNA and protein-level abundance suggested functional and cyto-architectural differences between brain regions; e.g., comparison of structurally similar cortical brain tissues showed important differences in abundance between the receptor-associated proteins and resident plasma membrane protein family, which was not evident in the transcription analysis of these tissues (Fig. 4.2).

The basic approach is to select an extracellular matrix material, such as PEG4-MAL (Cruz-Acuna et al., 2017) that can provide the right scaffold for the stem cells to grow and combine appropriate iPSCs in a syringe configuration that may be controlled by microfluidic controls in bio-printing process to develop physiologically relevant brain organoid tissue in 3D.

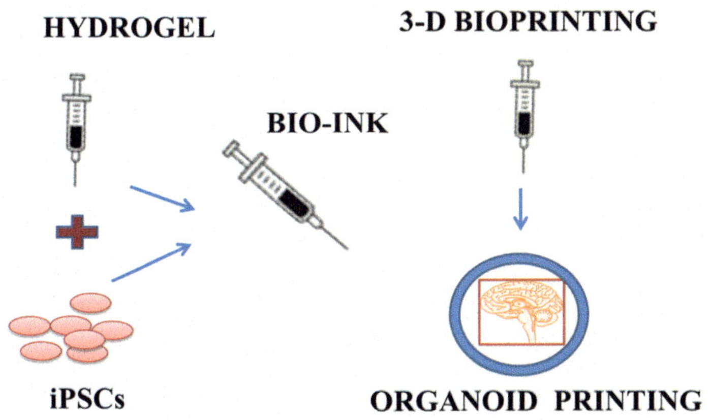

Fig. 4.2 Use of hydrogel and stem cells in bio-printing

Use of Hydrogel and Other Matrix in Brain Organoids

Generation of hPSC-derived human brain organoids has relied upon encapsulation of these brain cell aggregates using Matrigel—produced from biologically derived material—that are poorly characterized and hence show significant lot-based variability and the influence of their biophysical and biochemical attributes, and poor experimental control (Hughes et al., 2010; Miyoshi et al., 2013); Matrigel is obtained from transformed mouse cells and the complexity of these transformed mouse cells precludes its full translational potential (Cruz-Acuna et al., 2016; Gjorevski et al., 2016). Therefore, synthetic hydrogels that are fully defined and can be manipulated for biochemical and physiological properties have great potential alternatives to Matrigel and similar matrices to be used in brain organoid production (Gjorevski et al., 2016), as these could transduce innate cellular behavior by exposure to bioactive motifs, which facilitate cell-directed matrix degradation and cell–matrix-adhesive interactions (Gjorevski et al., 2016; Cruz-Acuna et al., 2017).

Synthesis of PEG-4MAL hydrogel has been described that supports the robust and highly reproducible in vitro generation of human brain organoids from human-induced pluripotent stem cells (hiPSCs) and human embryonic stem cell (hESC)-derived spheroids without the need for Matrigel encapsulation (Cruz-Acuna et al., 2017). Furthermore, PEG-4MAL hydrogel polymerization chemistry allows improved cellular compatibility when compared to free-radical-initiated polymer formation and increased efficiency in cross-linking over acrylate PEG4-A and PEG-4VS (vinylsulfone)-mediated polymerization (Enemchukwu et al., 2016; Cruz-Acuna et al., 2017). The PEG4-MAL hydrogel preparation has an advantage as this hydrogel is not dependent on animal-derived factors such as laminin-111 (Cruz-Acuna et al., 2017), and can be used for both in vitro production and in vivo delivery of organoids, thereby providing an excellent platform for tissue engineering studies and potential therapeutic applications. In addition, mechanical properties of PEG4-MAL synthetic matrix can be manipulated by altering the polymer density, without affecting the hydrogel's adhesive peptide type or density and other biochemical properties (Phelps et al., 2012; Enemchukwu et al., 2016; Cruz-Acuna et al., 2017).

These flexibility features of synthetic hydrogels are important as this allows manipulation of hydrogel properties and can be adapted to promote generation and culturing of a variety of human brain organoids. Hydrogel can also be cost-effective, as it is approximately half the cost of materials when synthetic hydrogels are used as compared to similar amounts of biologically derived Matrigel, which may cost almost twice the cost of synthetic hydrogels (Cruz-Acuna et al.,). In brief, PEG-4 MAL hydrogels can be utilized as in vitro scaffold, which can be manipulated suitable for a variety of developmental requirements, for example, the human brain organoid 3D culture, and not restricted with the limitations of materials that require the preparation from biological sources, such as the Matrigel (Cruz-Acuna et al., 2018).

Increasing the Scalability and Traceability of Organoids

Currently popular techniques to produce mouse pluripotent stem cells-generated organoids are expensive, intensive labor requiring and very difficult to scale, especially by utilizing robotic manipulations. Decembrini et al. (2020) have addressed the issue of scaling up and reproducibility of retinal organoid micro-fabrication, by culturing mouse embryonic stem cells in microenvironments with optimized physical and chemical properties, by using round bottomed milliwells fabricated from biomimetic scaffolds (hydrogels), combined with titrated medium components, resulting in rapid development of retinal organoids from mouse embryonic stem cells in a highly reproducible and efficient manner, such that more than 90% of the cellular aggregates consisted of retinal organoids. These retinal organoids beyond day 26 comprised about 80% of photoreceptor cells, of which about 22% showed GNAT2 marker-positive cone cells that is a critical and rare retinal sensory cell type that is hard to investigate in a mouse model. Thus, the ability to partitioning of retinal organoids into predetermined positions on a 2D array permitted most aggregates into retinal organoids and, furthermore, captured the dynamics of individual organoid that could facilitate for high-throughput screens for drugs and biochemical studies. This protocol combines two key positive developments, to increase the scalability and the ability to trace single retinal organoid, and could permit screens for small molecules that are neuroprotective and a possible source for transplantation of organoids in clinical studies. Decembrini et al. (2020) are credited with an improved technique in that it is based on an a simple one-step handling and manipulation to produce retinal organoids without the need for successive interventions, permitting automation of the 3D culturing process from cell inoculation and seeding to routine medium changes, and also the characterization of retinal organoid growth and differentiation.

Brain-on-a-Chip

The main purpose for developing 3D cell culture systems differs considerably—and ranges from engineering tissues for clinical studies of drug delivery through to the development of drug screening model. The development of the brain-on-a-chip technology primarily has to basically ask which a human brain model can be engineered by cell line assemblies to develop an organ-level model? Hence, it is critical to select appropriate cell lines for such organoids, since brain tissue comprises many different and distinct neuronal types, and additionally a wide variety of supporting cells such as astrocytes and glial cells. The chip design process should also take into account the neural network and brain architecture that varies throughout neural network and brain regions, in a significant and critical manner (Alepee et al., 2014; Park et al., 2014; Bhatia & Ingber, 2014, Jo et al., 2016; Killic et al., 2016; Bouyer et al., 2016; Haring et al., 2017; Qiao et al., 2018, Dolmetsch & Geschwind, 2011; Ducker et al., 2020).

Brain-on-a-chip is a platform, which is engineered to resemble the physiological microenvironment and tissue composition of a specific region of the brain. Therefore, such brain chips have an advantage in their capacity to reconstitute brain microenvironments in vivo, such as cell–cell interaction and scaffold composition, i.e., extracellular matrix and hemodynamics that can be manipulated according to the specific need of the researcher. In contrast, brain organoids investigate the developmental processes to recapitulate the early stages of fetal brain development, such as cell subtype heterogeneity, polarized neuroepithelium and compartmentalization of specific brain regions; furthermore, brain organoid culture has little control over physiological and biochemical factors in the three-dimensional microenvironment, whereas the brain-on-a-chip constructs have limits on the reconstitution of complexity and the temporal and spatial control as seen during the brain development. Thus, to combine the strengths of the brain organoid and the brain-on-a-chip, the organoid on a chip strategy serves as a useful new model synthesizing structural and physiological aspects of the in vivo brain region and the corresponding microenvironment in a 3D space (Park et al., 2014; Moreno et al., 2015; Killic et al., 2016; Skardal et al., 2016, Wang et al., 2018).

Thus in brief, three considerations are important for the brain chip design: first, composition and availability of the cell type; second, the cell maturity; and the third the cyto-architecture, i.e., structural organization of different cell types and their scaffold or matrix for cellular growth within the model. The brain-on-a-chip technique to be functional requires that all required cell types must be available and part of the engineered chip, combining the microsystem platform with hiPSC-derived cell subtypes. Such hiPSC-derived neuronal cell assemblies are useful as these neurons can build a given tissue architectural network depending on their differentiation stage. For a specific application, differentiation can be induced within a specific microenvironment of a compartment, within the physiological constraints of the desired experimental application.

In a brain organoid, in addition to the different varieties of neurons, supporting glial cells such as astrocytes, Schwann cells, oligodendrocytes and microglial cells are also part of the neural tissue, and due consideration has to be made to incorporate the appropriate cell type. These complex multicellular assemblies of brain organoids are necessary for studying the functional nervous system and required for investigating underlying basic developmental processes, including axonal growth and pathfinding, synaptogenesis and neural function. Thus, incorporation of glial and associated cells is critical as these support cells can function as mediators of chemically induced tissue damage and targets of the injury (Alepee et al., 2014). A number of studies have shown that the glial cells could modulate or affect the chemo-toxicity of chemicals for neurons (Schildknecht et al., 2009; Zurich et al., 2002; Vivani et al., 1998), or glial cells may cause neuroinflammatory response of the brain tissue (Falsig et al. 2004; Park et al., 2014; Henn et al., 2011; Boillee et al., 2006, Jasmine et al., 2010, Dolmetsch & Geschwind, 2011, Killic et al., 2016; Bouyer et al., 2016; Jo et al., 2016, Qiao et al., 2018, Wang et al., 2018; Achberger et al., 2019; Ducker et al., 2020). Thus, setting up a 3D brain organoid requires consideration of different types of glial cells and of course the choice of neuronal cells (Dolmetsch, & Geschwind, 2011).

With the availability of human-induced pluripotent stem cells (hiPSC), the complexity of brain tissue and neural network can be pursued, depending on the differentiation stage of the hiPSCs. The availability of specific stem cells has allowed design of compartmentalized micro-environments within the tissue culture with physiological attributes to attain specific requirements of a specific experimental application. Combining artificial intelligence and bioinformatics with in vitro tissue culture methods can enhance the speed and rate of the drug discovery and drug development process, allow improved pharmacokinetics and toxicological risk assessment and provide better understanding of the neural disease process. Since most of these disease models are organotypic, in which the main purpose is to recapitulate the primary function of an organ such as brain, more than one cell type need to be incorporated in the 3D organoid culture, and the scaffold or the cellular matrix factors should also be considered in the chip design (Fuchs et al., 2004; Morrison & Spradling, 2008, Achberger et al., 2019). Similar approach has been used in evaluation of dental pulp stem cells with different materials to study dental pulp injury (Youssef et al., 2019). It is hoped that the use of such models will increase experimental success by reducing errors and incorrect predictions from small molecule screens for therapeutic development.

Another key consideration in 3D brain organoid technology is formation of cell niches that are specific in the brain developmental process, as they arise by interaction of specific cell types and gradient of signaling factors and stimuli to produce the desirable cellular milieu for the optimum function and development of brain cells. Glial cells are often critical in the formation and modeling of such cellular niches, e.g., niches in the retinal model and niche generated by satellite glial cells in the trigeminal ganglia in association with the pain neuronal cell bodies (Jasmin et al., 2010). Such brain niches and gradients and signaling and trophic factors are required during the neurite growth and guidance and neuronal cell differentiation and thereby in shaping the brain architecture. Thus, such three-dimensional brain organoids are clearly superior to the 2D tissue culture setup when the purpose is to model brain-specific cellular niches (e.g., Fuchs et al., 2004; Morrison & Spradling, 2008, Zuhang et al., 2018; Ducker et al., 2020) (Fig. 4.3).

The architecture includes a flow of medium mimicking the BBB, enriched with soluble factors and peripheral immune cells, which are key players in neuro-inflammation and neuro-degeneration. The migration of peripheral immune cells through the BBB has been implicated in the pathogenesis of several neurodegenerative diseases. The role of infiltrating peripheral immune cells has been investigated in detail for MS, which involves the breakdown of the BBB and multifocal inflammation caused by the innate and adaptive immune systems. However, BBB impairment and the infiltration of peripheral immune cells also correlate with the pathogenesis of other neurodegenerative diseases, such as AD and PD. Adding a fluidic system to mimic the BBB is therefore necessary to investigate the pathological mechanisms of neurodegenerative diseases and eventually to study the transport of drugs across the BBB (Adapted from Slanzi et al., 2020).

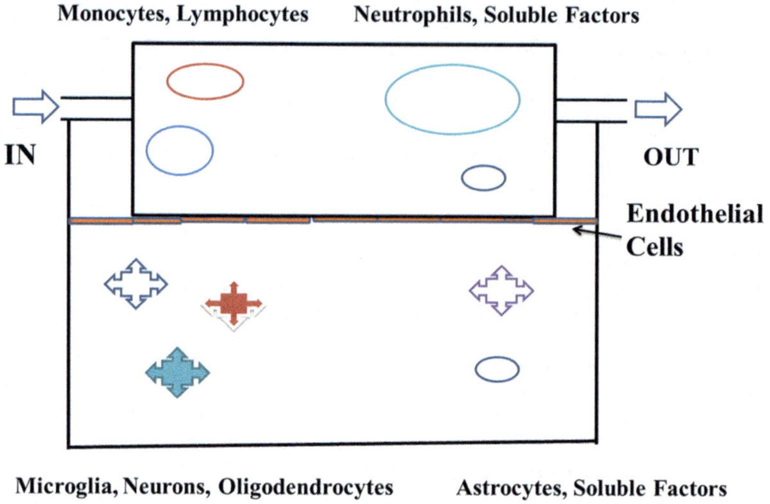

Fig. 4.3 Schematic representation of 3D in vitro models of the nervous system

The third basic requirement for the functional brain organoids is cell maturity, depending on if a organoid is being used to measure toxicity to the mature brain tissue, also known as neurotoxicity assay, or if the experimental design seeks to study perturbations in the development of brain that can be defined as developmental neurotoxicity, thereby requiring different brain organoids to investigate different questions on toxicity or developmental biological studies, respectively (Ducker et al., 2020). Thus, cell maturity in a given model will depend on the type of studies that such toxicity experiments are planned to assess, and these questions have been addressed using, in addition to traditional cell biological features, new technologies such as transcription profiling and the determination of epigenetic state of the cellular genome (Balmer et al., 2012; Waldmann et al., 2014; Qiao et al., 2018).

Stem Cells for Brain Research, HiPSCs from Patients

One alternative to the isolation of neural cells from fetal brain is to generate neural stem cells (NSCs) from pluripotent stem cells, but culturing such cell lines is challenging and requires long periods to generate and propagate; furthermore, the gene expression profile of these cells grown in two-dimensional traditional culture and NSC grown in three-dimensional organoids or chips results in alterations in the gene expression profile, and the cell function also shows distinct differences between the 2D and the 3D neuronal cultures (Koch et al., 2009; Birgersdotter et al., 2005; Zahir & Weaver, 2004).

Since human-induced pluripotent stem cells can generate from somatic cells, brain organoids using hiPSCs can be profitably used in brain organoids and brain-on-chip technology. One of the main advantages of hiPSCs over primary animal brain or immortalized neuronal cell lines is their constant availability (Takahashi et al., 2007), and that these cells can be differentiated into different neuronal and support cell types. Differentiation of human embryonic stem cells (hESCs), human-induced pluripotent stem cells (hiPSCs) or fetal neural progenitor cells (NPCs) into glial cells and neurons has been reported (Bal-Price et al., 2012; Fritsche et al., 2017). Furthermore, hiPSCs can be derived from human patients and thus can be genetically matched with a desired source (Dolmetsch, & Geschwind, 2011). Human iPSCs (NTERA2) can be differentiated into neural cell aggregates, consisting of astrocytes that interfaced with microvascular endothelial cells derived from human brain, which exhibited the characteristics of blood-brain barrier (Killic et al., 2016). Recently, Goz et al. (2020) have reported such cells derived from glioneuronal tumor, showing that BRAFV600E variants have a cell autonomous effect and the mutation changes several electrophysiological characteristics in neocortical neurons; in contrast, similar neuronal excitability changes were not observed in cells adjacent to BRAFV600E—expressing neurons, showing that BRAFV600E affects a cell autonomous, distinct and highly excitable neuronal electrophysiological response when somatically introduced into neocortical progenitor cells.

Similarly, in another report, Flaherty et al. (2019) showed that hiPSCs generated from NRXN1-alpha the plurality of alternate splicing seen in the human brain tissue, reporting 123 high confidences and in correct reading frame human NRXN1-alpha isoforms. Heterozygous NRXN1-alpha ± hiPSC-neural cells show more than twice inhibition in half of the wild-type NRXTN1-alpha isoforms and transcribe several novel isoforms from the NRXTN1-alpha mutant allele. The authors demonstrated that depending on the genotype, NRXN1-alpha ± mutations can affect the phenotype through the reduction of NRXN1-alpha isoform expression levels and also the presence of the mutant NRXN1-alpha variant isoforms.

In case of familial dysautonomia, where a single mutation can cause a disease, the iPSCs harboring a point mutation in IKBPKAP encoding gene, resulting in the depletion of sensory and autonomic neurons, have been used for wild-type versus diseased hiPSCs screens, for therapeutic drug discovery, and more personal patient-specific diagnosis (Lee et al., 2011, Dolmetsch & Geschwind, 2011). In a brain-on-a-chip variation, Achberger et al. (2019) have used these ideas on retina-on-a-chip (RoC), modeling human retina that combines seven different retinal cell subtypes generated from hiPSCs, demonstrating fluid perfusion similar to vasculature and mimicking in vitro and interaction of mature photoreceptor segments with retinal pigment epithelium (RPE). These authors showed that this interaction supports and increases the creation of outer segment-like networks and recapitulation of in vivo-like biochemical and physiological phenomena such as calcium dynamics and outer segment phagocytosis. This retina-on-a-chip can be used for drug screens such as the antibiotic gentamicin and the retinopathic injury of anti-malaria drug chloroquine, thereby showing promise for drug discovery and a platform to study retinal physiology and pathology of retinal disorders (Achberger et al., 2019).

The Blood-Brain Barrier

As a neurovascular component, the blood-brain barrier (BBB) provides a physical and chemical barrier against intrusion of blood cells, plasma factors and various pathogens for the protection of the central nervous system (CNS). Brain microvascular endothelial cells are the main components of the BBB, together with neurons, astrocytes, pericytes, and the scaffold extracellular matrix (ECM) consisting of type IV collagen, laminin, fibronectin, perlecan and heparin sulfate (Page et al., 2018). Many acute and chronic neural diseases and disorders such as Parkinson's disease, Huntington's disease, ischemic stroke and Alzheimer's disease have been attributed to the malfunction of the BBB (Sweeney et al., 2019).

The blood-brain barrier (BBB) controls the exposure of brain cells in a significant manner; therefore, various in silico or in vitro BBB models should incorporate the choice of cell subtypes, the transport properties and the extracellular matrix to recapitulate the features of human BBB, and a variety of such models has been reported, for example (Vandenhaute et al., 2012). With an increasing sophistication in BBB platform technology (e.g., see Frimat & Luttge, 2019; Hai, et al., 2010; Sweeney et al., 2019; Sun et al., 2020), organoids of different brain regions, such as the cerebral cortex layers, model that were developed by intercalating hydrogel–neuron layers with plain hydrogel layers (Kunze et al., 2011); these cortical layers displayed both the chemical gradient of trophic and growth factors and the differential synaptic density distributed in different layers (Choi et al., 2010).

Pericytes are important for the structure and function of the BBB, and their degeneration is related to neural disorders, with poorly understood mechanisms, due to the paucity of obtaining sufficient pericytes for investigations. Sun et al. (2020) describe pericytes-like cells (PCs) from human pluripotent stem cells (hPSCs) via the intermediate developmental stage of cranial neural crest (CNC) cells and show that CNC-derived pericyte-like cells express specific molecular markers such as NG2, CD146, CD13 and PDGFR-beta, with Vimentin and Caldesmon, and exhibit typical contractile features, endothelial barrier function and potential to form vessels; interestingly implanted into a model transient middle cerebral artery occlusion (tMCAO), with blood-brain barrier disruption hPSC-CNS-PCs are capable of improving functional recovery in the tMCAO mouse model by enhancing the integrity of the BBB and inhibiting neuronal cell death through apoptosis and may provide a model to study BBB function in a variety of neurological disorders (Sun et al., 2020).

Microfluidic engineering has been used to generate BBB models in 3D organoids. In such designs, intersecting channels are separated with a porous membrane (polycarbonate) upon which microvascular endothelial cells and astrocytes (brain) are grown on opposite a section, which in a way recapitulates the BBB and allows for the measurements of trans-endothelial electric resistance (TEER) to evaluate endothelial barrier function (Van Der Helm, 2016). Such BBB models have helped how various drugs and toxins may cross the BBB and find entry to the brain microenvironment.

Another important issue in blood-brain barrier models is to develop innovative drug delivery routes, as the BBB has special requirements for molecular passage across the barrier. Developing novel drug delivery vehicles is important for drug development of basic physiological studies. Nanotechnology is one of the emerging drug delivery strategies and could have enormous therapeutic potential and translational efficacy; however, there are some problems that remain to be solved, such as the removal of nanoparticles after the drug release and non-specific adverse effects on non-intended tissues and organs, and related toxicity. This will require examining the properties of each nanoparticle design, their intended target and pharmacokinetic properties of this drug delivery (Siddiqui et al., 2020). Thus, developing physiologically relevant models using stem cells can be very useful for drug development, drug delivery and elucidating molecular and structural mechanisms of both acute and chronic neurological disorders.

Neuronal Disorders and Disease Models

A very important use of stem cell technology to mimic brain function is to study neuronal disorders and human neural disease, such as Parkinson's disease and Alzheimer's disease. These disorders affect the integrity of synaptic connections and result in reduction and degradation of these connections, and other ailments such as epilepsy or autism have been attributed to abnormal neural architecture and network responses. For Parkinson's disease (PD) and Alzheimer's disease (AD), brain-on-a-chip approach has been applied and summarized below (Choi et al., 2013; Hai et al., 2010; Lu et al., 2012; Slanzi et al., 2020).

Alzheimer's Disease (AD)

The late onset chronic neurodegenerative Alzheimer's disease is devastating as the dementia grows slowly and develops into irreversible worse outcome over time. Early detection of AD is critical for disease monitoring and management, but conventional methods do not meet these challenges. In addition, animal models that are both expensive and labor- and time-intensive do not allow real-time studies on biological processes underlying the disease, and human and animal species differences also preclude extrapolation of animal studies for the progression of disease in humans. These limitations have prompted investigators to experiment on microfluidic brain-on-a-chip that may mimic the neuroanatomical and physiological features of AD. In Alzheimer's disease (AD), the traditional view is that synaptic abnormalities arise due to accumulation of proteins, such as amyloid-beta and tau protein; hence, some AD studies have focused on production and function of these two proteins on synaptogenesis and communication with the supporting glial cells (Hai et al., 2010); 3D models such as neurospheres have also been employed for AD research (Choi et al., 2010), particularly investigating the amyloid-beta protein expression and synapse formation (Choi et al., 2013). Furthermore, microfluidics technology was used to show the role of amyloid-beta in synapse formation and in the glia, including the phosphorylation of tau proteins (Cho et al., 2013; Choi et al., 2013; Deleglise et al., 2014; Kunze et al., 2011; Li et al., 2020). In addition, brain-on-a-chip model was also used to show wild-type tau protein transfer across neuron via trans-synaptic pathway (Dujardin et al., 2014).

In another study, fibroblasts from Alzheimer patients who have familial disease (FAD) with mutations in PS1 (A246E) and PS2 (N141I) have been used to generate iPSCs to study neuronal differentiation (Yagi et al., 2011) and showed that FAD-iPSC-generated neurons have higher amyloid-beta42 secretion, mimicking the biochemical pathology of mutant presenilins, and that secretion of amyloid-beta42 from these generated neurons responded well to the gamma secretase modulators and inhibitors, suggesting the possibility of drug screening and validation for high-throughput analysis (Yagi et al., 2011). Thus, such stem cells from hiPSCs derived from patients can provide very useful models to study diagnostic and therapeutic pursuits.

Parkinson's Disease (PD) Model

Parkinson's disease (PD) is progressive neural degeneration disease accompanied by loss of dopaminergic neuronal projections of the ventral forebrain, causing abnormalities in cognitive and motor functions. In spite of considerable efforts in studying PD abnormalities, no drug that can reverse the neurodegenerative process of PD has been discovered (Son et al., 2017; Kouroupi et al., 2020). To investigate mitochondrial transport on single dopaminergic axon, a microfluidic chip of the Parkinson disease (PD) was studied, in which axonal extension was investigated and mitochondria that were labeled were observed (Lu et al., 2012). The chip allowed oriented axonal extension into separate axonal compartments for visualization; in addition, this construction could also allow monitoring microtubule fragmentation and transport of vesicles on microtubules, processes that contribute to the severity of the PD, including the loss of dopaminergic neurons. This provides a great advantage from the conventional 2D culture studies to study the physiological aspects of the PD malformations and other neurodegenerative diseases. In another study, using human neuroepithelial stem cells differentiated into dopaminergic neurons in the microfluidic chip cell culture at a large scale it was shown that this technology could be harnessed to characterize dopaminergic neuron degeneration's marker substantia nigra, which is a specific marker for the progression of Parkinson's disease (Moreno et al., 2015).

Neural Disease Models

A number of other neurological disorders such as amyotrophic lateral sclerosis (ALS or Lou Gehrig's disease), Dravet syndrome, microcephaly, hyper-excitability, epilepsy, autism spectrum disorders and Zika virus-mediated brain malformation are all ready for the use of hiPSCs from the patients in 2D and 3D brain cultures, and organoids and brain-on-a-chip technology for improved understanding of physiological and structural brain studies, and drug screens for a possible therapeutic potential (Morin et al., 2006; Selmer et al., 2009; Gullo et al., 2014; Costamagna et al., 2019; Frimat & Luttge, 2019) (Table 4.2).

Table 4.2 Modeling of neural disease using 3D organoids from human iPSCs

Neural disease	Remarks	Selected references
Alzheimer's disease	Adult onset disease affecting cognition and memory function	Raja et al. (2016) Arber et al. (2017) Gonzalez et al. (2018) Lin et al. (2018) Ranjan et al. (2018) Fan et al. (2019) Ghatak et al. (2019) Meyer et al. (2019) Tzekaki et al. (2019) Choi et al. (2020)
Parkinson's disease	Neurodegenerative disease affecting motor functions, tremor, rigidity and stiffness	Monzel et al. (2017) Ho et al. (2018)
Macrocephaly or the autism	Social interaction disorder, with early onset	Mariani et al., (2015) Ho et al. (2018)
Primary microcephaly	Significantly small head of the newborn, where brain growth is impaired	Kelava et al. (2016) Dang et al. (2016) Lancaster et al. (2013, 2014) Li et al. (2017)
Congenital brain defects and Zika virus-associated malformations	Birth defects in the brain development associated with Zika virus infection	Dang et al. (2016) Cugola et al. (2016) Kelava et al. (2016) Garcez et al. (2016)
SARS-CoV2 virus-associated brain defects	Blood supply to the brain is compromised, with hemorrhage and strokes, loss of taste and smell	Ramani et al. (2020) Song et al. (2020) Mesci et al. (2020) Shpichka et al. (2020) Zimmerling and Chen (2020)
Retinal neuropathy	In diabetes, it is a complication that affects eyes; caused by damage to the blood vessels of the light-sensitive tissue at the back of the eye (retina)	Slembrouck-Brec et al. (2019) Rabesandratana et al. (2020)

Spinal Cord Injury (SCI) and Stem Cell Transplantation

The spinal cord injury (SCI) is a highly common neurological disorder resulting from the destruction of long axis of spinal cord and affects a very large number of young and old people every year, and this is not accessible to simple therapeutic treatments, necessitating combinatory approach to treat SCI, and regeneration of the spinal cord. The SCI results in a cascade of tissue damage, starting with the death of the cells in the central nervous system (CNS), affecting astrocytes, endothelial cells, microglia, oligodendrocytes and, most importantly, neurons. More specifically, long axonal projections are damaged that inhibits descending and ascending axonal pathways that communicate stimuli between the brain and the rest of the body. Subsequently, vascular deterioration causes neuro-inflammation, demyelination, acute injury-associated signaling activation, tissue degeneration and remodeling of the extracellular matrix, enhancing the initial cord injury-associated pathology (Griffin & Bradke, 2020; Hilton & Bradke, 2017; Hilton et al., 2017).

Thus, SCI unfolds a series of physiological and anatomical alterations that can extend from months to years following the injury (Donnelly & Popovich, 2008; Griffin & Bradke, 2020). The key advances required in treating SCI are in nerve regeneration and limiting the tissue damage. For the regeneration of the nerve, tissue engineering and transplantation of appropriate cell subtypes to provide neural protection, axonal growth and path-finding, immune response regulation, myelin regeneration, and neuronal circuitry establishment, to allow a neuron to regenerate and form neural circuitry. The use of induced pluripotent cells (iPSCs) is an emerging technology in treating SCI, and the use of such stem cells also bypasses the ethical problems associated with the embryonic cells or cells from the fetus; thus, neural progenitor cells (NPCs) derived from iPSCs have proven useful after transplantation in animal models of SCI (Nagoshi, & Okano, 2018). One critical bottleneck in using the iPSC-NPCs is the incidence of tumor formation after the cellular transplantation, although some results in marmosets show that iPSC-NPCs mostly differentiated into neural cells around the transplant site, without tumor formation and facilitated axonal regrowth and exhibited vascularization as angiogenesis and protected myelin formation (Griffin & Bradke, 2020; Kobayashi et al., 2012) (Table 4.3).

Transplantation of Stem Cells in Model Metazoans

Therapeutic transplantation application of stem cells, specifically iPSCs, has made great strides. But, it is important to resolve issues concerning immunogenicity and immunological dynamics after transplantation of iPSC-derived cells in such transplantation studies (Itakura et al., 2017). Neural stem cells and neural progenitor cells generated from human and rodent iPSCs (iPSC-NPSCs) can be transplanted in spinal cord injury in animal models (Nori, 2011; Tsuji, 2010), since iPSCs technology allows autologous transplantation. Nevertheless, the limitations are a long waiting

Table 4.3 Spinal cord injury trials—summary of included studies

Trial phase and trial identifier	Name of the trial and the type of cells used in the trial	Intervention and type of the cells used in the trial	Transplantation route
NCT01328860 Phase 1	Autologous stem cells for spinal cord injury (SCI) in children	Autologous bone marrow progenitor cells	Intravenous
NCT01162915 Phase 1	Transfer of bone marrow-derived stem cells for the treatment of spinal cord injury	Autologous bone marrow-derived mesenchymal stem cells	Intrathecal
NCT03308565 Phase 1	Adipose stem cells for traumatic spinal cord injury (CELLTOP)	Autologous, adipose-derived mesenchymal stem cells	Intrathecal
NCT01772810 Phase 1	Safety study of human spinal cord-derived neural stem cell transplantation for the treatment of chronic SCI	Human spinal cord-derived neural stem cell	Intramedullary
NCT03225625 Phase NA	Stem cell spinal cord injury exoskeleton and virtual reality treatment study (SciExVR)	Autologous bone marrow-derived stem cells	Intravenous
NCT02163876 Phase 2	Study of human central nervous system (CNS) stem cell transplantation in cervical spinal cord injury	Human central nervous system stem cell	Intrathecal
NCT03979742 Phase 2	Umbilical cord blood cell transplant into injured spinal cord with lithium carbonate or placebo followed by locomotor training	Umbilical cord blood mononuclear stem cells	Intrathecal
NCT02302157 Phase 1/2a	Dose escalation study of AST-OPC1 in spinal cord injury	Human embryonic stem cell-derived oligodendrocyte progenitor cells	Intramedullary

Adapted from: Platt et al. (2020); Stem Cell Clinical Trials in Spinal Cord Injury: A Brief Review of Studies in the United States, Medicines (Basel) 0.2020;7(5):27

period of several months necessary to induce iPSC to differentiate into the desired mature cell subtype and added cost of scaling-up (Theodorou,). In addition, these autologous iPSC lines require determination of the safety and efficacy of each line; thereby, allogeneic transplantation used in combination with iPSC banks is a better alternative. However, the problem of immune rejection of allograft transplantation still remains. Cells derived from iPSC show low immunogenicity (Liu et al., 2013), but little is known about immunogenicity or immune rejection of iPSC-generated cells in vivo. One caveat of Itakura et al. (2017) study is that transplantation experiments were done in allogeneic and syngeneic mouse models, which are certainly different from the human immune dynamics. Their data suggest that fetus-NPSCs and iPSC-NPSCs display similar immunogenicity, and that therapeutic cell transplantation into the spinal cord may immunologically better tolerated than transplantation into other organs, which may have some clinical therapeutic potential.

Future Directions

Brain Regeneration: Reversing the Vision Loss in Mouse Model by Reprogramming Stem Cells

A remarkable study published by Lu et al. (2020) reported a huge breakthrough finding in which the authors were able to use genetic reprogramming in old cells to return to their youthful stage and restore vision in a mouse model of glaucoma in aged mice. Since aging gradually degenerates tissue causing cell death and dysfunction and has been postulated to be associated with the epigenetic status of the genome (such as DNA methylation pattern), Lu et al. (2020) investigated whether older animals keep the genetic information needed to restore the epigenetic signature of the young adults, and if so restoring those epigenetic patterns may lead to improvement in the tissue function?

Since there is a gradual loss of function of the central nervous system (CNS) and its capacity to proliferate, Lu et al. (2020) ectopically expressed *Oct4, Sox2, and Klf4 (OSK)* in mouse retinal ganglion cells and demonstrated that transcription pattern and the DNA methylation pattern of the youthful mouse can be restored. Furthermore, they showed that expression of these transcription factors enhanced capacity of axons to regenerate after injury, and the vision loss caused by glaucoma in mouse model can be reversed and vision is restored. The genetic construct used in this study was based on adeno-associated virus (AAV), to allow expression of *Oct4, Sox2* and *Klf4* genes that are expressed in early embryogenesis during early development, and these transcription factors were previously discovered by Shinya Yamanaka's group in Kyoto, Japan (2007, 2012), as the key to induce somatic cells into pluripotent stem cells. The reprogramming of cells by expression of OSK transcription factor to regenerate axons and restore vision was dependent on expression of two DNA demethylases TET1 and TET2. The important improvement in the protocol was to delete the use of c-Myc in the transcription factor cocktail and use only three (*Oct4, Sox2* and *Klf4)* of the four OSKM factors, as no tetratomic growth or cancer was observed in these experiments, as the development of cancerous cells is a huge bottleneck in reprogramming of the iPSCs Fig. 4.4.

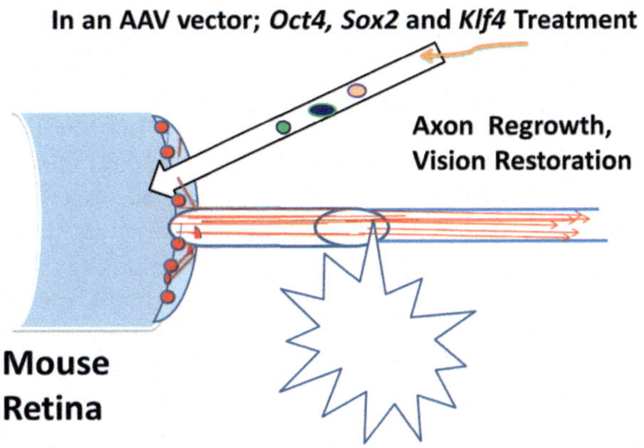

Fig. 4.4 Axonal outgrowth and restoration of vision in a mouse glaucoma model

The retinal ganglion cells (RGCs) communicate visual input from the eye to the brain though axonal connections. Injury to the retinal ganglion axons blocks transmission of this visual information to the brain for processing, causing blindness and loss of vision. Remarkably, Lu et al., 2020, report that damaged retinal ganglion cells (RGCs) can be injected with a cocktail of three transcription factors: *Oct4, Sox2,* and *Klf-4*, also known as OSK factors in an adeno-associated virus (AAV) vector restoring the RGCs to a youthful stage, regrowth of axonal projections and a gain in eyesight. Schematic cartoon is adapted from Lu et al. (2020) and Huberman (2020).

These observations strongly suggest that mouse tissues retain an epigenetic pattern memory of youthful methylation status that pattern may allow designing experiments to facilitate axonal regeneration and improve tissue physiology and function in vivo. The highlights of this important work are that it shows axonal regeneration can be achieved after injury to the optic nerve in mice with injured optic nerves, it restored vision loss in mice with a glaucoma-like condition, and more importantly the technique reversed the loss of vision in older aging animals without glaucoma and in human cells grown in Petri dish. These important observations indicate that aging clocks may be reversed by appropriate transcription control and epigenetic memory recapitulation. The technology is being licensed by Harvard University to a Boston-based company, to try the technique in humans. How the memory of youthful epigenetic state is retained still remains unknown.

CRISPR and iPSC

Another technology that has been by the Nobel Committee and has transformed the biological science landscape is the CRISPR technique and was awarded the Nobel Prize, 2020, in chemistry to Emmanuelle Charpentier and Jennifer Doudna for discovering one of gene technology's critical tools: the Clustered Regularly Interspersed Palindromic Repeats (CRISPR/Cas9) with Cas9 enzyme, providing genetic scissors for genomic editing (Jinek et al., 2012). It is now possible to use these to change the DNA of animals, plants and microorganisms with extremely high precision. Focusing on neural disease, such as Parkinson's disease, CRISPR technology can potentially allow genome editing, and review of PD patients' iPSCs has been published (Safari et al., 2019; Anzalone et al., 2019; Iarkov et al., 2020). One of the major issues using the CRISPR technology has been the gene alterations in non-specific genome region due to the double-strand breaks in the target DNA or the off-site effect, giving rise to the unintended mutations. However, Rees et al. (2019) have reported a modification in the CRISPR/Cas9 technique in which the Cas9 hybridizes to the target gene site (DNA) using a guide-engineered RNA with a spacer sequence that is complementary; the transfer of this genetic sequence information from these designed guide RNAs helps genomic DNA nicking only in one strand, thereby precluding or greatly reducing the possibility of unwanted DNA nicks in both stands and generation of mutations (Anzalone et al. 2019, 2020). Such approach may revolutionize the therapy of Parkinson's disease and other disorders linked to single-gene mutations.

Conclusions

How do genes control the structure and function of the nervous system is an age-old question that is the key to understand the working of neurons at different levels of complexity and organization (Brenner, 1974; Siddiqui, 1990). New methods and techniques of molecular genetics and cell biology in the last quarter century have given an unprecedented access to the working of brain, such as the discovery of hybridoma technology for generating mono-specific antibodies as a marker for neurons, sequencing of the human genome, labeling cells with green fluorescence protein (GFP) for live imaging of neurons. Similarly, inducing somatic cells to acquire stem cells like pluripotency (iPSCs), including neurons and support cells, using a cocktail of specific transcription factors, the use of CRISPR technology to edit genome at will are great discoveries that promise novel technologies for the mankind. Most recently, the ability to turn the aging clock backward in an old mouse by introduction of OSK transcription factors and reverse the DNA methylation to recapitulate the epigenome of youthful period and in doing so restoring vision in an old mouse and in a mouse with glaucoma and restoration of axonal growth in retinal ganglion cells is a paradigm changing and may be used to reverse not only aging and disease in nervous system, but most likely in other tissues and organs. Stay tuned; there is a lot of good science and discovery that the human brain will continue to contribute, and stem cells and their genetic manipulation will provide new answers to old questions.

Acknowledgements We regret that due to the limitation of space, many important publications were not included in the text for review and any omission is due to constraints of page limits. SSS is grateful for the award of the research grant, 19-MED-1-01-0010, by the Department of Scientific Research, U. Q., in 2020, for which the funding is still awaited. Mr. Azeem Omar Siddiqui provided excellent stimulatory discussion during the COVID-19 lockdown period of 2020, for which SSS and ZKS remain obliged.

References

Achberger, K. et al. (2019, August 27) Merging organoid and organ-on-a-chip technology to generate complex multi-layer tissue models in a human retina-on-a-chip platform. eLife (Vol. 8, p. e46188).

Alépée, N., Bahinski, A., Daneshian, M., et al. (2014). State-of-the-art of 3D cultures (organs-on-a-chip) in safety testing and pathophysiology. *Altex, 31*(4), 441–477.

Anzalone, A. V., Randolph, P. B., Davis, J. R., Sousa, A. A., Koblan, L. W., Levy, J. M., Chen, P. J., Wilson, C., Newby, G. A., Raguram, A., & Liu, D. R. (2019, December). Search-and-replace genome editing without double-strand breaks or donor DNA. *Nature, 576*(7785), 149–157.

Anzalone, A. V., Koblan, L. W., & Liu, D. R. (2020 July). Genome editing with CRISPR-Cas nucleases, base editors, transposases and prime editors. *Nature Biotechnology, 38*(7), 824–844.

Arber, C., Lovejoy, C., & Wray, S. (2017, June 13). *Alzheimers Res Ther., 9*(1), 42.

Balmer, N. V., Weng, M. K., & Zimmer, B. (2012, September 15). Epigenetic changes and disturbed neural development in a human embryonic stem cell-based model relating to the fetal valproate syndrome. *Human Molecular Genetics, 21*(18), 4104–4114.

Bal-Price, A. K., Coecke, S., & Costa, L. (2011). Conference report: Advancing the science of developmental neurotoxicity (dnt) testing for better safety evaluation. *Altex, 2012*(29), 202–215.

Bershteyn, M. et al. (2017). Human iPSC-derived cerebral organoids model cellular features of lissencephaly and reveal prolonged mitosis of outer radial glia. *Cell Stem Cell, 20*(4), 435–449.e4.

Bhatia, S. N., & Ingber, D. E. (2014, August). Microfluidic organs-on-chips. *Nature Biotechnology, 32*(8), 760–772.

Birgersdotter, A., & Sandberg, R. (2005, October). Ernberg. Gene expression perturbation in vitro—A growing case for three-dimensional (3D) culture systems. *Seminars in Cancer Biology, 15*(5), 405–412.

Boillée, S., Yamanaka, K., Lobsiger, C. S., et al. (2006, June 2). Onset and progression in inherited ALS determined by motor neurons and microglia. *Science, 312*(5778), 1389–1392.

Bordoni, M., Rey, F., Fantini, V., Pansarasa et al. (2018, December). From neuronal differentiation of iPSCs to 3D neuro-organoids: modelling and therapy of neurodegenerative diseases. *International Journal of Molecular Science, 10, 19*(12), 3972.

Bouyer, C., Chen, P., Güven, S., et al. (2016, January 6). A bio-acoustic levitational (BAL) assembly method for engineering of multilayered, 3D brain-like constructs, using human embryonic stem cell derived neuro-progenitors. *Advanced Materials, 28*(1), 161–167.

Brenner, S. (1974). The genetics of Caenorhabditis elegans. *Genetics, 77*(1), 71–94.

Cakir, B., et al. (2019). Engineering of human brain organoids with a functional vascular-like system. *Nature Methods, 16*, 1169–1175.

Camp, G. J., Badsha, F., Florio, M., et al. (2015, December). Cerebral organoids model neocortex gene networks. *Proceedings of the National Academy of Sciences, 112*(51), 15672–15677.

Carlyle, B. C., Kitchen, R. R., Kanyo, J. E., et al. (2017, December). A multiregional proteomic survey of the postnatal human brain. *Nature Neuroscience, 20*(12), 1787–1795.

Chambers, S. M., et al. (2009). Highly efficient neural conversion of human ES and iPS cells by dual inhibition of SMAD signaling. *Nature Biotechnology, 27*(3), 275–280.

Cho, H., Hashimoto, T., Wong, E., Hori, Y., Wood, L. B., Zhao, L., et al. (2013). Microfluidic chemotaxis platform for differentiating the roles of soluble and bound amyloid-beta on microglial accumulation. *Science and Reports, 3*, 1823.

Choi, Y. Y., Chung, B. G., Lee, D. H., Khademhosseini, A., Kim, J. H., & Lee, S. H. (2010). Controlled-size embryoid body formation in concave microwell arrays. *Biomaterials, 31*, 4296–4303.

Choi, Y. J., Park, J., & Lee, S. H. (2013). Size-controllable networked neurospheres as a 3D neuronal tissue model for Alzheimer's disease studies. *Biomaterials, 34*, 2938–2946.

Choi, H., Kim, H. J., Yang, J., et al. (2020, January). Acetylation changes tau interactome to degrade tau in Alzheimer's disease animal and organoid models. *Aging Cell, 19*(1), e13081.

Costamagna, G., Andreoli, L., Corti, S., & Faravelli, I. (2019, November 14). iPSCs-based neural 3D systems: A multidimensional approach for disease modeling and drug discovery. *Cells, 8*(11), 1438.

Cruz-Acuña, R., et al. (2018). PEG-4MAL hydrogels for human organoid generation, culture, and in vivo delivery. *Nature Protocols, 13*(9), 2102–2119.

Cruz-Acuña, R., & García, A. J. (2016). Synthetic hydrogels mimicking basement membrane matrices to promote cell-matrix interactions. *Matrix Biology* 57–58, 324–333.

Cruz-Acuña, R., Quirós, M., Farkas, A. E., et al. (2017, November). Synthetic hydrogels for human intestinal organoid generation and colonic wound repair. *Nature Cell Biology, 19*(11), 1326–1335.

Cugola, F. R., Fernandes, I. R., Russo, F. B., Freitas, B. C., Dias, J. L., Guimaraes, K. P., Benazzato, C., Almeida, N., Pignatari, G. C., Romero, S., et al. (2016). The Brazilian Zika virus strain causes birth defects in experimental models. *Nature, 534*, 267–271.

Dang, J., Tiwari, S. K., Lichinchi, G., Qin, Y., Patil, V. S., Eroshkin, A. M., & Rana, T. M. (2016). Zika virus depletes neural progenitors in human cerebral organoids through activation of the innate immune receptor TLR3. *Cell Stem Cell, 19*, 258–265.

Decembrini, S., Hoehnel, S., Brandenberg, N., et al. (2020). Hydrogel-based milliwell arrays for standardized and scalable retinal organoid cultures. *Science and Reports, 10*, 10275.

Deleglise, B., Magnifico, S., Duplus, E., Vaur, P., Soubeyre, V., Belle, M., et al. (2014). β-Amyloid induces a dying-back process and remote trans-synaptic alterations in a microfluidic–based reconstructed neuronal network. *Acta Neuropathologica Communications, 2*, 145.

Dolmetsch, R., & Geschwind, D. H. (2011). The human brain in a dish: The promise of iPSC-derived neurons. *Cell, 145*, 831–834.

Donnelly, D. J., & Popovich, P. G. (2008). Inflammation and its role in neuroprotection, axonal regeneration and functional recovery after spinal cord injury. *Experimental Neurology, 209*, 378–388.

Ducker, M., Millar, V., Ebner, D., & Szele, F. G. (2020, September 8). A semi-automated and scalable 3D spheroid assay to study neuroblast migration. *Stem Cell Reports, 15*(3), 789–802.

Dujardin, S., Lecolle, K., Caillierez, R., Begard, S., Zommer, N., Lachaud, C., et al. (2014). Neuron-to-neuron wild-type Tau protein transfer through a trans-synaptic mechanism: Relevance to sporadic tauopathies. *Acta Neuropathologica Communications, 30*, 14.

Eiraku, M., & Sasai, Y. (2012). Self-formation of layered neural structures in three-dimensional culture of ES cells. *Current Opinion in Neurobiology, 22*, 768–777.

Elkabetz, Y., et al. (2008). Human ES cell-derived neural rosettes reveal a functionally distinct early neural stem cell stage. *Genes & Development, 22*(2), 152–165.

Enemchukwu, N. O., Cruz-Acuña, R., Bongiorno, T., et al. (2016, January 4). Synthetic matrices reveal contributions of ECM biophysical and biochemical properties to epithelial morphogenesis. *Journal of Cell Biology, 212*(1), 113–124.

Falsig, J., Latta, M., & Leist, M. (2004, January). Defined inflammatory states in astrocyte cultures: Correlation with susceptibility towards CD95-driven apoptosis. *Journal ofNeurochemistry, 88*(1), 181–193.

Fan, W., Sun, Y., Shi, Z., Wang, H., & Deng, J. (2019). Mouse induced pluripotent stem cells-derived Alzheimer's disease cerebral organoid culture and neural differentiation disorders. *Neuroscience Letters, 15, 711*, 134433. 4.

Fernandopulle, M. S., Prestil, R., Grunseich, C., Wang, C., Gan, L., & Ward, M. E. (2018). Transcription-factor mediated differentiation of human iPSCs into neurons. *Current Protocols in Cell Biology*, e51. https://doi.org/10.1002/cpcb.51

Flaherty, E., Zhu, S., Barretto, N., et al. (2019, December). Neuronal impact of patient-specific aberrant NRXN1α splicing. *Nature Genetics, 51*(12), 1679–1690.

Frimat, J. P., & Luttge, R. (2019, May). The need for physiological micro-nanofluidic systems of the brain. *Frontiers in Bioengineering and Biotechnology, 7*(7), 100.

Fritsche, E., Crofton, K. M., Hernandez, A. F., Hougaard Bennekou, S., Leist, M., Bal-Price, A., Reaves, E., Wilks, M. F., Terron, A., Solecki, R., Sachana, M., & Gourmelon, A. (2017). OECD/EFSA workshop on developmental neurotoxicity (DNT): The use of non-animal test methods for regulatory purposes. *Altex, 34*(2), 311–315.

Fuchs, E., Tumbar, T., & Guasch, G. (2004, March 19). Socializing with the neighbors: Stem cells and their niche. *Cell, 116*(6), 769–778.

Garcez, P. P., Loiola, E. C., Madeiro da Costa, R., Higa, L. M., Trindade, P., Delvecchio, R., Nascimento, J. M., Brindeiro, R., Tanuri, A., Rehen, S. K. (2016). Zika virus impairs growth in human neurospheres and brain organoids. *Science* 352:3.

Ghatak, S., Dolatabadi, N., Trudler, D., Zhang, X., Wu, Y., Mohata, M., Ambasudhan, R., Talantova, M., & Lipton, S. A. (2019, November 29). Mechanisms of hyperexcitability in Alzheimer's disease hiPSC-derived neurons and cerebral organoids versus isogenic controls. *Elife, 8*, e50333.

Gjorevski, N., Sachs, N., Manfrin, A., Giger, S., Bragina, M. E., Ordóñez-Morán, P., Clevers, H., & Lutolf, M. P. (2016, November 24). *Nature, 539*(7630), 560–564.

González, F., Boué, S., & Izpisúa Belmonte, J. C. (2011, April). Methods for making induced pluripotent stem cells: Reprogramming à la carte. *Nature Reviews Genetics, 12*(4), 231–242.

Gonzalez, C., Armijo, E., Bravo-Alegria, J., Becerra-Calixto, A., Mays, C. E., & Soto, C. (2018, December). Modeling amyloid beta and tau pathology in human cerebral organoids. *Molecular Psychiatry, 23*(12), 2363–2374.

Goz, R. U., Akgül, G., & LoTurco, J. J. (2020, June 1). BRAFV600E expression in neural progenitors results in a hyperexcitable phenotype in neocortical pyramidal neurons. *Journal of Neurophysiology, 123*(6), 2449–2464. https://doi.org/10.1152/jn.00523.2019 (Epub 2020 May 13. Erratum in: J Neurophysiol, 2020 Sep 1;124(3):1005).

Griffin, J. M., & Bradke F. (2020). Therapeutic repair for spinal cord injury: Combinatory approaches to address a multifaceted problem. *EMBO Molecular Medicine, 12*, e11505.

Gullo, F., Manfredi, I., Lecchi, M., Casari, G., Wanke, E., & Becchetti, A. (2014). Multi-electrode array study of neuronal cultures expressing nicotinic beta2-V287L subunits, linked to autosomal dominant nocturnal frontal lobe epilepsy. An in vitro model of spontaneous epilepsy. *Frontiers in Neural Circuits, 8*, 87.

Hai A., Shappir J., & Spira, M. E. (2010). In-cell recordings by extracellular micro-electrodes. *Natural Methods, 7*, 200–202

Haring, A. P., Sontheimer, H., & Johnson, B. N. (2017, June). Microphysiological human brain and neural systems-on-a-chip: Potential alternatives to small animal models and emerging platforms for drug discovery and personalized medicine. *Stem Cell Revised Report, 13*(3), 381–406.

Haycock, J. W. (2011). 3D cell culture: A review of current approaches and techniques. *Methods in Molecular Biology, 695*, 1–15.

Henn, A., Kirner, S., & Leist, M. (2011, Mar 1). TLR2 hypersensitivity of astrocytes as functional consequence of previous inflammatory episodes. *The Journal of Immunology, 186*(5), 3237–3247.

Hilton, B. J., & Bradke, F. (2017). Can injured adult CNS axons regenerate by recapitulating development? *Development, 144*, 3417–3429.

Hilton, B. J., Moulson, A. J., & Tetzlaff, W. (2017). Neuroprotection and secondary damage following spinal cord injury: Concepts and methods. *Neuroscience Letters, 652*, 3–10.

Ho, B. X., Pek, N. M. Q., & Soh, B. S. (2018, March 21). Disease modeling using 3D organoids derived from human induced pluripotent stem cells. *International Journal of Molecular Sciences, 19*(4), 936.

Huberman, A. D. (2020, December). Sight restored by turning back the epigenetic clock. *Nature, 588*(7836), 34–36.

Hughes, C. S., Postovit, L. M., & Lajoie, G. A. (2010, May). Matrigel: A complex protein mixture required for optimal growth of cell culture. *Proteomics, 10*(9), 1886–1890.

Iarkov, A., Barreto, G. E., Grizzell, J. A., & Echeverria, V. (2020, January). Strategies for the treatment of Parkinson's disease: Beyond dopamine. *Frontier Aging Neuroscience, 31*(12), 4.

Itakura, G., Ozaki, M., Nagoshi, N., et al. (2017). Low immunogenicity of mouse induced pluripotent stem cell-derived neural stem/progenitor cells. *Science and Reports, 7*, 12996.

Jasmin, L., Vit, J. P., Bhargava, A., & Ohara, P. T. (2010, February). Can satellite glial cells be therapeutic targets for pain control? *Neuron Glia Biology, 6*(1), 63–71.

Jinek, M., Chylinski, K., Fonfara, I., Hauer, M., Doudna, J. A., Charpentier, E. (2012, August 17). A programmable dual-RNA-guided DNA endonuclease in adaptive bacterial immunity. *Science, 337*(6096), 816–821. https://doi.org/10.1126/science.1225829. Epub 2012 Jun 28. PMID: 22745249; PMCID: PMC6286148.

Jo, J., Xiao, Y., Sun, A. X., & Cukuroglu, E. (2016, August 4). Midbrain-like organoids from human pluripotent stem cells contain functional dopaminergic and neuromelanin-producing neurons. *Cell Stem Cell, 19*(2), 248–257.

Kadoshima, T., et al. (2013). Self-organization of axial polarity, inside-out layer pattern, and species-specific progenitor dynamics in human ES cell-derived neocortex. *Proceedings of National Academy of Science USA, 110*(50), 20284–20289.

Kelava, I., & Lancaster, M. A. (2016). Dishing out mini-brains: Current progress and future prospects in brain organoid research. *Developmental Biology, 420*, 199–209.

Kilic, O., Pamies, D., Lavell, E., Schiapparelli, P., Feng, Y., Hartung, T., et al. (2016). Brain-on-a-chip model enables analysis of human neuronal differentiation and chemotaxis. *Lab on a Chip, 16*, 4152–4162.

Kobayashi, Y., Okada, Y., Itakura, G., Iwai, H., Nishimura, S., Yasuda, A., Nori, S., Hikishima, K., Konomi, T., Fujiyoshi, K., et al. (2012). Pre-evaluated safe human iPSC-derived neural stem cells promote functional recovery after spinal cord injury in common marmoset without tumorigenicity. *PLoS One, 7*, e52787.

Koch, P., Kokaia, Z., Lindvall, O., & Brüstle, O. (2009, September). Emerging concepts in neural stem cell research: Autologous repair and cell-based disease modelling. *Lancet Neurology, 8*(9), 819–829.

Kouroupi, G., Antoniou, N., Prodromidou, K., Taoufik, E., & Matsas, R. (2020). Patient-derived induced pluripotent stem cell-based models in Parkinson's disease for drug identification. *International Journal of Molecular Sciences, 21*(19), 7113.

Kunze, A., Meissner, R., Brando, S., & Renaud, P. (2011). Co-pathological connected primary neurons in a microfluidic device for Alzheimer studies. *Biotechnology and Bioengineering, 108*, 2241–2245.

Lancaster, M. A., & Knoblich, J. A. (2014). Organogenesis in a dish: Modeling development and disease using organoid technologies. *Science, 345*, 1247125.

Lancaster, M. A., Renner, M., Martin, C. A., Wenzel, D., Bicknell, L. S., Hurles, M. E., Homfray, T., Penninger, J. M., Jackson, A. P., & Knoblich, J. A. (2013). Cerebral organoids model human brain development and microcephaly. *Nature, 501*, 373–379.

Lancaster, M. A., Corsini, N. S., Wolfinger, S., Gustafson, E. H., Phillips, A. W., Burkard, T. R., et al. (2017). Guided self-organization and cortical plate formation in human brain organoids. *Nature Biotechnology, 35*, 659–666.

Li, R., Sun, L., Fang, A., Li, P., Wu, Q., & Wang, X. (2017). Recapitulating cortical development with organoid culture in vitro and modeling abnormal spindle-like (aspm related primary) microcephaly disease. *Protein & Cell, 8*, 823–833.

Li, Y., Li, D., Zhao, P., Nandakumar, K., Wang, L., & Song, Y. (2020). Microfluidics-based systems in diagnosis of Alzheimer's disease and biomimetic modeling. *Micromachines (Basel), 11*(9), 787. Published 2020 August 19.

Lin, Y. T., Seo, J., Gao, F., et al. (2018, June 27). APOE4 causes widespread molecular and cellular alterations associated with Alzheimer's disease phenotypes in human iPSC-derived brain cell types. *Neuron, 98*(6), 1141-1154.e7.

Liu, P., et al. (2013). Low immunogenicity of neural progenitor cells differentiated from induced pluripotent stem cells derived from less immunogenic somatic cells. *PloS One, 8*, e69617.

Lu, X., Kim-Han, J. S., O'Malley, K. L., & Sakiyama-Elbert, S. E. (2012). A microdevice platform for visualizing mitochondrial transport in aligned dopaminergic axons. *Journal of Neuroscience Methods, 209*, 35–39.

Lu, Y., Brommer, B., Tian, X., Krishnan, A., et al. (2020, December). Reprogramming to recover youthful epigenetic information and restore vision. *Nature, 588*(7836), 124–129.

Madhavan, M., et al. (2018). Induction of myelinating oligodendrocytes in human cortical spheroids. *Nature Methods, 15*, 700–706.

Mariani, J., Coppola, G., Zhang, P., Abyzov, A., Provini, L., Tomasini, L., Amenduni, M., Szekely, A., Palejev, D., Wilson, M., et al. (2015). FOXG1-dependent dysregulation of GABA/glutamate neuron differentiation in autism spectrum disorders. *Cell, 162*, 375–390.

Marton, R. M., et al. (2019). Differentiation and maturation of oligodendrocytes in human three-dimensional neural cultures. *Nature Neuroscience, 22*, 484–491.

Mesci, P., et al. (2020). Preprint at bioRxiv. https://doi.org/10.1101/2020.05.30.125856

Meyer, K., Feldman, H. M., Lu, T., et al. (2019, January 29). REST and neural gene network dysregulation in iPSC models of Alzheimer's disease. *Cell Reports, 26*(5), 1112-1127.e9.

Miyoshi, H., & Stappenbeck, T. S. (2013, December). In vitro expansion and genetic modification of gastrointestinal stem cells in spheroid culture. *Nature Protocol, 8*(12), 2471–2482.

Monzel, A. S., Smits, L. M., Hemmer, K., Hachi, S., Moreno, E. L., van Wuellen, T., Jarazo, J., Walter, J., Bruggemann, I., Boussaad, I., et al. (2017). Derivation of human midbrain-specific organoids from neuroepithelial stem cells. *Stem Cell Rep., 8*, 1144–1154.

Moreno, E., Hachi, S., Hemmer, , et al. (2015). Differentiation of neuroepithelial stem cells into functional dopaminergic neurons in 3D microfluidic cell culture. *Lab on a Chip, 15*, 2419–2428.

Morin, F., Nishimura, N., Griscom, L., Le Pioufle, B., Fujita, H., Takamura, Y., et al. (2006). Constraining the connectivity of neuronal networks cultured on microelectrode arrays with microfluidic techniques: A step towards neuron-based functional chips. *Biosensors & Bioelectronics, 21*, 1093–1100.

Morrison, S. J., & Spradling, A. C. (2008, February 22). Stem cells and niches: Mechanisms that promote stem cell maintenance throughout life. *Cell, 132*(4), 598–611.

Nagoshi, N., & Okano, H. (2018). iPSC-derived neural precursor cells: Potential for cell transplantation therapy in spinal cord injury. *Cellular and Molecular Life Sciences, 75*, 989–1000.

Nakatake, Y., et al. (2020). Generation and profiling of 2,135 human ESC lines for the systematic analyses of cell states perturbed by inducing single transcription factors. *Cell Reports, 31*, 107655.

Ng, A. H. M., Khoshakhlagh, P., Rojo Arias, J. E., et al. (2020). A comprehensive library of human transcription factors for cell fate engineering. *Nature Biotechnology*.

Nori, S., et al. (2011). Grafted human-induced pluripotent stem-cell-derived neurospheres promote motor functional recovery after spinal cord injury in mice. *Proceedings of the National Academy of Sciences of the United States of America, 108*, 16825–16830.

Ormel, P. R., Vieira de Sá, R., van Bodegraven, E. J., et al. (2018, October 9). Microglia innately develop within cerebral organoids. *Nature Communications, 9*(1), 4167.

Page, S., Patel, R., Raut, S., & Al-Ahmad, A. (2018). Neurological diseases at the blood-brain barrier: Stemming new scientific paradigms using patient-derived induced pluripotent cells. *Biochimica Biophysical Acta Molecular Basis Disease, 1866*, 165358.

Park, J., Kim, S., Park, S. I., et al. (2014). A microchip for quantitative analysis of CNS axon growth under localized biomolecular treatments. *Journal of Neuroscience Methods, 221*, 166–174.

Paşca, A. M., et al. (2015). Functional cortical neurons and astrocytes from human pluripotent stem cells in 3D culture. *Nature Methods, 12*(7), 671–678.

Pfrieger, F. W., & Barres, B. A. (1997). Synaptic efficacy enhanced by glial cells in vitro. *Science, 277*(5332), 1684–1687.

Phelps, E. A., Enemchukwu, N. O., & Fiore, V. F. (2012, January 3). Maleimide cross-linked bioactive PEG hydrogel exhibits improved reaction kinetics and cross-linking for cell encapsulation and in situ delivery. *Advanced Materials, 24*(1), 64–70, 2.

Platt, A., David, B. T., & Fessler, A. R. G. (2020). Stem cell clinical trials in spinal cord injury: A brief review of studies in the United States. *Medicines (Basel), 7*(5), 27. [Table 4]

Qiao, H., Zhang, Y. S., & Chen, P. (2018). Commentary: Human brain organoid-on-a-chip to model prenatal nicotine exposure. *Frontiers on Bioengineering and Biotechnology, 6*, 138. Published 2018 October 4.

Quadrato, G., Nguyen, T., Macosko, E. Z., Sherwood, J. L., Min Yang, S., Berger, D. R., Maria, N., Scholvin, J., Goldman, M., Kinney, J. P., Boyden, E. S., Lichtman, J. W., Williams, Z. M., McCarroll, S. A., & Arlotta, P. (2017, May 4). Cell diversity and network dynamics in photosensitive human brain organoids. *Nature, 545*(7652), 48–53.

Rabesandratana, O., Chaffiol, A., Mialot, A., et al. (2020). Generation of a transplantable population of human iPSC-derived retinal ganglion cells. *Frontiers in Cell Deviation Biology, 8*, 585675. https://doi.org/10.3389/fcell.2020.585675

Raja, W. K., Mungenast, A. E., Lin, Y. T., Ko, T., Abdurrob, F., Seo, J., & Tsai, L. H. (2016). Self-organizing 3D human neural tissue derived from induced pluripotent stem cells recapitulate Alzheimer's disease phenotypes. *PLoS ONE, 11*, e0161969.

Ramani, A., Müller, L., Ostermann P. N., et al. (2020). SARS-CoV-2 targets neurons of 3D human brain organoids. *EMBO Journal, 39*(20), e106230.

Ranjan, V. D., Qiu, L., Tan, E. K., et al. (2018, September). Modelling Alzheimer's disease: Insights from in vivo to in vitro three-dimensional culture platforms. *Tissue Engineering and Regenerative Medicine, 12*(9), 1944–1958.

Rees, H. A., Wilson, C., Doman, J. L., & Liu, D. R. (2019, May 8). Analysis and minimization of cellular RNA editing by DNA adenine base editors. *Science Advances, 5*(5), eaax5717.

Reynolds, B. A., & Weiss, S. (1992). Generation of neurons and astrocytes from isolated cells of the adult mammalian central nervous system. *Science, 255*(5052), 1707–1710.

Safari, F., Zare, K., Negahdaripour, M., Barekati-Mowahed, M., Ghasemi, Y. (2019, May 9). CRISPR Cpf1 proteins: structure, function and implications for genome editing. *Cell Bioscience, 9*, 36. https://doi.org/10.1186/s13578-019-0298-7. PMID: 31086658; PMCID: PMC6507119.

Sartore, R. C., Cardoso, S. C., Lages, Y. V., et al. (2017). Trace elements during primordial plexiform network formation in human cerebral organoids. *Peer Journal, 5*, e2927.

Sasai, Y. (2013). Cytosystems dynamics in self-organization of tissue architecture. *Nature, 493*, 318–326.

Schildknecht, S., Pöltl, D., Nagel, D. M., et al. (2009, November 15). Requirement of a dopaminergic neuronal phenotype for toxicity of low concentrations of 1-methyl-4-phenylpyridinium to human cells. *Toxicology and Applied Pharmacology, 241*(1), 23–35.

Selmer, K. K., Eriksson, A. S., Brandal, K., et al. (2009). Parental SCN1A mutation mosaicism in familial Dravet syndrome. *Clinical Genetics, 76*, 398–403.

Shpichka, A., Bikmulina, P., Peshkova, M., et al. (2020, August 28). Engineering a model to study viral infections: Bioprinting, microfluidics, and organoids to defeat coronavirus disease 2019 (COVID-19). *International Journal Bioprint, 6*(4), 302.

Siddiqui, S. S. (1990). Mutations affecting axonal growth and guidance of motor neurons and mechanosensory neurons in the nematode Caenorhabditis elegans. *Neuroscience Research Supplements, 13*, S171–S190.

Siddiqui, S. S. (2002, January). Metazoan motor models: Kinesin superfamily in C. elegans. *Traffic, 3*(1), 20–28.

Siddiqui, S., Al-Qahtani, M., Allaf, F., Sivakumar, L., & Siddiqui, Z. (2020). Application of nanomaterials in cancer diagnosis. *Drug Delivery, and Therapy*. https://doi.org/10.1007/978-981-15-4802-4_8

Skardal, A., Shupe, T., & Atala, A. (2016, September). Organoid-on-a-chip and body-on-a-chip systems for drug screening and disease modeling. *Drug Discovery Today, 21*(9), 1399–1411.

Slanzi, A., Iannoto, G., Rossi, B., Zenaro, E., & Constantin, G. (2020). In vitro models of neurodegenerative diseases. *Frontiers Cell Deviation Biology, 8*, 328. Published 2020 May 13. https://doi.org/10.3389/fcell.2020.00328

Slembrouck-Brec, A., Rodrigues, A., Rabesandratana, O., et al. (2019, July). Reprogramming of adult retinal Müller glial cells into human-induced pluripotent stem cells as an efficient source of retinal cells. *Stem Cells International, 15*(2019), 7858796.

Sloan, S. A., Darmanis, S., Huber, N., Khan, T. A., et al. (2017, August 16). Human astrocyte maturation captured in 3D cerebral cortical spheroids derived from pluripotent stem cells. *Neuron, 95*(4), 779-790.e6.

Son, M. Y., Sim, H., Son, Y. S., Jung, K. B., et al. (2017 December). Distinctive genomic signature of neural and intestinal organoids from familial Parkinson's disease patient-derived induced pluripotent stem cells. *Neuropathology and Applied Neurobiology, 43*(7), 584–603.

Song, E., et al. (2020). *Preprint at bioRxiv*. https://doi.org/10.1101/2020.06.25.169946

Sun, J., Huang, Y., Gong, J., et al. (2020). Transplantation of hPSC-derived pericyte-like cells promotes functional recovery in ischemic stroke mice. *Nature Communications, 11*, 5196.

Sweeney, M. D., Zhao, Z., Montagne, A., Nelson, A. R., & Zlokovic, B. V. (2019). Blood-brain barrier: From physiology to disease and back. *Physiological Review, 99*, 21–78.

Takahashi, K., & Yamanaka, S. (2006). Induction of pluripotent stem cells from mouse embryonic and adult fibroblast cultures by defined factors. *Cell, 126*, 663–676.

Takahashi, K., & Yamanaka, S. (2006, August 25). Induction of pluripotent stem cells from mouse embryonic and adult fibroblast cultures by defined factors. *Cell, 126*(4), 663–676.

Takahashi, K., Tanabe, K., Ohnuki, M., Narita, M., Ichisaka, T., Tomoda, K., & Yamanaka, S. (2007, November 30). Induction of pluripotent stem cells from adult human fibroblasts by defined factors. *Cell, 131*(5), 861–872.

Theodorou, E., et al. (2009). A high throughput embryonic stem cell screen identifies Oct-2 as a bifunctional regulator of neuronal differentiation. *Genes & Development, 23*, 575–588.

Thomson, J. A., Itskovitz-Eldor, J., Shapiro, S. S., et al. (1998, November 6). Embryonic stem cell lines derived from human blastocysts. *Science, 282*(5391), 1145–1147.

Tsuji, O., et al. (2010). Therapeutic potential of appropriately evaluated safe-induced pluripotent stem cells for spinal cord injury. *Proceedings of the National Academy of Sciences of the United States of America, 107*, 12704–12709.

Tzekaki, E., Tsolaki, M., & Pantazaki, A. (2019, Jan–Apr). Hell technical characteristics of Alzheimer model based on organ technology (organoid). *Journal of Nuclear Medicine*, 22 Suppl, 195–208.

Ullian, E. M., et al. (2001). Control of synapse number by glia. *Science, 291*(5504), 657–661.

Van der Helm, M., Odijk, M., Frimat, J.-P., van der Meer, A., Eijkel, J., Van den Berg, A., & Segerink, L. (2016). Direct quantification of transendothelial electrical resistance in organs-on-chips. *Biosensors and Bioelectronics* 85. https://doi.org/10.1016/j.bios.2016.06.014

Vandenhaute, E., Sevin, E., Hallier-Vanuxeem, D., Dehouck, M. P., & Cecchelli, R. (2012, April). Case study: Adapting in vitro blood-brain barrier models for use in early-stage drug discovery. *Drug Discovery Today, 17*(7–8), 285–290.

Viviani, B., Corsini, E., Galli, C. L., & Marinovich, M. (1998, June). Glia increase degeneration of hippocampal neurons through release of tumor necrosis factor-alpha. *Toxicology and Applied Pharmacology, 150*(2), 271–276.

Waldmann, T., Rempel, E., & Balmer, N. V. (2014, March 17). Design principles of concentration-dependent transcriptome deviations in drug-exposed differentiating stem cells. *Chemical Research in Toxicology, 27*(3), 408–420.

Wang, Y., Wang, L., Zhu, Y., & Qin, J. (2018, March 13). Human brain organoid-on-a-chip to model prenatal nicotine exposure. *Lab on a Chip, 18*(6), 851–860.

Xiang, Y., Tanaka, Y., Patterson, B., et al. (2017, September 7). Fusion of regionally specified hPSC-derived organoids models human brain development and interneuron migration. *Cell Stem Cell, 21*(3), 383-398.e7.

Yagi, T., Ito, D., Okada, Y., Akamatsu, W., Nihei, Y., Yoshizaki, T., Yamanaka, S., Okano, H., & Suzuki, N. (2011). Modeling familial Alzheimer's disease with induced pluripotent stem cells. *Human Molecular Genetics, 20*, 4530–4539.

Yamanaka, S. (2007). Strategies and new developments in the generation of patient-specific pluripotent stem cells. *Cell Stem Cell, 1*, 39–49.

Yamanaka, S. (2012, June 14). Induced pluripotent stem cells: past, present, and future. *Cell Stem Cell, 10*(6), 678–684. https://doi.org/10.1016/j.stem.2012.05.005. PMID: 22704507

Youssef, A. R., Emara, R., Taher, M. M., Al-Allaf, F. A., Almalki, M., Almasri, M. A., & Siddiqui, S. S. (2019, July 2). Effects of mineral trioxide aggregate, calcium hydroxide, biodentine and Emdogain on osteogenesis, Odontogenesis, angiogenesis and cell viability of dental pulp stem cells. *BMC Oral Health, 19*(1), 133.

Zahir, N., & Weaver, V. M. (2004, Febraury). Death in the third dimension: Apoptosis regulation and tissue architecture. *Current Opinion in Genetics & Development, 14*(1), 71–80.

Zhang, S. C., Wernig, M., Duncan, I. D., Brustle, O., & Sloan, J. A. (2001). In vitro differentiation of transplantable neural precursors from human embryonic stem cells. *Nature Biotechnology, 19*, 1129–1133.

Zhang, Y., et al. (2013). Rapid single-step induction of functional neurons from human pluripotent stem cells. *Neuron, 78*, 785–798.

Zhang, T., Lin, Y., & Liu, J. (2016, July). Rbm24 regulates alternative splicing switch in embryonic stem cell cardiac lineage differentiation. *Stem Cells, 34*(7), 1776–1789.

Zhuang, P., Sun, A. X., An, J., Chua, C. K., et al. (2018, Febraury). 3D neural tissue models: From spheroids to bioprinting. *Biomaterials, 154*, 113–133.

Zimmerling, A., & Chen, X. (2020, December). Bioprinting for combating infectious diseases. *Bioprinting, 20*, e00104.

Zurich, M. G., Eskes, C., Honegger, P., Bérode, M., & Monnet-Tschudi, F. (2002, October 1). Maturation-dependent neurotoxicity of lead acetate in vitro: Implication of glial reactions. *Journal of Neuroscience Research, 70*(1), 108–116.

Chapter 5
Differentiation of Stem Cells into Cardiomyocyte Lineage: In Vitro Cell Culture, In Vivo Transplantation in Animal Models

Sumira Malik and Archna Dhasmana

Abstract Globally, cardiovascular disease is a significant threat responsible for the higher death rate in the current scenario. Myocardial infarction causes ischemic injury to irreversibly damages cardiomyocytes, making them non-functional and leading to heart failure due to the lack of regeneration capacity. Clinically, organ transplantation and autologous cell-based therapies are potentially used to replace and restore damaged and unhealthy tissues. Nevertheless, the deficit of donor cells and lack of cell potency to differentiate into cardiac cells is another major problem for repairing damaged tissue. Thus, identifying the mechanism of cell differentiation, proliferation, and specification into the cardiac cells is a crucial step. This process can be accelerated by utilizing stem cells like pluripotent embryonic stem (ES) cells, which have the potency to differentiate into different cell lines and infinite probability to act as a source of cardiovascular cells under the influence of unspecified regulatory elements. This pluripotent ES differentiates into cardiomyocytes through the complex cellular pathways, controlled under gene regulation, expression, specific signaling molecules, and physiological parameters. To understand the diverse molecular machinery and regulatory pathway of stem cell differentiation is one of the difficult conundrums for our researcher. Although, to date, many innovations have been made to resolve this uncertainty to the formation of novel cures such as induced pluripotent stem cells (iPSCs). The chapter reviews the concepts of stem cell differentiation into cardiomyocytes through various in vitro cell culture and in vivo therapies at the pre-clinical to clinical level to evaluate the therapeutic application for the regeneration of the bio-functional heart.

Keywords Myocardial infarction · Pluripotent stem cells · IPScs · Cell culture · In vivo model organisms

S. Malik (✉)
Amity Institute of Biotechnology, Amity University Jharkhand, Ranchi, Jharkhand 834001, India
e-mail: smalik@rnc.amity.edu

A. Dhasmana
Department of Biotechnology, School of Applied and Life Sciences, Uttaranchal University, Uttarakhand 248007, India

F. A. Khan (ed.), *Advances in Application of Stem Cells: From Bench to Clinics*, Stem Cell Biology and Regenerative Medicine 69,
https://doi.org/10.1007/978-3-030-78101-9_5

Differentiation of Stem Cells into Cardiomyocytes Lineage: In Vitro Cell Culture

Introduction to Pluripotent Embryonic Cells

The stem cells that have unlimited capability to undergo self-renewal and the ability to undergo differentiation for the formation of ectoderm, endoderm, and mesoderm are called Pluripotent embryonic stem cells. These stem cells can arise into various cells that can form a complete human body (Ameen et al., 2008). The pluripotent stem cell-derived, in vitro, differentiated cardiomyocytes retain the functional properties of the pluripotent stem cell and own the phenotype of cardiac cells that are more stabilized and reproducible on both clinical and physiological aspects. Further, these pluripotent stem cell-derived, in vitro differentiated cardiomyocytes are the important models for in dissection of molecular events involved in cardiogenesis. These pluripotent stem cell-derived cardiomyocytes may serve as significant in vitro tools for developing and generating safe drugs in the pharmaceutical industry. The previous studies have reported the numerous methodologies in the formation of functional cardiomyocytes through the differentiation of human Embryonic Stem cells (hESCs) and induced pluripotent stem (iPS) cell technology (Zhang et al., 2009; Zwi et al., 2009; Haase et al., 2009).

There is a significant contribution of pluripotent stem cell-derived, in vitro differentiated cardiomyocytes from the future health perspective. Therefore, it is a critical requirement to validate standardized assays of such cells in the in vitro models to ensure the safety potential and efficacy of new drugs in the pharmaceutical industry (Vidarsson et al., 2010; Bram et al., 2009). However, there are impediments to generate the pluripotent stem cell-derived, in vitro differentiated cardiomyocytes cellular preparations that may cause cancer risk. The successful transplantation may reduce the chances of immune rejection (Braam et al., 2009). Furthermore, the advancement of research in the field of stem cell technology has generated the possibilities for the treatment of damaged and degenerated cardiac tissue through the repairing of damaged myocardium using stem cell-derived cardiomyocytes in preclinical studies in translational medicine (Nelson et al., 2009; van Laake, et al., 2007).

Development of Cardio Myocytes from Pluripotent Stem Cells

There are efficient methods that induce differentiation in cardiomyocytes to generate the qualitative and quantitative homogenous populations of cardiomyocytes for futuristic cell-based therapeutic applications. Perhaps, these futuristic cell therapies will urge well-structured methodologies and protocols for repetitive results fulfilling the regulatory requirements. The process of cardiogenesis is an extraordinarily vigorous and well-coordinated process that involves sequential expression of signaling molecules such as signal transduction molecules and transcription factors. The process of early differentiation of early mesoderm through cardiac mesoderm and committed cardiac progenitors to functional beating cardiomyocytes with the expression of markers is explained in Fig. 5.1.

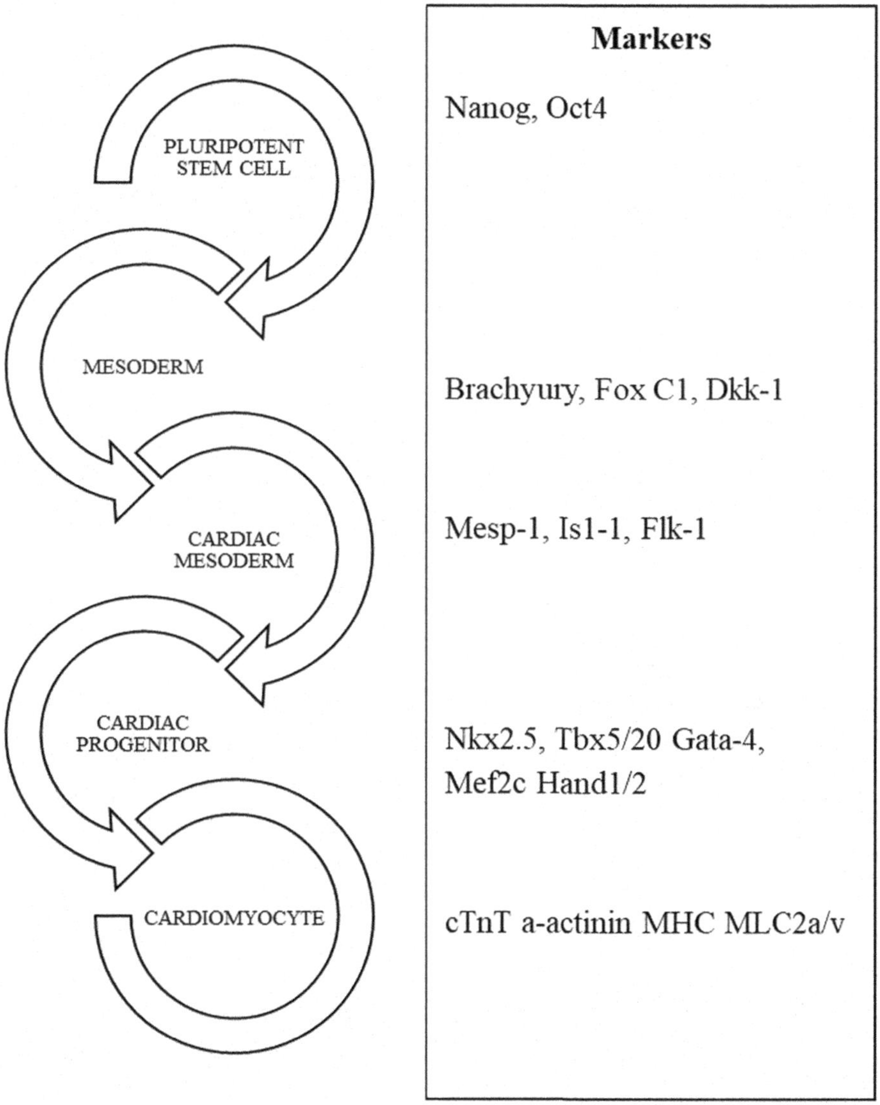

Fig. 5.1 Sequential steps in the formation of cardio myocytes from pluripotent stem cells depicting the markers expressed

Table 5.1 Transcription factors, signaling pathways, growth factors and microRNAs involved in the cardio genesis

S. No.	Factors and pathways involved in cardio genesis	Names of Factors and pathways involved in cardio genesis	References
1	Signaling pathways	Wnts/Nodal BMPs FGFs	• Peterkin, et al. (2003) • Plageman and Yutzey (2004) • Riley and Cross (1998) • Watt et al., (2004) • Hiroi et al., (2001)
2	Growth factors	T/Brachyury, Mesp1, Nkx2.5, Tbx5/20, Gata4, Mef2c, and Hand1/2	• Watt et al., (2004) • Marvin and et al., (2001) • Mima et al. (1995) • Winnier et al., (1995)
3	MicroRNAs	miR-1, miR-133, miR-206, miR-143 and -145 miR-133	• Zhang and Bradley (1996) • Pasquinelli et al., (2005) • Zhao and Srivastava (2007) • Calliset al., (2008) • van Rooij et al., (2008) • Wang et al., (2008) • Ivey et al., (2008)

There are four major steps in the generation of cardio myocytes from pluripotent stem cells:

1. Mesoderm formation which requires T/Brachyury, Mesp1 and Nkx2.5, Tbx5/20, Gata4, Mef2c, and Hand1/2 transcription factors
2. Pattern formation of mesoderm toward cardiogenic mesoderm
3. Formation of cardiac mesoderm
4. Maturation of early cardio myocytes.

The transcription factors, signaling pathways, growth factors, and microRNAs involved in developing specialized cardiac subtypes and differentiation process during the cardiogenesis are summarized in Table 5.1 and Fig. 5.1.

Steps of Pluripotent Stem Cells Differentiation into Cardiomyocytes

1. Formation of embryoid body and spontaneous cardiomyocyte differentiation- The embryoid body consists of derivatives of the three germ layers (ectoderm, endoderm, and mesoderm), which develop spontaneously. These populations of cells are the mixed cells with functional properties of cardiomyocytes. These cardiomyocytes are the first cell types induced from pluripotent stem cells in embryoid bodies, which induce the stimulation of expression of markers for

mesodermal and early cardiac cell lineages through the process of cell to cell communication (Tran et al., 2009).

2. Co-culture of Pluripotent Stem Cells and cardio inductive Cell- The widely used for cardiomyocyte in vitro differentiation approach of co-culture involves the crucial function of anterior endoderm in the cardiac induction of adjacent mesodermal structures (Synnergren et al., 2008; Cao et al., 2008a, b; Xu et al., 2009). In this method, co-culture of the visceral endoderm-like cell line (END-2), a derivative of mouse P19 embryonic carcinoma (EC) cells and pluripotent stem cells which forms the beating clusters of cells that also demonstrate the characteristics of cardiomyocytes (Passier, 2008).
3. Guided cardiomyocyte differentiation with Specific Factors-In this method, signaling pathways that are responsible for the regulation of cardiogenesis are mimicked in the form of cell culture. To this system, growth factors such as FGFs, BMPs, and Wnts are supplemented that has the capability to induce mesoderm or endoderm development in pluripotent stem cells (Xu et al., 2006; Rust et al., 2009; Kolossov et al., 2005).
4. Cardiac progenitor cells-The three major cardiac cell lineages may arise from a common multipotent cardiovascular progenitor cell population that has the capability to display the specific expression of markers (Bu et al., 2009; Kattman et al., 2006; Moretti et al., 2006).

Methods in Cardiomyocyte Differentiation

The methods in cardiomyocyte differentiation from pluripotent stem cells are explained in Fig. 5.2.

Concept and Methods of In Vitro Differentiation of ESCs and iPSCs Cells into Cardiomyocyte In Vitro Differentiation of ESCs

Embryonic stem cells (ESCs) or ES cells act as the best source for genetic manipulation through cultivation in vitro in the form of 3D aggregates, which are called embryoid bodies. ES cells can differentiate into derivatives of all three primary germ layers, including cardiomyocytes.

Several parameters specifically influence ES cells' differentiation potency to form cardiomyocytes in culture (Wobus et al., 2002), as summarized in Fig. 5.3.

(1) The initial number of cells in the EBs.
(2) Supplements in media, FBS, growth factors
(3) ES cell lines, and
(4) The time of EB plating.

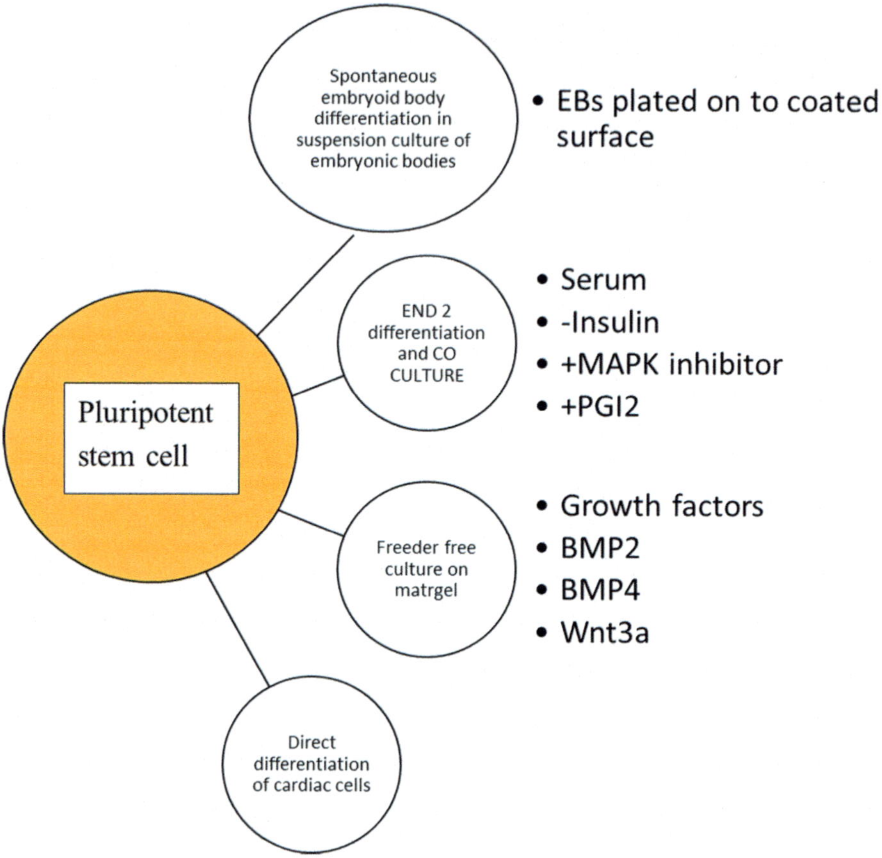

Fig. 5.2 Methods in cardiomyocyte differentiation from pluripotent stem cells

Differentiation of iPSCs into Human Cardio Myocytes

Induced pluripotent stem cells iPSCs are derived from patients who had different cardiac diseases or disorders, such as congenital heart disease, and become essential tools for studying the mechanisms underlying the disease pathogenesis and for the development of new treatment opportunities (Itzhaki et al., 2011; Moretti et al., 2010). However, on the basis of the methodology using embryonic stem cells, human iPSCs are capable of differentiating into beating cardiomyocytes through exposure to a variety of stimuli (Shiba et al., 2009; Yoshida & Yamanaka, 2011) (Table 5.2). ESCs, methods for producing embryoid bodies using iPSCs, have been successfully differentiated into cardiomyocytes (Zhang et al., 2009).

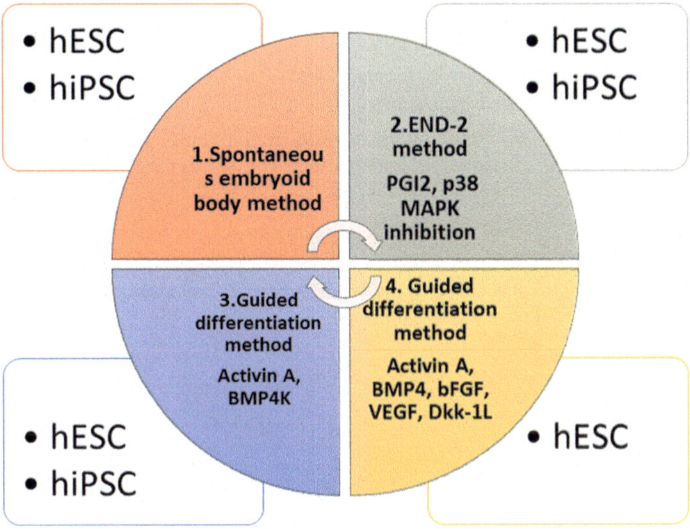

Fig. 5.3 Cardiac differentiation methods of heSCs and hiPSCs into cardiomyocytes

Methods for Differentiation of iPSCs and ESCs into Cardiomyocytes

Figure 5.3 and Table 5.2 summarizes the differentiation approaches currently used for cardiomyocyte differentiation from human embryonic stem cells, and human-induced pluripotent stem cells involve different growth factors as well as signaling molecules to form cardiomyocytes.

Future Trends of iPSCs in Stem Cell Research with Regard to Cardio Myocytes

- Although iPSCs are good candidates for drug screening and disease modeling applications, they are not without limitations. The following reasons make them unfavorable promising entities.
- The late-onset nature of many diseases likely shows a failure in reiterating the disease development accurately.
- There is no assurance of result without in vivo studies as complex cellular interactions in human metabolic processes often cannot be recapitulated in these in vitro culture systems.
- A large panel of patient-derived iPSCs needs to be evaluated to consider when the wide range of immune responses may elicit when applied to a larger heterogeneous

Table 5.2 Differentiation approaches currently used for cardiomyocyte differentiation from human embryonic stem cells and human-induced pluripotent stem cells involves different growth factors as well as signaling molecules to form cardiomyocytes

Method	Pluripotent culture	Mesoderm induction factors	Cardiac specification factors	Cardiac differentiation factors	References
Suspension EB in StemPro34	Knock-out Serum Replacement (KSR)/FGF2	Activin A, BMP4, FGF2	VEGFA, DKK1	VEGFA, FGF2	Yang and Soonpaa (2008)
			VEGFA, DKK1, SB431542, dorsomorphin	VEGFA, FGF2	Kattman et al. (2011)
			IWR1	Tri-iodothyronine	Willems et al. (2011)
Forced aggregation EB	Colonies on MEFs	Activin A, FGF2	20% FBS and DMEM	20% FBS and DMEM	Burridge et al. (2012)
	Monolayer on Geltrex	BMP4, FGF2	RPMI and FBS or RPMI-INS	RPMI-INS	Burridge and Thompson (2011)
	KSR/FGF2 on MEF	Activin A, BMP4, FGF2, VEGFA, SCF	LI-BEL	LI-BEL	Elliott et al. (2011)
Monolayer differentiation	Monolayer on Matrigel with MEFs	Activin A, BMP4	RPMI plus B27	RPMI plus B27	Laflamme et al. (2007)
		BMP4, FGF2, Activin A	NOGGIN, RAi, DKK1	DKK1	Zhang and Jiang (2011)
	Monolayer on Matrigel with MEFs	Activin A, BMP4, FGF2	VEGFA, DKK1	VEGFA, FGF2	Uosaki et al. (2011)
	KSR/FGF2 on MEFs	Activin A, BMP4, FGF2, VEGFA, SCF	LI-BEL	LI-BEL	Elliott and Braam (2011)
	mTeSR1	Activin A, BMP4	IWR1 or IWP4	RPMI plus B27	Hudson et al. (2012)
	mTeSR1 plus Y27632	CHIR99021	IWR2 or IWP4	RPMI plus B27	Lian and Hsiao (2012)

(continued)

Table 5.2 (continued)

Method	Pluripotent culture	Mesoderm induction factors	Cardiac specification factors	Cardiac differentiation factors	References
	Chemically defined E8 medium on a synthetic vitronectin peptide matrix	CHIR99021, Wnt-C59	CDM3	CDM3	Burridge and Matsa (2014)

human population. It requires clinical trials of a specific drug (Grskovic et al., 2011).

- For these reasons, and despite their considerable promise, patient-specific iPSC models failed to substitute animal models. The persistent issue with maturity and accurate recapitulation of onset disease phenotypes has led to an explosion of bioengineering strategies to improve engineered cardiac tissue development in vitro. Such techniques seek to impersonate multiple aspects of the cardiac micro environmental niche.

Applications of In-Vitro Culture in Cardiogenesis

Cardiac transplantation is currently the treatment of choice for end-stage heart failure; however, the number of available donor organs limits this treatment to a minority of patients. In-vitro Cell-based therapies have recently emerged as an innovative approach for the treatment of degenerative heart diseases. Significant challenges remain to be overcome before this therapy can be practiced in clinics. Other critical problems include inefficient differentiation, tumorigenicity, immunogenicity, as well as complicated ethical issues surrounding the isolation of cells from in vitro fertilized human embryos (Blin et al., 2010). Human ES cells obviously represent a potentially valuable and renewable source of cells that could be used for transplantation therapy. Human ES cell lines are immortal and pluripotent, and consequently, derivatives of these cell lines can theoretically be used to treat a wide range of devastating diseases whose underlying pathology involves cell degeneration, death, or acute injury. Significant challenges to the therapeutic application are the sustenance against the host system immune rejection. The development of large stem cell banks that represent a wide array of histocompatibility backgrounds is a suggested trial to overcome immune rejection challenge.

Differentiation of Stem Cells into Cardiomyocyte Lineage: In Vivo Transplantation in Animal Models

Introduction to Embryonic Cells Involved in In-Vivo Development of Cardiomyocyte Lineage

In invertebrates, the heart is the first functional organ developed after the gastrulation phase of embryogenesis. The intercalating anterior mesodermal cell between the ectoderm and endoderm germ layer forms the primary mid-streak act as progenitor cardiomyocytes. These cardiac progenitors' cells specifically differentiate into cardiomyocyte lineages to form heart muscle cells (Aguilar-Sanchez et al., 2018). Besides that, the endocardia cells form the endothelial cell lining, and vascular smooth muscle forms the vascular system. The cardiomyocytes have a low division rate, and their division rate subsequently reduces or stops until the postnatal time reached. This results in a less or restricted number of proliferated cells, but the heart grows with enlarged cell size to perform more activity (Senyo et al., 2013; Ali et al., 2014). Therefore, in the case of adults, the cardiomyocytes' self-renewal or proliferation rate is significantly much lesser than the endothelial and mesenchymal cell; as a result heart injury or damage cannot be cured selfheal themselves (Kajstura et al., 2010). The turnover rate of the human cardiomyocytes is very poor; only the cardiac progenitor cells present in the heart possess the self-renewing capacity and have multipotent stem cell that could differentiate into cardiac cell lineages (Bang et al., 2016; Bergmann et al., 2009). In a developing embryo body, the early mesodermal cell differentiated to form cardiovascular progenitors, and these progenitors potentially differentiate into cardiac cells, i.e., cardiomyocytes, smooth muscle cells, and endothelial cells, respectively (Brade et al., 2013). These cardiac progenitor matures and results in the separated chambered heart formation. Thus, embryonic stem (ES) cells could be utilized as sources for the repair and regeneration of cardiac incision or injury.

- Embryo stem cell → Ectoderm, Endoderm, Mesoderm
- Mesoderm → Cardiovascular progenitor → Smooth muscle cells, Endothelial cells, Cardiac progenitors
- Cardiac progenitors → Mature cardiomyocytes (MC)
- Mature ***cardiomyocytes (MC)***, Smooth muscle cells, endothelial cells → Bio-functional heart.

5. These progenitors, embryonic pluripotent stem cells differentiated into differentiated cell lineages, make a complex cardiac system. Other than it different cell lines such as adult stem cells, cardiac cells, vascular endothelial cells, mesenchymal cells act as the progenitor to cardiac tissue / whole organ generation (Brade et al., 2013; Wysoczynski & Bolli, 2020). However, during the post-natal period, the differentiation or capacity of cardiac cells to regenerate the tissue or replace the damaged one reduces; stem cells are the only opportunity.

6. In cardiac damage, grafting or whole organ transplantation is the only clinical treatment to save a life. In clinical studies, it was reported that patients own stem cells (multipotent cell or bone-marrow) used for repair of damaged tissue, but due to lack of potency they are less effective and have a short life-span (Keller, 2005). ES cells have pluripotency, differentiate into any cell lineage, and make a better opportunity for clinical studies and tissue regeneration purposes. ES cells derived from the inner blastomeres of a developing embryo, subsequently cultured, maintained into cardiomyocytes lineages and function of different factors of signal transduction pathways involved in ES cell differentiation (Guo et al., 2016). It was reported that the ES cell obtained from the mouse demonstrated pluripotency and regenerated in all types of tissue cell under in vivo conditions (Morey et al., 2015). The potential of ES cells to differentiate and form cardiac lineage, can be identified on the basis of their physiological properties, such as the contracting nature of cardiomyocytes (Bartosh et al., 2008). The development of cardiomyocyte lineage form ES cells results in the formation of hematopoietic and vascular systems as same as the in vivo development (Arabadjiev et al., 2014). The applicability of ES cell differentiated cardiomyocytes used for the transplantation and to cure the cardiovascular problems. Therefore, ES cells have vast potential in the area of therapeutics and tissue regeneration, such as to treat cardiovascular diseases and restore cardiac function.

Function of Different Factors of Signal Transduction Pathways Involved in ES Cell Differentiation

ES cells are isolated from the embryonic inner cell mass self-renewal and pluripotent, but they need specific signaling to differentiate and proliferate into specific cell lineage. Besides that, to improve the cell functioning and survivability of ES cells derived cardiomyocyte, there is a need for extrinsic and intrinsic factors. (Arabadjiev et al., 2014). The progenitor cell lineage commitment and differentiation into the cardiomyocyte require the specific intrinsic factors such as zinc finger transcription factor GATA4, Nkx2.5 for the activation of myocardial differentiation genes and plays a role in the activation of many myocardial differentiation genes (Arabadjiev et al., 2014; Chen et al., 2008). The specific differentiation of genes expressed at different myocardial differentiation stages is listed in the figure below (Table 5.3).

Thus, in 2002, Yang and co-worker designed VEGF incorporated cardiomyocytes to treat myocardial infarction in mice model (Yang et al., 2002). The transplanted VEGF-expressing cells result in neo-vascularization and regain functioning of damaged hearts compared to non-VEGF-expressed transplanted cells. Besides the intrinsic factors, extrinsic factors i.e., growth factors, and chemicals induced the signaling and gene expression for the ES cells differentiation into the myocardial lineages (Bartosh, 2008; Gude et al., 2018; Pal et al., 2012). In cardiomyogenesis, these factors (Table 5.4) play an essential role in heart functioning and be a potential therapeutic agent used for cardiomyocyte transplantation.

Table 5.3 Gene expressed by specific cell lines for cardiomyogenesis

Cell linages	Expressed genes
ES cell	Oct 4, Nanog
Mesoderm	Brachyury
CVP	Flk-1, Nkx2.5, Isl1, c-kit
CP	Nkx2.5, GATA-4, Mef2c
MC	Nkx2.5, GATA-4, αMHC, βMHC, ANF, MLC-2V
Functional heart	Nkx2.5, GATA-4, αMHC, ANF, SM-MHC, CD-144

Table 5.4 Mode of action of extrinsic factors in cardiomyogensis and cardiac repair

Extrinsic factor	Mode of actions/effects
Growth factor	
BMPs (bone morphogenetic proteins)	BMPs play an essential role in morphogenetic development. Provides signaling for the mesodermal induction cardiac lineage differentiation and expression of cardiac markers for cardiomyocyte contraction
FGF (fibroblast growth factor)	In the case of mesodermal cell differentiation and tissue formation, FGF plays a vital role. During the gastrulation phase, FGF released from the endoderm induces the mesodermal lineage for cardiomyogenesis. The FGF receptors regulate the heart's signal induction and development by gene expression regulation such as Nkx2.5
HGF (hepatocyte growth factor)	HGF upregulates the expression of cardiac markers and cardiomyogenesis through the PI3 kinase/Akt pathway. MSCs differentiated cardiomyocytes enhance the functioning of infracted myocardium by repairing and regenerating the neo-vascularized cardiac tissue
IGF (insulin growth factor)	IGF regulates the autocrine and paracrine pathway in the cardiac development
PDGF-BB (platelet-derived growth factor-BB)	PDGF-BB have potential to differentiate the ES cells into the cardiogenesis
Wnts	Cysteine-rich glycoprotein found in vertebrates plays an essential role in cell–cell communication, determination, and organogenesis. It plays a biphasic role in early gastrulation phase induction Wnt/b-catenin enhances the cardiac differentiation from ES cells and inhibition of hematopoietic and vascular cell lineages differentiation
OT (oxytocin)	Female reproductive hormone OT plays an essential role in the development of heart and prevent the cardiac deformities at early embryonic stage
EPO (erythropoietin)	EPO and its receptors helps to prevent the cardiac defects such as ventricular hyperplasia, interventricular septum. They regulates the ES cell differentiation into erythrocytes, megakaryocytes, and cardiomyocytes, and promotes vascularization of heart

(continued)

Table 5.4 (continued)

Extrinsic factor	Mode of actions/effects
Chemicals	
DMSO (dimethyl sulfoxide)	This cryoprotectant induces the ES cells differentiation into cardiomyocytes, expression of cardiac factors, enhances intracellular calcium ion level
ROS (reactive oxygen species)	In myocardial cell generation, ROS regulates the cell growth, activates the repairing pathways and differentiation of ES cells
Opioid	Opioid receptor ligand along with the protein kinase helps to induce the ES cell differentiation into the cardiac cell lineage differentiation
5 Azacytidine	This drug have potential to differentiate the embryonic and adult stem cell into cardiomyocytes
Retinoic acid (vitamin A)	Vitamin A and its receptor helps to differentiate the ES cell into the cardiomyocytes lineage, cardiovascular morphogenesis in fetal development
Ascorbic acid (vitamin C)	Vitamin C induces the ES cell differentiation into the cardiomyocytes. In case myocardial injury, Vit. C acts healing agent to repair and regain the damaged cardiac tissue functioning

Evaluation of Efficacy of Embryonic Cells in Functional and Anatomical Cardiac Repair in Animal Model

The embryonic and cardiac progenitor cells in the heart have self-renewing capability; thus, they will differentiate into the neo myocardial cell lineages to repair the cardiac incision or injury. It was reported that the mesenchymal cells used for the treatment of myocardial infarction have paracrine effects but lack the differentiation property into the cardiomyocytes (Brade et al., 2013). Besides that, the adult cardiac cell lineages in vivo differentiation result in enhancement of sarcomere organization and high beating rate (Bang et al., 2016). Hence, embryonic cells must be the focused cell line for cardiac repair.

In 1996, the first in vivo study in mice model having dystrophy to demonstrate the role of ES cell-derived cardiomyocytes for expressing α-cardiac MHC and improve the survivability (Klug et al., 1996). The survivability rate of in vivo transplanted ES-cell-derived cardiac cells up to 32 weeks and cured the myocardial infraction (Min et al., 2003). In the twentieth century, in vivo studies revealed the human ES cells (hESC) differentiation into cardiomyocytes for cardiac injury and diseases (Nir et al., 2003; Wysoczynski & Bolli, 2020). They observed both mouse ES and hES cells have the same in vivo maturation rate, but the cardiomyocytes derived from hES cells lack complete maturation and conduction.

After that, it was shown that the clinical applicability of hES cell-derived cardiomyocytes for the treatment of bradycardia in pig model by making as biological pace-makers' for xenogeneic transplantation (Kehat et al., 2004; Shiba et al., 2012). The regenerated, repaired heart showed the spontaneous rhythmic contradiction of the transplanted cells. In the animal model study, in mouse treated with stem cell injection, results in regeneration of ventricular ejection fractions. However, in another study, the myocardial infraction was repaired with the engraftment of the matrix associated with the allogeneic stem cells (Cheraghi et al., 2016).

The ES cells differentiated into the cardiomyocyte lineages is mimic the in vivo development of the cardiomyocyte. The histopathological analysis studies reveal that the ES differentiated cardiac cell lineages require the incorporation of another signaling factor such as VEGF for the regeneration of functional cardiac tissue with better vascularization (Cao et al., 2008a, b).

However, it was found that the injected mature differentiated cardiomyocyte has a lesser therapeutic effect as compared to the immature ES cell lineage (Cheraghi et al., 2016; Nir et al., 2003). Thus, the identification, selection, and utilization of ES cells at specific cell dividing stage still query that has been focused by the researcher at pre-clinical for the evaluation of ES cell treatment.

Evaluation of Embryonic Cell Therapy for the Repair and Regeneration of Cardiac Tissue at the Clinical Level

ES cells is the most focused and novel source for the treatment of any kind of tissue injury, to repair, regenerate the neo-tissue or organ, and to cure different diseases or deformities such as cancer, cardiovascular, neural damages, etc. (Cheraghi et al., 2016). However, the in vivo success rate of the ES cell therapy demonstrates the limited functional potential and requires standard cell therapy parameters (Terashvili & Bosnjak, 2019). Thus, for the application of ES cell cardiac tissue repair and regeneration, there are some merits and demerits that we have to focus on the better outstanding. The challenges for the clinical application of ES cells are the identification of optimized cell lineages, therapy duration, chances of host rejection, the susceptibility of evocation of immune response, optimization of dosages, and enhancement of cell signaling for cardiac repair infarction (Wysoczynski & Bolli, 2020). Thus, there is a need to develop the methodologies, protocols for the enhancement and propagation of injected stem cells in the host body of myocardial regeneration.

The isolation of human ES cell (hESC) was focused by the researcher, as well as their applicability as a therapeutic agent has been done. The application of donor allogeneic embryonic cells for the treatment of cardiac injury is mocktail of cell population involves the MSCs, CPCs, pluripotent cells. At the clinical level, the ES cell therapy for the repairing of myocardial infraction results in an improved healing rate with better cardiac efficiency. The maturation rate of hES cells is not the same as the mES cells. Therefore, it shows more effectiveness as compared to the other cell lineages (Hoshino et al., 2007; Terashvili & Bosnjak, 2019). However, the standard dosage and treatment parameter for the cardiomyocyte is still a question.

Cardiac stem cell therapies may also result in inflammation and graft rejection by the host body; thus, clinical cell therapy has no significant outcome (Tang et al.,

2018). The selection of the specific immature cell lineages and to provide a micro-environment to differentiate them into specific cell lineage under in vivo conditions is an ongoing issue to be resolved by the researcher. The success rate of the ES cell lines for cardiac repair has been going on, and research focused on the cell lineages associated with matrix or grafts to overcome the problem of graft rejection, fibrosis, and bio-functional tissue regeneration (Cheraghi et al., 2016). In situ injection of the bio-engineered hydrogel enriched with ECM protein and seeded with cells results in enhanced cardiac differentiation at in vitro level as well as repair of myocardial injury (Bai et al., 2019). Matrix associated cell grafts open the new era for embryonic cell-based therapies to study their synergistic effect for the complete neo-tissue regeneration. At the worldwide level, scientists and medical experts work together to cure millions of patients suffering from myocardial infarctions, congenital disabilities and improve the survival rate.

References

Aguilar-Sanchez, C., Michael, M., & Pennings, S. (2018). Cardiac stem cells in the postnatal heart: Lessons from development. *Stem Cells International.*

Ali, S. R., Hippenmeyer, S., Saadat, L. V., Luo, L., Weissman, I. L., & Ardehali, R. (2014, June 17). Existing cardiomyocytes generate cardiomyocytes at a low rate after birth in mice. *Proceedings of the National Academy of Sciences, 111*(24), 8850–8855.

Ameen, C., Strehl, R., Bjorquist, P., Lindahl, A., Hyllner, J., & Sartipy, P. (2008). Human embryonic stem cells: current technologies and emerging industrial applications. *Critical Reviews in Oncology/Hematology, 65*(1), 54–80.

Arabadjiev, A., Petkova, R., Chakarov, S., Pankov, R., & Zhelev, N. (2014). We heart cultured hearts. A comparative review of methodologies for targeted differentiation and maintenance of cardiomyocytes derived from pluripotent and multipotent stem cells. *BioDiscovery, 14*, p. e8962.

Bai, R., Tian, L., Li, Y., Zhang, J., Wei, Y., Jin, Z., Liu, Z., & Liu, H. (2019). Combining ECM hydrogels of cardiac bioactivity with stem cells of high cardiomyogenic potential for myocardial repair. *Stem Cells International.*

Bang, O. Y., Kim, E. H., Cha, J. M., & Moon, G. J. (2016). Adult stem cell therapy for stroke: Challenges and progress. *Journal of Stroke, 18*(3), 256.

Bartosh, T. J. (2008). Molecular regulation of cardiac stem cell growth and differentiation by extrinsic factors and novel intracellular signaling pathways. University of North Texas Health Science Center at Fort Worth.

Bergmann, O., Bhardwaj, R. D., Bernard, S., Zdunek, S., Barnabé-Heider, F., Walsh, S., Zupicich, J., Alkass, K., Buchholz, B. A., Druid, H., & Jovinge, S. (2009). Evidence for cardiomyocyte renewal in humans. *Science, 324*(5923), 98–102.

Blin, G., Nury, D., Stefanovic, S., Neri, T., Guillevic, O., Brinon, B., & Bellamy, V., et al. (2010). A purified population of multipotent cardiovascular progenitors derived from primate pluripotent stem cells engrafts in postmyocardial infarcted nonhuman primates. *The Journal of Clinical Investigation, 120*(4), 1125–1139.

Braam, S. R., Passier, R., & Mummery, C. L. (2009). Cardiomyocytes from human pluripotent stem cells in regenerative medicine and drug discovery. *Trends in Pharmacological Sciences, 30*(10), 536–545.

Brade, T., Pane, L. S., Moretti, A., Chien, K. R., & Laugwitz, K. L. (2013). Embryonic heart progenitors and cardiogenesis. *Cold Spring Harbor Perspectives in Medicine, 3*(10), a013847.

Bu, L., Jiang, X., Martin-Puig, S., et al. (2009). Human ISL1 heart progenitors generate diverse multipotent cardiovascular cell lineages. *Nature, 460*(7251), 113–117.
Burridge, P. W., Keller, G., Gold, J. D., & Wu, J. C. (2012). Production of de novo cardiomyocytes: human pluripotent stem cell differentiation and direct reprogramming. *Cell Stem Cell, 10*(1), 16–28. https://doi.org/10.1016/j.stem.2011.12.013.
Callis, T. E., Deng, Z., Chen, J. F., & Wang, D. A. Z. (2008). Muscling through the microRNA world. *Experimental Biology and Medicine, 233*(2), 131–138.
Cao, F., Wagner, R. A., Wilson, K. D., et al. (2008). Transcriptional and functional profiling of human embryonic stem cell-derived cardiomyocytes. *PLoS One, 3*(10), Article ID e3474.
Cao, F., Wagner, R. A., Wilson, K. D., Xie, X., Fu, J. D., Drukker, M., Lee, A., Li, R. A., Gambhir, S. S., Weissman, I. L., & Robbins, R. C. (2008). Transcriptional and functional profiling of human embryonic stem cell-derived cardiomyocytes. *PloS One, 3*(10), e3474.
Chen, K., Wu, L., & Wang, Z. Z. (2008). Extrinsic regulation of cardiomyocyte differentiation of embryonic stem cells. *Journal of Cellular Biochemistry, 104*(1), 119–128.
Cheraghi, M., Namdari, M., Eatemadi, A., & Negahdarib, B. (2016). Recent advances in cardiac regeneration: Stem cell, biomaterial and growth factors. *Biomedicine & Pharmacotherapy, 87*(2017).
Elliott, D. A., Braam, S. R., Koutsis, K., Ng, E. S., Jenny, R., Lagerqvist, E. L., & Biben, C., et al. (2011). NKX2-5(eGFP/w) hESCs for isolation of human cardiac progenitors and cardiomyocytes. *Nature Methods, 8*(12), 1037–1040.
Grskovic, M., Javaherian, A., Strulovici, B., & Daley, G. Q. (2011). Induced pluripotent stem cells–opportunities for disease modelling and drug discovery. *Nature Reviews. Drug Discovery, 10*, 915–929. [PubMed: 22076509].
Gude, N. A., Broughton, K. M., Firouzi, F., & Sussman, M. A. (2018). Cardiac ageing: Extrinsic and intrinsic factors in cellular renewal and senescence. *Nature Reviews Cardiology, 15*(9), 523–542.
Guo, G., von Meyenn, F., Santos, F., Chen, Y., Reik, W., Bertone, P., Smith, A., & Nichols, J. (2016). Naive pluripotent stem cells derived directly from isolated cells of the human inner cell mass. *Stem Cell Reports, 6*(4), 437–446.
Haase, A., Olmer, R., Schwanke, K., et al. (2009). Generation of induced pluripotent stem cells from human cord blood. *Cell Stem Cell, 5*(4), 434–441.
Hiroi, Y., Kudoh, S., Monzen, K., et al. (2001). Tbx5 associates with Nkx2-5 and synergistically promotes cardiomyocyte differentiation. *Nature Genetics, 28*(3), 276–280.
Hoshino, K., Ly, H. Q., Frangioni, J. V., & Hajjar, R. J. (2007). In vivo tracking in cardiac stem cell-based therapy. *Progress in Cardiovascular Diseases, 49*(6), 414–420.
Hudson, J., Titmarsh, D., Hidalgo, A., Wolvetang, E., & Cooper-White, J. (2012). Primitive cardiac cells from human embryonic stem cells. *Stem Cells and Development, 21*(9), 1513–1523.
Itzhaki, I., Maizels, L., Huber, I., et al. (2011). Modelling the long QT syndrome with induced pluripotent stem cells. *Nature, 471*(7337), 225–229.
Ivey, K. N., Muth, A., Arnold, J., et al. (2008). MicroRNA regulation of cell lineages in mouse and human embryonic stem cells. *Cell Stem Cell, 2*(3), 219–229.
Kajstura, J., Urbanek, K., Perl, S., Hosoda, T., Zheng, H., Ogórek, B., Ferreira-Martins, J., Goichberg, P., Rondon-Clavo, C., Sanada, F., & D'Amario, D. (2010). Cardiomyogenesis in the adult human heart.
Kattman, S. J., Huber, T. L., & Keller, G. (2006). Multipotent Flk-1+ cardiovascular progenitor cells give rise to the cardiomyocyte, endothelial, and vascular smooth muscle lineages. *Developmental Cell, 11*(5), 723–732.
Kattman, S. J., Witty, A. D., Gagliardi, M., Dubois, N. C., Niapour, M., Hotta, A., Ellis, J., & Keller, G. (2011). Stage-specific optimization of activin/nodal and BMP signaling promotes cardiac differentiation of mouse and human pluripotent stem cell lines. *Cell Stem Cell, 8*(2), 228–240.
Kehat, I., Khimovich, L., Caspi, O., Gepstein, A., Shofti, R., Arbel, G., Huber, I., Satin, J., Itskovitz-Eldor, J., & Gepstein, L. (2004). Electromechanical integration of cardiomyocytes derived from human embryonic stem cells. *Nature Biotechnology, 22*(10), 1282–1289.

Keller, G. (2005). Embryonic stem cell differentiation: Emergence of a new era in biology and medicine. *Genes & Development, 19*(10), 1129–1155.
Klug, M. G., Soonpaa, M. H., Koh, G. Y., & Field, L. J. (1996). Genetically selected cardiomyocytes from differentiating embronic stem cells form stable intracardiac grafts. *The Journal of Clinical Investigation, 98*(1), 216–224.
Kolossov, E., Lu, Z., Drobinskaya, I., et al. (2005). Identification and characterization of embryonic stem cell-derived pacemaker and atrial cardiomyocytes. *FASEB Journal, 19*(6), 577–579.
Laflamme, M. A., Chen, K. Y., Naumova, A. V., Muskheli, V., Fugate, J. A., Dupras, S. K., & Reinecke, H., et al. (2007). Cardiomyocytes derived from human embryonic stem cells in pro-survival factors enhance function of infarcted rat hearts. *Nature Biotechnology, 25*(9), 1015–1024.
Marvin, M. J., Di Rocco, G., Gardiner, A., Bush, S. M., & Lassar, A. B. (2001). Inhibition of Wnt activity induces heart formation from posterior mesoderm. *Genes and Development, 15*(3), 316–327.
Mima, T., Ueno, H., Fischman, D. A., Williams, L. T., & Mikawa, T. (1995). Fibroblast growth factor receptor is required for in vivo cardiac myocyte proliferation at early embryonic stages of heart development. *Proceedings of the National Academy of Sciences of the United States of America, 92*(2), 467–471.
Min, J. Y., Yang, Y., Sullivan, M. F., Ke, Q., Converso, K. L., Chen, Y., Morgan, J. P., & Xiao, Y. F. (2003). Long-term improvement of cardiac function in rats after infarction by transplantation of embryonic stem cells. *The Journal of Thoracic and Cardiovascular Surgery, 125*(2), 361–369.
Moretti, A., Caron, L., Nakano, A., et al. (2006). Multipotent embryonic isl1+ progenitor cells lead to cardiac, smooth muscle, and endothelial cell diversification. *Cell, 127*(6), 1151–1165.
Moretti, A., Bellin, M., Welling, A., et al. (2010). Patient-specific induced pluripotent stem-cell models for long-QT syndrome. *New England Journal of Medicine, 363*(15), 1397–1409.
Morey, L., Santanach, A., & Di Croce, L. (2015). Pluripotency and epigenetic factors in mouse embryonic stem cell fate regulation. *Molecular and Cellular Biology, 35*(16), 2716–2728.
Nelson, T. J., Martinez-Fernandez, A., Yamada, S., PerezTerzic, C., Ikeda, Y., & Terzic, A. (2009). Repair of acute myocardial infarction with induced pluripotent stem cells induced by human stemness factors. *Circulation, 120*(5), 408–416.
Nir, S. G., David, R., Zaruba, M., Franz, W. M., & Itskovitz-Eldor, J. (2003). Human embryonic stem cells for cardiovascular repair. *Cardiovascular Research, 58*(2), 313–323.
Pal, R., Mamidi, M. K., Das, A. K., & Bhonde, R. (2012). Diverse effects of dimethyl sulfoxide (DMSO) on the differentiation potential of human embryonic stem cells. *Archives of Toxicology, 86*(4), 651–661.
Pasquinelli, A. E., Hunter, S., & Bracht, J. (2005). MicroRNAs: A developing story. *Current Opinion in Genetics and Development, 15*(2), 200–205.
Passier, R., Ward-Van Oostwaard, D., Snapper, J., et al. (2005). Increased cardiomyocyte differentiation from human embryonic stem cells in serum-free cultures. *Stem Cells, 23*(6), 772–780.
Peterkin, T., Gibson, A., & Patient, R. (2003). GATA-6 maintains BMP-4 and Nkx2 expression during cardiomyocyte precursor maturation. *EMBO Journal, 22*(16), 4260–4273.
Plageman, T. F., Jr., & Yutzey, K. E. (2004). Differential expression and function of Tbx5 and Tbx20 in cardiac development. *Journal of Biological Chemistry, 279*(18), 19026–19034.
Riley, L. A.-C., & Cross, J. C. (1998). The Hand1 bHLH transcription factor is essential for placentation and cardiac morphogenesis. *Nature Genetics, 18*(3), 271–275.
Rust, W., Balakrishnan, T., & Zweigerdt, R. (2009). Cardiomyocyte enrichment from human embryonic stem cell cultures by selection of ALCAM surface expression. *Regenerative Medicine, 4*(2), 225–237.
Senyo, S. E., Steinhauser, M. L., Pizzimenti, C. L., Yang, V. K., Cai, L., Wang, M., Wu, T. D., Guerquin-Kern, J. L., Lechene, C. P., & Lee, R. T. (2013). Mammalian heart renewal by pre-existing cardiomyocytes. *Nature, 493*(7432), 433–436.
Shiba, Y., Hauch, K. D., & Laflamme, M. A. (2009). Cardiac applications for human pluripotent stem cells. *Current Pharmaceutical Design, 15*, 2791–2806. [PubMed: 19689350].

Shiba, Y., Fernandes, S., Zhu, W. Z., Filice, D., Muskheli, V., Kim, J., Palpant, N. J., Gantz, J., Moyes, K. W., Reinecke, H., & Van Biber, B. (2012). Human ES-cell-derived cardiomyocytes electrically couple and suppress arrhythmias in injured hearts. *Nature, 489*(7415), 322–325.
Synnergren, J., Akesson, K., Dahlenborg, K., et al. (2008). Molecular ° signature of cardiomyocyte clusters derived from human embryonic stem cells. *Stem Cells, 26*(7), 1831–1840.
Tang, J. N., Cores, J., Huang, K., Cui, X. L., Luo, L., Zhang, J. Y., Li, T. S., Qian, L., & Cheng, K. (2018). Concise review: Is cardiac cell therapy dead? embarrassing trial outcomes and new directions for the future. *Stem Cells Translational Medicine, 7*(4), 354–359.
Terashvili, M., & Bosnjak, Z. J. (2019). Stem cell therapies in cardiovascular disease. *Journal of Cardiothoracic and Vascular Anesthesia, 33*(1), 209–222.
Tran, T. H., Wang, X., Browne, C., et al. (2009). Wnt3a-induced mesoderm formation and cardiomyogenesis in human embryonic stem cells. *Stem Cells, 27*(8), 1869–1878.
Uosaki, H., Fukushima, H., Takeuchi, A., Matsuoka, S., Nakatsuji, N., Yamanaka, S., & Yamashita, J. K. (2011). Efficient and scalable purification of cardiomyocytes from human embryonic and induced pluripotent stem cells by VCAM1 surface expression. *PloS One, 6*(8), e23657.
van Laake, L. W., Passier, R., Monshouwer-Kloots, J., et al. (2007). Human embryonic stem cell-derived cardiomyocytes survive and mature in the mouse heart and transiently improve function after myocardial infarction. *Stem Cell Research, 1*(1), 9–24.
van Rooij, E., Liu, N., & Olson, E. N. (2008). MicroRNAs flex their muscles. *Trends in Genetics, 24*(4), 159–166.
Vidarsson, H., Hyllner, J., & Sartipy, P. (2010). Differentiation of human embryonic stem cells to cardiomyocytes for in vitro and in vivo applications. *Stem Cell Reviews and Reports, 6*(1), 108–120.
Wang, Z., Luo, X., Lu, Y., & Yang, B. (2008). miRNAs at the heart of the matter. *Journal of Molecular Medicine, 86*(7), 771–783.
Watt, A. J., Battle, M. A., Li, J., & Duncan, S. A. (2004). GATA4 is essential for formation of the proepicardium and regulates cardiogenesis. *Proceedings of the National Academy of Sciences of the United States of America, 101*(34), 12573–12578.
Willems, E., Spiering, S., Davidovics, H., Lanier, M., Xia, Z., Dawson, M., Cashman, J., & Mercola, M. (2011). Small-molecule inhibitors of the Wnt pathway potently promote cardiomyocytes from human embryonic stem cell-derived mesoderm. *Circulation Research, 109*(4), 360–364.
Winnier, G., Blessing, M., Labosky, P. A., & Hogan, B. L. M. (1995). Bone morphogenetic protein-4 is required for mesoderm formation and patterning in the mouse. *Genes and Development, 9*(17), 2105–2116.
Wobus, A. M., Guan, K., Yang, H.-T., & Boheler, K. R. (2002). Embryonic stem cells as a model to study cardiac, skeletal muscle, and vascular smooth muscle cell differentiation. *Methods in Molecular Biology, 185*, 127–156.
Wysoczynski, M., & Bolli, R. (2020). A realistic appraisal of the use of embryonic stem cell-based therapies for cardiac repair. *European Heart Journal, 41*(25), 2397–2404.
Xu, C., Police, S., Hassanipour, M., & Gold, J. D. (2006). Cardiac bodies: A novel culture method for enrichment of cardiomyocytes derived from human embryonic stem cells. *Stem Cells and Development, 15*(5), 631–639.
Xu, X. Q., Soo, S. Y., Sun, W., & Zweigerdt, R. (2009). Global expression profile of highly enriched cardiomyocytes derived from human embryonic stem cells. *Stem Cells, 27*(9), 2163–2174.
Yang, Y., Min, J. Y., Rana, J. S., Ke, Q., Cai, J., Chen, Y., Morgan, J. P., & Xiao, Y. F. (2002). VEGF enhances functional improvement of postinfarcted hearts by transplantation of ESC-differentiated cells. *Journal of Applied Physiology, 93*(3), 1140–1151.
Yoshida, Y., & Yamanaka, S. (2011). iPS cells: A source of cardiac regeneration. *Journal of Molecular and Cellular Cardiology, 50*, 327–332. [PubMed: 21040726].
Zhang, H., & Bradley, A. (1996). Mice deficient for BMP2 are nonviable and have defects in amnion/chorion and cardiac development. *Development, 122*(10), 2977–2986.

Zhang, J., Wilson, G. F., Soerens, A. G., Koonce, C. H., Yu, J., Palecek, S. P., et al. (2009). Functional cardiomyocytes derived from human induced pluripotent stem cells. *Circulation Research, 104*, e30–e41. [PubMed: 19213953].

Zhao, Y., & Srivastava, D. (2007). A developmental view of microRNA function. *Trends in Biochemical Sciences, 32*(4), 189–197.

Zwi, L., Caspi, O., Arbel, G., et al. (2009). Cardiomyocyte differentiation of human induced pluripotent stem cells. *Circulation, 120*(15), 1513–1523.

Chapter 6
Differentiation of Stem Cells into Hepatocyte Lineage: In Vitro Cell Culture, In Vivo Transplantation in Animal Models

Munther Alomari

Abstract Liver failure is one of the life-threatening illnesses, and it accounts for 3.5% of all deaths worldwide. Liver transplantation is the solution, but due to the long waiting list for an available donor, patients die before they get the chance to get a healthy liver. Shortage supply of donor organs, lifelong need for immunosuppression, and the adult hepatocytes culturing difficulty are the serious limitations for liver transplantation. Therefore, hepatocyte transplantation of stem cells with tissue engineering is the alternative therapeutic approach to liver transplantation. Differentiation of stem cells from many origins into hepatocytes has been reported for therapeutic and research purposes. This book chapter summarizes the differentiation process into hepatocytes from different stem cell types, including mesenchymal, pluripotent, hematopoietic, and umbilical cord stem cells. In addition, it discusses the use of hepatocytes in drug discovery and clinical studies, as well as, in 2D and 3D conformation and liver formation.

Keywords Hepatocytes · Hepatocytes like stem cells · Differentiation · Induced pluripotent stem cells (iPSCs) · Embryonic stem cells (ESCs) · Hematopoietic stem cells transplantation · Umbilical cord · 2D · 3D

Introduction

Liver is the body's biggest organ, which plays an essential role in detoxification, metabolism, protein synthesis, regulation of glucose levels, and control of blood homeostasis. The liver consists of many types of cells. Generally, they are classified as non-parenchymal and parenchymal cells (hepatocytes). The hepatocytes form 70% of the liver mass that is derived from endoderm during embryonic development along with biliary epithelial cells (cholangiocytes). After resection (hepatectomy) or injury, the liver is able to regrow 70% of its mass, by multiplying the hepatocyte

M. Alomari (✉)
Department of Stem Cell Biology, Institute for Research and Medical Consultations, Imam Abdulrahman Bin Faisal University, Dammam 31441, Saudi Arabia
e-mail: maomari@iau.edu.sa

F. A. Khan (ed.), *Advances in Application of Stem Cells: From Bench to Clinics*, Stem Cell Biology and Regenerative Medicine 69,
https://doi.org/10.1007/978-3-030-78101-9_6

cells (Fausto et al., 2006; Michalopoulos, 2007). However, the capability of liver regeneration is insufficient in many diseases, such as advanced cirrhosis and hepatitis that end up in liver failure and thus life-threatening (Alqahtani, 2012). For such problem, liver transplantation is the best solution, but due to the long waiting list for available liver, patients die before they get the chance to get healthy liver of donor. Shortage supply of donor organs, lifelong need for immunosuppression, and the adult hepatocytes culturing difficulty are the serious limitations for liver transplantation (Bodzin & Baker, 2018; Iansante et al., 2018; Ibars et al., 2016; Langer & Vacanti, 2016). Therefore, the use of bio-artificial liver devices or transplantation of isolated hepatocytes is the option to provide a limited liver function (Dan & Yeoh, 2008; Demetriou et al., 2004; Horslen & Fox, 2004; McKenzie et al., 2008). Hepatocyte transplantation of stem cells with tissue engineering is the promising approach of unlimited sources for transplantation (Iansante et al., 2018; Muraca et al., 2002). Stem cells can differentiate into diverse cell progenies, such as hepatocytes, which could be used in liver tissue engineering and hepatocyte transplantation. Hepatocyte primary cell cultures and the use of animal models, such as chicken, mouse, and zebra fish, have identified many of the genes and molecular pathways that regulate the embryonic development of the liver. This scientific data has encouraged scientist to generate practical hepatocytes of stem cells including, embryonic stem cells (ESC), induced pluripotent stem cells (iPSC), hepatic progenitor/stem cells (HPC), and mesenchymal stem cells (MSC) (Ang et al., 2018; Corbett & Duncan, 2019; Ghosheh et al., 2020). Transplantation of these stem cells to generate functional hepatocytes in liver may lead to cure liver diseases. The lack of knowledge of differentiation hindered the differentiation process of stem cells into functional hepatocytes. Commonly, stem cells are differentiated by small molecules or growth factors to induce the cells into becoming hepatocyte-like cells, usually via a stepwise strategy in 2D or 3D (Zhao et al., 2020). Transplantation of human hepatocytes has been used to treat a number of liver disorders such as glycogen storage disease type 1 (Muraca et al., 2002), phenylketonuria (Stéphenne et al., 2012), urea cycle disorders (Meyburg et al., 2009; Mitry et al., 2004; Soltys et al., 2017; Stéphenne et al., 2006), factor VII deficiency (Dhawan et al., 2004), infantile Refsum's disease (Sokal et al., 2003), acute liver failure (Bilir et al., 2000; Habibullah et al., 1994; Khan et al., 2004; Schneider et al., 2006), and severe infantile oxalosis (Beck et al., 2012).

Isolation and Culture of Hepatocytes

Isolation of hepatocytes was performed firstly in 1957 (Branster & Morton, 1957), followed by advanced techniques to achieve high-quality hepatocytes. Hepatocytes harvesting is based on using two-step collagenase perfusion technique. In general, the first perfusion of liver with buffers containing ethylene glycol tetra acetic acid (EGTA) or ethylene diamine tetraacetic acid (EDTA) is to wash blood out of tissue into waste and to avoid coagulation of leftover blood in the tissue. The second perfusion buffer contains collagenase to dissolve the tissue collagen. Following perfusion with collagenase, the liver is cut into halves and opens the capsule that leads to free the cells. Then a gauze filter is used to separate the cells from the connective tissue and debris. Followed by several washes and the hepatocytes finally separated by centrifugation at 50–100 g and 4 °C (Berry & Friend, 1969; Charni-Natan & Goldstein, 2020; Knobeloch et al., 2012; Lee et al., 2020a, b; Strom et al., 1982; Seglen, 1976). One of the limitations of this method that the ability of each collagenase lot to release hepatocytes must be assessed. As the collagenase, lots vary in their composition, and consequently, its activity will vary among batches. In addition, individual collagenase lots activity may vary between animal models.

Two ways can be followed to access the isolated hepatocyte, one is the trypan blue exclusion, which is considered a poor guide for hepatocyte engraftment and function. Hepatocytes viability has not been shown to correlate with engraftment using this method (Akhter et al., 2007; Matsumura et al., 2019; Mitry et al., 2003). The second way is the adherence of hepatocytes to tissue culture plates after 24 h of seeding that showed a better correlation with engraftment (Holzman et al., 1993). After isolation, the cells will be subjected to direct transplantation, primary cell culture, or storage (cryopreserved) for later use. Cell culture or cryopreserved options has some possible advantages including increasing number of cells of donors for transplantation, tissue matching, and immunological modulation of donor cells if needed. Even though cryopreservation of cells is vital for urgent transplantation needs, this option is damaging the hepatocytes through caspases activation during freezing and thawing steps (Baust et al., 2001; Yagi et al., 2001). Generally, the hepatocytes are placed in liquid nitrogen after resuspended in cryopreservation freezing medium, which contains culture medium, 10% dimethyl sulfoxide, and 10% fetal calf serum (Aoki et al., 2005; Hang et al., 2010; Kusano et al., 2008). Even after thawing this suspension provides valid cell viability and function, it is not reliable for clinical transplantation. Recent methods were introduced to preserve the cells during cryopreservation that includes the modification of freezing medium, such as CryoStor CS10 (Woods et al., 2009) and University of Wisconsin solution, which contains 5% glucose, 10% DMSO, and a cytoprotectant such as the pan-caspase inhibitor benzyloxycarbonyl-Val-Ala-DL-Asp-fluoromethylketone (ZVAD). This freezing medium showed better cell attachment, viability, and function after thawing (Jitraruch et al., 2017). Post-cryopreservation thawing hepatocytes were maintained in culture for 24 h in order to access the cellular function and morphology (Jitraruch et al., 2017).

The best relevant in vitro model to human liver is the primary human hepatocytes culture, which can show drug metabolism profile that is very similar to liver drug metabolism profile (Gómez-Lechón et al., 2003). Thus, it can be used for pharmacological and toxicological experiments, such as drug clearance and hepatotoxicity (Gómez-Lechón et al., 2008; Hewitt et al., 2007; Leist et al., 2017; Soldatow et al., 2013; Vinken & Hengstler, 2018), that will reduce the cost and reduce the number of animal models used for drug discovery and development. Therefore, the researchers tried to establish long-lasting primary human hepatocytes cell culture relevant to in vivo situation using the extracellular matrix (ECM), hormones, growth factors, and cytokines (Clause & Barker, 2013; Michalopoulos et al., 2001; Navarro-Alvarez et al., 2006; Pediaditakis et al., 2001; Tanaka et al., 2006). The updated culture modifications significantly maintained the hepatocytes function and morphology in the culture. The function and morphology of hepatocytes were investigated through, gene expression profiles, levels of cytochrome P450 activity, and functional apical and basal polarity (Jindal et al., 2009; Kidambi et al., 2009).

Hepatocyte Morphological and Functional Characterization

Isolation of pure hepatocytes required functional and morphological evaluation after cryopreservation or primary culture to be suitable for clinical application.

Hepatocytes filters and process blood nutrients, drugs, metabolites and hormones, and synthesis and secret the bile (Treyer & Müsch, 2013). To mediate these functions, the hepatocytes are highly polarized with multiple apical membranes forming bile canaliculi and multiple basolateral membranes facing the sinusoids (Schulze et al., 2019; Slim, 2014; Treyer & Müsch, 2013). Within this particular cell morphology, protein secretion, membrane trafficking, cell signaling, and bile transport are highly organized (McNiven et al., 2009; Schulze et al., 2019; Thi et al., 2020). The examination of cell morphology in cell culture accessed using light microscopy and electron microscopy (SEM and TEM), which presented viable hepatocytes with smooth membrane, precise nuclear membranes, and intact cristae in mitochondria, while the unhealthy or apoptotic hepatocytes showed irregular plasma membranes, condensation of nuclear chromatin, and swelling in mitochondria (Jitraruch et al., 2017; Lillegard et al., 2011; Wu et al., 2014).

Specific characterization of differentiated HLC in comparison with adult hepatocytes is an important indicator of successful functional hepatocyte differentiation. This characterization is required after pharmacological research or transplantation in vivo. Hepatocytes functional activity can be evaluated using many ways, including

1. Gene expression profile of differentiated HLC in compared to mature liver hepatocytes (Gao et al., 2017; Hewitt et al., 2007; Li et al., 1990).
2. Evaluating the enzymatic functions, such as glycogen storage, LDL uptake, and cytochrome P450 (Snykers et al., 2007; Yin et al., 2015; Zheng et al., 2015).
3. Accessing drug metabolism function and other materials such as ammonia and hormones (Duncan et al., 1998; Hewitt et al., 2007; Yoon et al., 2010).
4. Screening for different protein synthesis and/or secretion: albumin, alpha fetoprotein, CK-18, complements, clotting factors, and transporter proteins (Hasan et al., 2017; Raoufil et al., 2015; Wei et al., 2008; Yin et al., 2015; Zhang et al., 2009).
5. Assessment of mitochondrial dehydrogenase activity and cell adherence (Ho et al., 2012; Jitraruch et al., 2017).
6. Measuring the ability of urea production, bile acids clearance, and lipids and lipoproteins secretion (Hasan et al., 2017; Hewitt et al., 2007; Mita et al., 2006; Yoon et al., 2010).

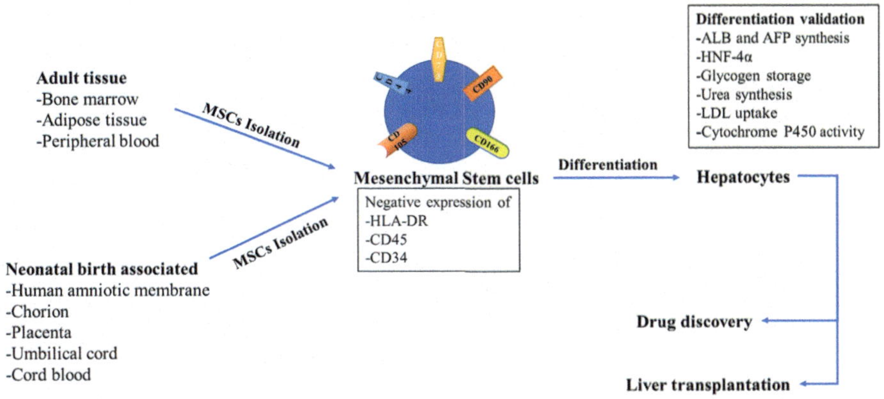

Fig. 6.1 Schematic figure of differentiation of mesenchymal stem cells into hepatocytes

Differentiation of Mesenchymal Stem Cells into Hepatocytes

Functional hepatocytes can be generated by inducing the differentiation of mesenchymal stem cells (MSCs) (Fig. 6.1). MSCs were firstly isolated from bone marrow in 1968 (Friedenstein et al., 1968). There are many resources to isolate MSCs from, including adult tissues (bone marrow, adipose tissue, and peripheral blood) (Chen et al., 2009; Hass et al., 2011; Jung et al., 2013; Kim et al., 2011; Kolanko et al., 2019; Liang & Sun, 2015) and neonatal birth-associated tissues (cord blood, umbilical cord, placenta, chorion, and human amniotic membrane). MSCs are characterized by positive cell surface expression of CD90, CD73, CD44, CD166, and CD105 and negative expression of HLA-DR, CD45, and CD34 (Dominici et al., 2006; Kholodenko et al., 2019; Maleki et al., 2014). MSCs are derived from neonatal birth-associated tissues well known for their differentiation and proliferation capabilities in vitro (Ullah et al., 2015) and in vivo (Danielyan et al., 2014).

Differentiation of MSCs was successfully reported to be used for diseases treatment (Afshari et al., 2020; Hu et al., 2013; Lim et al., 2017; Phan et al., 2018; Wu et al., 2020), especially liver disorders. Table 6.1 summarizes some of these examples of approaches in order to treat liver disorders by differentiating MSCs into functional hepatocytes and then liver transplantation.

Differentiation of Pluripotent Stem Cells into Hepatocytes

Human pluripotent stem cells (PSCs, including human embryonic stem cells (ESCs) and induced pluripotent stem cells (iPSCs)) differentiation into HLCs are being widely investigated for generation of functional hepatocyte, because of their potential ability to avoid the immune system, limitless proliferation quantity, and reflect a

Table 6.1 Protocols for MSCs differentiation toward hepatocytes, including hepatocyte characterization and protocol results

MSC source	Growth factors and cytokines used for differentiation	Time (days)	Functional characterization and differentiation markers	Results	References
Human adipose tissue	– HDM, FGF (20 ng/ml), DEX (10^{-8} M), OSM (40 ng/ml), EGF (20 ng/ml), HGF (40 ng/ml)	–14	Cell morphology, ALB, LDL uptake, and store glycogen	TSA promotes the MSC differentiation into hepatocyte, which may relieve the symptoms of mouse injured liver (CCl4)	Yin et al. (2015)
	– HDM, DEX (10^{-8} M), FGF (20 ng/ml), HGF (40 ng/ml), TSA (1 μM), EGF (20 ng/ml), OSM (40 ng/ml)	–14			
Human adipose tissue	– HCM, HGF (150 ng/ml), FGF4 (25 ng/ml), ascorbic acid (2 mM), Dex (10^{-8} M), EGF (20 ng/ml), FGF1 (300 ng/ml)	–21	Cell morphology, ALB, LDL uptake, and ammonia detoxification. CYP1A1, CYP3A4, CYP2C9, and NAPDH P-450	Differentiation of functional and transplantable hepatocytes from MSCs	Banas et al. (2007)
	– Dex (2×10^{-5} mol/l), OSM (30 ng/ml)	–14			
	– HCM alone	–14–35			
Umbilical cord vein	– DMEM-LG, EGF (10 ng/ml), bFGF (20 ng/ml)	–60–80% confluence	Expression of transthyretin, glucose 6-phosphatase, CK-8,18, AFP, ALB, and HNF-3β. Glycogen storage, ICG, and LDL uptake	MSCs isolated from human umbilical cord vein can differentiate into functional HLS	Raoufil et al. (2015)
	– ITS + Premix (1x), Dex (0.5 μmol/l), HGF (50 ng/ml), EGF (10 ng/ml), bFGF (20 ng/ml)	–14			
	– bFGF (20 ng/ml), EGF (10 ng/ml), OSM (50 ng/ml), Dex (0.5 μmol/L), ITS + Premix (1x)	–14			

(continued)

Table 6.1 (continued)

MSC source	Growth factors and cytokines used for differentiation	Time (days)	Functional characterization and differentiation markers	Results	References
Umbilical cord blood	– DMEM, HGF (20 ng/mL), FGF4 (10 ng/ml)	−28	AFP, ALB, and urea production	Derived MSCs from umbilical cord blood can differentiate into HLS, which may be useful for liver treatments	Hasan et al. (2017)
Umbilical cord	– ADMEM, hHGF (20 ng/mL)	−7	ALB, HNF-4α, Glycogen storage, urea synthesis, LDL uptake	Wharton's jelly derived MSCs derived from all the parts of the whole cord are viable cells sources and beneficial in regenerative medicine	Bharti et al. (2018)
	– ADMEM, Dex (10 nmol/L), OSM (10 ng/ml) and ITS mix (1%)	−15			
Umbilical cord	– IMDM, bFGF (10 ng/mL), EGF (20 ng/ml)	−2	TAT, CYP3A4, G6P, AFP ALB, A1AT	MSCs derived from umbilical cord is better than that derived from bone marrow in hepatic differentiation	Yu et al. (2018)
	– IMDM, bFGF (10 ng/mL), nicotinamide (0.61 g/mL), HGF (20 ng/mL) – OSM (20 ng/ml), Dex (1 μmol/l), ITS (50 mg/ml)	−7			
Human amniotic membrane	−DMEM/LG, HGF (20 ng/ml), Dex 1(0^{-7} M), OSM (10 ng/ml) and ITS	−21	AFP, CK18	cKit enrichment in amniotic cells enhances differentiation and stemness possibility	Resca et al. (2015)
Chorion of human term placenta	−α-MEM, HGF (20 ng/ml), OSM (10 ng/ml), Dex (10^{-7} M), FGF-4 (10 ng/ml), N-acetyl-L-cysteine (10 Mm) and ITS	−21	A1AT, AFP, ALB	human placenta (Chorion) is valuable source of stem cells, which may useful for cell-based therapies	Nazarov et al. (2012)

hepatic differentiation medium (HDM), trichostatin A (TSA), cytokeratin (CK), insulin-transferrin-selenium mix (ITS), hepatocyte nuclear factor (HNF) potential renewable source. Differentiation of PSCs to HLCs is a complex process that involves growth factors or cytokines (Hannan et al., 2013; Tolosa et al., 2015; Wang et al., 2020), small molecules (Asumda et al., 2018; Gao et al., 2020), or microRNAs (Jaafarpour et al., 2020; Jung et al., 2016; Li et al., 2018; Viiri et al., 2019) over different time intervals. All these used materials affect the signaling pathway in the cells to direct it into hepatocyte specifically.

Functional hepatocytes are derived from hiPSCs and hESCs in three stages. First stage, generally involves the usage of Activin A only or along, LY294002, BMP4, and Wnt3a to differentiate hiPSCs and hESCs into definitive endoderm (DE) cells (Cameron et al., 2015; Danoy et al., 2020; Hannan et al., 2013; Hay et al., 2008). These cells keep expressing pluripotent markers along with DE markers that require further differentiation (Carpentier et al., 2016). In the second stage, using DMSO alone or in combination with FGF4, BMP4, and FGF2, resulting in suppression of pluripotency markers and differentiation of the DE cells into hepatoblast (hepatic specification) (Asgari et al., 2013; Cameron et al., 2015; Czysz et al., 2015; Gao et al., 2020; Si-Tayeb et al., 2010). In addition, in this stage other researchers use more materials such as TTNPB, Forskolin, A8301, and C59 to efficiently generate high numbers of hepatocytes (Ang et al., 2018; Loh et al., 2019). The third stage includes the differentiation of hepatoblast into functional hepatocyte using Leibovitz's L-15 or HC-HepatoZYME or HCM mediums supplemented with HGF, oncostatin M (OSM), hydrocortisone (HC), and dexamethasone (Asgari et al., 2013; Cameron et al., 2015; Medine et al., 2011; Si-Tayeb et al., 2010; Tolosa et al., 2015) (Table 6.2; Fig. 6.2). There is a shift of growth factors between the stages depending on the procedure followed by researchers, some examples are summarized in Table 6.1. On the other hand, some researchers used small molecules only to generate mature hepatocytes of hiPSCs and hESCs (Table 6.1). That showed effective differentiation with cost reduction and better reproducibility overgrowth factors (Gao et al., 2020; Siller et al., 2015; Tasnim et al., 2015; Varghese et al., 2019). The differential stages were validated by quantifying the definitive endoderm, hepatoblast, and mature hepatocytes expression markers and measuring the hepatocytes functions (Table 6.2).

The third used application for hepatocyte differentiation is microRNAs, which regulate the gene expression. Overexpression of specific microRNAs such as microRNA-375 (miR-375) and miR-122 (Jaafarpour et al., 2020), microRNA-194 (Jung et al., 2016), miR-192 and miR-372-3p (Li et al., 2018) resulted in changing the cell fate decision to hepatocytes (Table 6.2; Fig. 6.2). This way of differentiation may have a high concern regarding inherent virus genome mutations in the host genome.

Hematopoietic Stem Cells Differentiation into Hepatocyte

HSCs are found in very low numbers in whole bone marrow, umbilical cord, and peripheral blood, and HSCs normally are located in the niche of bone marrow at quiescent mood and will respond to intrinsic or extrinsic signals (Tümpel & Rudolph, 2019)

Table 6.2 Protocols for pluripotent stem cells differentiation toward hepatocytes, including three-stage differentiation

Source	iPSCs -derived cells	Definitive endoderm differentiation	Hepatoblast	Hepatic maturation	Characterization and biomarkers	References
Growth factors						
Human blastocyst	H9, Man11, and Man12 (hESCs)	– RPM1-1640 + B27, Wnt3A (50 ng/ml), Activin A (100 ng/ml), **3 days**	– KO-DMEM + KO(20%) SR, DMSO (1%) **5 days**	– HepatoZYME, HC (10 μM), OSM (20 ng/ml), HGF (100 ng/ml) **18–20 days**	ALB, TTR, ECAD, CEBPA, CYP3A4, CYP1A2, PROX1, CYP2D6, FOXA2, MRP1, HNF4A, AFP, GATA4, TBX3, SOX17	Cameron et al. (2015)
Human blastocyst	H9 (hESCs)	– Medium alone **1 day** – Activin (A100 ng/ml), BMP4 (10 ng/ml), bFGF (100 ng/ml), CHIR99021 (3 μM), LY294002 (10 μM), **1 day** – bFGF (100 ng/ml), Activin A (100 ng/ml), **1 day** and Activin A (50 ng/ml)	– FGF10 (10 ng/ml), BMP4 (20 ng/ml) **4 Days**	– OSM (30 ng/ml), HGF (50 ng/ml), **15 days** – VEGF (10 ng/ml), γ secretase inhibitor (1.5 μM), FGF2 (10 ng/ml), Dex (0.1 μM), HGF(100 ng/ml), DMSO (1%), EGF (10 ng/ml), **6 days** – Vitamin K (6 μg/ml), OSM (20 ng/ml), HGF (100 ng/ml), Dex (0.1 μM), **6 days**	ATT, TTR, A1AT, ALB, TAT, TDO2, APOF, HNF4 and CYP3A7, CP450 activity, glycogen storage, LDL uptake, asialoglycoprotein receptor, urea secretion	Hannan et al. (2013)
Human foreskin fibroblasts	– Val9 (hESCs) – hiPSCs	– CHIR99021 (3 μM), Activin A (100 ng/ml), LY294002 (10 μM), BMP4 (10 ng/ml) FGF (100 ng/ml), **1 day** – Same cytokines in the absence of CHIR, **1 day** – bFGF (100 ng/ml), Activin A (100 ng/ml), **1 day**	– Activin A (50 ng/ml) **3 days** – FGF10 (10 ng/ml), BMP4 (10 ng/ml), **4 days**	– HGF (50 ng/ml), **1–2 days** – HCM in the presence of HGF and OSM, **3 days** – HCM, HGF (20 ng/ml), **15–16 days**	Urea secretion, UGT1A1, CYP450 activity	Tolosa et al. (2015)

(continued)

Table 6.2 (continued)

Source	iPSCs -derived cells	Definitive endoderm differentiation	Hepatoblast	Hepatic maturation	Characterization and biomarkers	References
– Human dermal fibroblasts – Blastocysts	– Royan hiPSC1 – hESC line, Royan H5	– RPMI-1640, Activin A (100 ng/ml), **3 days**	– ITS (0%) **1 day** – ITS (0.5%) **1 day** – ITS (1.5%) **1 day** – Basal media, KOSR(10%), HGF (10 ng/ml), FGF-4 (10 ng/ml), **8 days** – RPMI (50%) and 50% HCM, **2 days**	– HCM, OSM (10 ng/ml), KOSR (10%), Dex (0.1 μM), **10 days**	CYP450 activity, Glycogen storage, ALB, AFP, urea secretion, LDL and ICG uptake, lipid storage	Asgari et al. (2013)
– Rat bone marrow – Human blastocysts – Mouse embryonic fibroblasts	– Rat multipotent adult progenitor cells (rMAPCs) – HSF6 and H9 (hESC) – MEFs	– Wnt3A (50 ng/ml), Activin A (100 ng/ml), **6 days**	– BMP-4 (50 ng/ml), FGF-2 (10 ng/ml), **4 days** – FGF 4 (10 ng/ml), FGF1 (50 ng/ml), FGF8b (25 ng/ml), **4 days**	– HGF (20 ng/ml), Follistatin (100 ng/ml), **6 days**	ASGPR, FOXA2, SOX17, AFP, ALB, HNF4A, CYP3A4, CK18, CK19, A1AT, ALB, MRP2,, CYP1A, Glycogen storage, urea secretion, CYP450 activity, GST activity, Electron microscope	Roelandt et al. (2010)
Human fibroblast	– H9 – iPSC	– RPMI + B27, Activin A100 ng/ml, **5 days**	– RPMI + B27, FGF2 (10 ng/ml), BMP4 (20 ng/ml), **5 days** – HGF (20 ng/ml), **5 days**	– HCM, OSM (20 ng/ml), **5 days**	GATA4, SOX17, FOXA2, FOXA2, HNF4A, AFP, ALB	Si-Tayeb et al. (2010)
Moue GFP expressing fibroblast	Mouse iPSC	– Wnt3A (50 ng/ ml), Activin A (50 ng/ ml), **6 days** – DMSO (1%), bFGF (10 ng/ml), **3 days**	– HGF (10 ng/ml), DMSO (1%), **9 days**	– OSM (10 ng/ml), Dex, HGF (10 ng/ml), **7 days**	Glycogen storage, ICG uptake, CYP450 activity, LDL uptake	Gai et al. (2010)

(continued)

Table 6.2 (continued)

Source	iPSCs -derived cells	Definitive endoderm differentiation	Hepatoblast	Hepatic maturation	Characterization and biomarkers	References
Human dermal fibroblasts	hiPSCs (TKDN4,M clones)	– RPMI-1640 + B27, Activin A (100 ng/ml), **5 days**	– RPMI-1640 + B27, FGF2 (10 ng/ml), BMP4 (20 ng/ml), **5 days** – HGF (20 ng/ml), **7 days**	– Lonza HBM without hEGF, OSM (20 ng/mL), **14 days**	HNF1A, PPARA, ACYP1A2, CYP2B6, CYP2C8, CYP2D6, CYP3A4, ALB, bile acid metabolism and transport	Danoy et al. (2020)
Human blastocysts	H1, H7, H9 hESCs	– CDM2, PI103 (50 nM), CHIR99201 (3 μM), Activin (100 ng/ml), FGF2 (10 ng/ml), **1 day** – DM3189 (250 nM), PI103 (50 nM), **1 day**	– CDM3, BMP4 (30 ng/ml), TTNPB (75 nM), A83-01 (1 μM), FGF2 (10 ng/ml), **1 day** – Activin (10 ng/ml), Forskolin (1 μM), C59 (1 μM), BMP4 (30 ng/ml), **2 days** – BMP4 (30 ng/ml), CHIR99201 (1 μM), Activin (10 ng/ml), Forskolin (1 μM), **1 day** – CDM4, Insulin (10 μg/ml), AA2P (200 μg/ml), Ro4929097 (2 μM), Forskolin (10 μM), Dex (10 μM), OSM (10 ng/ml), BMP4 (10 μg/ml), **6 days**	– CDM4 or CDM5, Ro4929097 (2 μM), Forskolin (10 μM), Dex (10 μM), AA2P (200 μg/ml), Insulin (10 μg/ml), **6 days**	AFP, SOX17, FOXA2, TBX3, HNF4A, ALB	Loh et al. (2019)

(continued)

Table 6.2 (continued)

Source	iPSCs -derived cells	Definitive endoderm differentiation	Hepatoblast	Hepatic maturation	Characterization and biomarkers	References
Human blastocysts	H9 (hESCs)	– LY294002 (10 μM), bromo-indirubin-3′-oxime (2 μM), Activin A (100 ng/ml), **6 days** – Wnt3A (50 ng/ml), Activin A (100 ng/ml), **3 days**	– DMSO (0.5%), Sodium butyrate (250 Nm), **6–8 days** – FGF-2 (10 ng/ml), BMP-4 (50 ng/ml), **4 days**	– SB431542 (1 μM), DMSO (1%), **8 days** – Follistatin (100 ng/ml), HGF (20 ng/ml), **6 days**	Urea secretion, CYP1A2, CYP1A1and CYP3A4 activities, ALB, A1AT, ASGPR, MRP2,, HNF4A, CK18, CK19, FOXA2, AFP, SOX17	Tasnim et al. (2015)
Small molecules						
– Blastocysts – Human Skin Fibroblast	– H1 (hESCs) – 207 (hESCs) – Detroit 551 (hiPSCs)	– RPMI + B27, CHIR99021 (3 μM), **1 day**	– KOSR, DMSO (1%), **5 days**	L-15, DiHexa (100 nM), Dex (100 nM), **10 days**	Glycogen storage, ICG uptake, CYP450 activity, SOX17, GSC, FOXA2, MIXL1, Wnt signaling, AFP, CEBPA, GATA4, HNF4A PROX1, TBX3, TTR, ALB, HNF4A, A1AT, CYP3A4, CYP1A2, Fibronectin	Siller et al. (2015)

(continued)

Table 6.2 (continued)

Source	iPSCs -derived cells	Definitive endoderm differentiation	Hepatoblast	Hepatic maturation	Characterization and biomarkers	References
– Human blood – Endothelial progenitor cells	S06 (hiPSCs)	– RPMI 1640 + B27, CHIR99021 (3 μM), **1 day** – RPMI + B27, **1 day**	DMEM, DMSO (1%), **5 days**	– L-15, DiHexa (100 nM), Dex (100 nM), **17 days**	CYP1A, CYP2B6, MIXL1, HHEX, FGF17, GATA4, FOXA2, SOX17, SOX7, HNF4A, HNF4A,CER1, CYP2C9, CYP2D6, CYP3A, ALB, GSC, fibronectin, urea synthesis, CYP450 activity	Gao et al. (2020)
MicroRNAs						
Embryonic Fibroblasts	HiPSCs	Overexpression of miR-375, **7 days**	–	Overexpression of miR-122, **14 days**	FOXA2, SOX17 ALB, CK18, HNF4a, urea secretion, LDL uptake, qRT-PCR, immunofluorescence	Jaafarpour et al. (2020)
– Liver tumor associated with chronic hepatitis C – Human blastocysts	– HepaRG – H9	Continuous overexpression of miR-194 over **14 days**			ECT2, FGFR3, GALNT7, HDAC2, NOTCH2, RACGAP1, RHEB, THBS1, HNF4A, ALDOB, CYP3A4, ALB	Jung et al. (2016)

fibroblast growth factor (FGF); dimethyl sulfoxide (DMSO); vascular endothelial growth factor (VEGF); N-hexanoic-Tyr-Ile-(6) aminohexanoicamide (DiHexa); epidermal growth factor (EGF); hepatocyte growth factor (HGF); transforming growth factor-α (TGF-α); bone morphogenetic protein (BMP); dexamethasone (Dex); hydrocortisone (HC), oncostatin M (OSM); insulin-transferrin-selenite (ITS); basic fibroblast growth factor (bFGF); indocyanin green (ICG) uptake; 4-[(E)-2-(5,6,7,8-Tetrahydro-5,5,8,8-tetramethyl-2-naphthalenyl)-1-propenyl] benzoic acid (TTNPB); TGF-b inhibitor (A8301); Porcupine inhibitor Wnt-C59 (C59); aldolase B (ALDOB), hydrocortisone (HC)

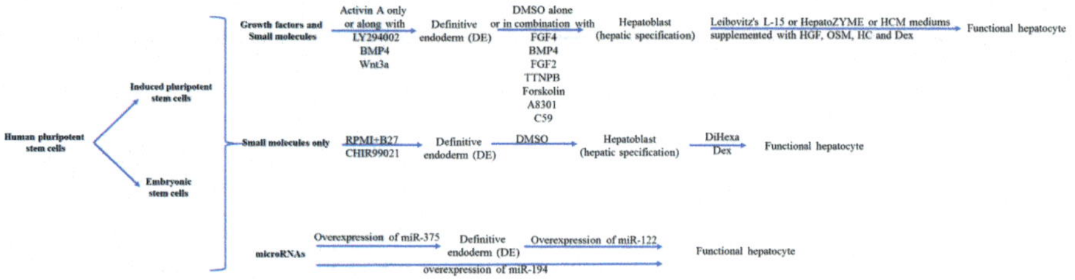

Fig. 6.2 Schematic of pluripotent stem cells differentiation toward hepatocytes

such as growth factors. With cell surface-specific markers in human CD45RA$^-$, CD90$^\pm$, CD38$^-$, CD34$^+$, Lin$^-$, CD49f$^+$, and RHO^l while in mouse the markers as follows Lin$^-$, CD34$^-$, cKit$^+$, Sca-1$^+$, FLK2$^-$, and Slamf1$^+$ (Chotinantakul & Leeanansaksiri, 2012).

HSCs are able to self-renew and differentiation into specialized blood cells (Doulatov et al., 2012) and another type of cells such as liver cells (Sellamuthu et al., 2011), osteochondrocytes (Mehrotra et al., 2010), adipocytes (Sera et al., 2009), endothelial cells (Elkhafif et al., 2011), and pancreatic cells (Minamiguchi et al., 2008). HSCs have been proposed as a replacement for hepatocyte transplantation in liver; as at the first stage of embryogenesis, the liver is considered hematopoietic organ; also, it is an erythrocytes source in the first trimester of pregnancy.

Some approaches were reported as potential cell replacement therapy of hepatocytes, such as inducing bone marrow HSCs differentiation into hepatocytes, which is the cell source for transplant procedures to treat hemophilia patients (Gabr et al., 2014). HSCs derived from umbilical cord, that differentiation in short duration (14 days) into hepatocytes using combination of growth factors (FGF 4 and HGF), this approach indicates that these stem cells have a potential to be used in liver replacement therapy (Sellamuthu et al., 2011). HSCs derived from bone marrow showed functional hepatocyte differentiation and engraftment (Khurana & Mukhopadhyay, 2008). In addition, these HSCs were found effective in curing the liver in the fumarylacetoacetate hydrolase (FAH)-deficient mouse model, the mutant mouse was intravenously injected with different numbers of HSCs, few months later the hepatocyte repopulated, FAH expression was detected, and the activity of hepatocytes was accessed by albumin and β-galactosidase expression (Lagasse et al., 2000). The recovery of animals transplanted with bone marrow-derived HSCs is probably due to HSCs-hepatocyte fusion and reprogramming and not due to their actual differentiation into hepatocytes (Vassilopoulos et al., 2003; Wang et al., 2003). Unfortunately, human liver regeneration by bone marrow HSCs is not clinically relevant at that time (Pilat et al., 2013). Recent reports were showed that human and mouse bone marrow HSCs are successfully incorporated into liver regeneration and transdifferentiating into hepatocytes (Lee et al., 2015). In addition, hematopoietic cell (autologous CD34$^+$) infusion in patients with decompensated cirrhosis of the liver showed significantly improved albumin expression and thus liver function, indicating the importance of HSCs transdifferentiation in liver transplantation (Sharma et al., 2015).

Umbilical Cord Stem Cell Differentiation into Hepatocytes

Umbilical cord (UC) veins, arteries, Wharton's jelly, perivascular, and lining membrane regions are considered the best source for mesenchymal stem cells (MSC) (Nagamura-Inoue & He, 2014). MSC collection of UC is painless and no tissue damage; in addition, this MSC is fast self-renewable and differentiation cells (Hsieh et al., 2010), promotes tissue repair, modulates immune response (Deuse et al., 2011), and can be used to autologous and allogeneic transplantation. UC-MSC was isolated and differentiated into hepatocytes in vitro using growth factors, small molecules, or microRNAs (Bharti et al., 2018; Campard et al., 2008; Raut & Khanna, 2016; Xue et al., 2016; Zhang et al., 2009). The differentiated hepatocytes were tested for cell activity, protein expression, and gene upregulation, which all showed successful hepatocyte differentiation. UC-MSC was used for the treatment of liver failure in many disease models. For examples, liver fibrosis, cirrhosis, and failure in rats (Chai et al., 2016; Zhang et al., 2017, 2018) and liver failure, chronic injury, ischemia/reperfusion injury in mice (Cui et al., 2017; Feng et al., 2015; Le et al., 2017; Yang et al., 2015; Zheng et al., 2019). In addition, UC-MSC was differentiated into hepatocytes and then transplanted in murine model of CCL4-induced liver injury (Cui et al., 2013; El Baz et al., 2020; Kao et al., 2015), that showed improvement in liver function and restored the liver injury.

UC-MSC is also used in human clinical trials for liver diseases treatments, such as transplantation of UC-MSC for treatment of newly onset type 1 diabetes mellitus (Hu et al., 2013), treatment of severe bronchopulmonary dysplasia in children (Wu et al., 2020), treatment of systemic lupus erythematosus (Liang & Sun, 2015), treatment of decompensated cirrhosis (Zhang et al., 2012), treatment of acute allograft rejection (Shi et al., 2017), and primary biliary cirrhosis (Wang et al., 2013). These clinical trials improved liver function and patients' life quality. One of the differentiation limitations of MSC derived from umbilical cord is the low number of MSC in the umbilical cord (Han et al., 2013).

Application in Drug Discovery, Preclinical, and Clinical Trials

The new drug discovery needs a lot of testing and validation, hepatocytes fasten this procedure and reduce the use of animal models. Hepatocytes function in drug metabolism (drug elimination of body) involves CYP-dependent oxidation, and CYP enzyme activation or inhibition is necessary for drug–drug interactions prediction (Riley & Grime, 2004). Therefore, hepatocytes were used as tool for testing drug clearance and toxicology, access drug uptake, and drug–drug interaction prediction (Andersson et al., 2012; Bernasconi et al., 2019; Hallifax et al., 2005; Louisse et al., 2020). Biopsy of primary liver cells is considered the best way for in vitro drug screening, as these cells are similar to healthy human cells (Guo et al., 2011).

Unfortunately, difficulty of cell supply and difficulty to get cell division for in vitro expansion, as well as their rapid dedifferentiation, which results in losing their function make them unsuitable for research or drug screening (Heslop et al., 2017). For regular supply of hepatocytes for drug screening purposes, researchers successfully differentiated many types of stem cells into functional hepatocytes such as, iPSCs, which are used for screening five drugs after differentiating them into hepatocytes (Choi et al., 2013). In addition, hepatocytes derived from iPSCs were also used to screen for drugs that improve mitochondrial function and reveal that nicotinamide adenine dinucleotide (NAD) is a potential treatment for mtDNA depletion syndrome 3 (Jing et al., 2018).

The liver failure is one of the life-threatening issues, and the liver transplantation is the solution, but many patients die before they get a liver from a donor. Due to that, hepatocyte transplantation was suggested for the replacement of liver transplantation. As a result, many preclinical trials on animal were carried on trying to replace liver cells with functional hepatocytes. These preclinical studies showed functional replacement of hepatocytes that reduce the effect of the liver disease in the animal models (Gilgenkrantz and l'Hortet, 2018; Gramignoli, 2016). These promising results encouraged human clinical trials establishment using hepatocytes like stem cells (HLSCs), which showed no side effect and no need to use immunosuppression, since HLSCs possess immunomodulatory activities (Spada et al., 2020). Some clinical trials examples are listed (Table 6.3).

Table 6.3 Application of differentiated hepatocytes in clinical trials

Disease	Cell therapy	Cell source	Patients number	Outcome	References
Chronic liver failure	Autologous $CD34^+$ stem cells	Peripheral blood	5	Improvement of albumin and bilirubin	Gordon et al. (2006)
End stage liver disease	Autologous $CD34^+$, $CD133^+$ stem cells	Bone marrow	90	Improvement of liver function	Salama et al. (2010)
Chronic liver failure-Hepatitis B	Mesenchymal stem cell and plasma exchange (PE)	Umbilical Cord	30	This treatment combination is safe but cannot significantly improve the condition of this disease	Xu et al. (2019)
Acute allograft rejection	Mesenchymal stem cell	umbilical cord	14	UC-MSCs transplantation is feasible and may mediate a therapeutic immunosuppressive effect	Shi et al. (2017)

2D and 3D Hepatocyte Confirmation, Liver Formation

Three-dimensional (3D) and Two-dimensional (2D) cell culture techniques were used to induce hepatocytes differentiation and then characterization by immunocytochemistry, gene expression, electron microscopy, and morphology. Two-dimensional (2D, Monolayer) culture is the most frequent technique used to coax differentiation of isolated stem cells into hepatocytes. Even though 2D provides useful observation of functional tests, low cost, and high efficacy as screening tool for hypertoxicity, there are disadvantages related to differentiated hepatocytes phenotype lasting period, environmental issues, biochemical processes, and loss of tissue-specific architecture (Duval et al., 2017). In order to improve the hepatocyte differentiation, morphology, and liver-specific functions, the researchers start thinking of adding more conditions to the culture that will be similar to in vivo environment (Bachmann et al., 2015). Using extracellular matrix (ECM), stromal cells, soluble factors, and macromolecules showed that the cells differentiate better in these conditions (Clause & Barker, 2013; Yu et al., 2011). In three-dimensional (3D) culture, the researchers try to reach the in vivo condition to give better prediction results and to overcome some disadvantages of 2D, such as maintaining the expansion, differentiation, and function capacity of stem cells for long period. 3D hepatic culture model involves three categories, spheroids, liver on a chip, and bioreactors.

Spheroids are formed in spherical shape in low attachment plates under optimized culture condition to produce uniform spheroid-like structures, such as differentiated HepaRG spheroid culture (Fig. 6.3). The cells in this 3D culture showed polarized and functional hepatocyte for more than 28 days (Ramaiahgari et al., 2017). In addition, extracellular matrix-based hydrogel model was used to generate differentiated hepatocyte spheroids of HepG2 cell line (Ramaiahgari et al., 2014). These spheroids grow to a certain diameter of 118 μm that allows enough oxygen diffusion to the core (Ramaiahgari et al., 2014). The spheroid technique is maintaining the

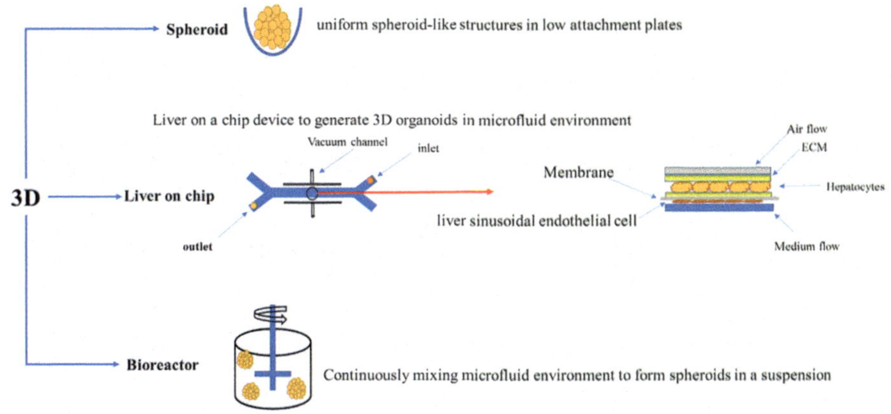

Fig. 6.3 Schematic of the in vitro 3D culture types

function and metabolism, easy to reproduce 3D form and differentiated hepatocytes showed high sensitivity to toxins, and also it is useful system for analyzing the drug-induced toxic or pathogenesis of diseases effects even though there are disadvantages for this technique such as dedifferentiation, varied physiological response, limited characterization, possible necrosis in the core (Ramaiahgari et al., 2014, 2017; Ryu et al., 2019; Sirenko et al., 2016). However, spheroids showed better differentiation efficiency when compared to 2D (Bratt-Leal et al., 2009).

Liver-on-a-chip device generates 3D organoids in microfluid environment, with preservation of hepatocytes viability, biological activity, and cellular phenotype (Moradi et al., 2020). This kind of culture is permeable to soluble growth factors and oxygen and allows nutrients and waste exchange that will avoid the hypoxia that could happen in spheroids (Ziolkowska et al., 2010) (Fig. 6.3). This system could be replicated in liver environment and offer homogenous distribution, consistent density of the cells in the culture, high cell viability, increased or stable liver enzyme levels, capacity for detoxification, in vivo-like dynamic flow, and chemical gradients even though there are some disadvantages of this system should be noted, such as high cost, limited evaluation tests, lack of scaffold (extracellular matrix), lack of biliary system, and low expression of structural proteins (Bovard et al., 2018; Corrado et al., 2019; Goral & Yuen, 2012; Grix et al., 2018; Lee et al., 2019; Rajan et al., 2020; Utech et al., 2015). Liver on a chip can be used in the application of cellular cytotoxicity, drug metabolism, liver disease model, and drug-induced liver injury.

Bioreactors, it is continuously mixing microfluid environment to form spheroids in a suspension (Lin & Chang, 2008) (Fig. 6.3). Fast stirring speed may damage the spheroids and slow stirring may inhibit spheroid formation as the cells will sink to the bottom (Achilli et al., 2012). This technique enables spheroids formation with the expression of relevant markers and enzymes, maintains cells phenotype and high seeding efficiency. In addition, it increases phases I and II enzyme activity and toxin sensitivity (Baudoin et al., 2011; Freyer et al., 2018; Ryu et al., 2019; Tostões et al., 2012). On the other hand, this technique is high cost and hard to standardize and reproduce, that make it less used (Tostões et al., 2012).

Overall, 3D reflects in vivo culture better than 2D; also, 3D cell culture enables the formation of organoids that result in the expansion of human primary tissues (Broutier et al., 2016; Xu et al., 2018). In addition, 3D is more effective in differentiation process than 2D (Afshari et al., 2020) and provides more realistic physical and biochemical environment than 2D (Chan et al., 2016; Duval et al., 2017). For example, 3D culture is better than 2D culture in lowering transaminase (El Baz et al., 2020), in drug-induced phospholipidosis sensitivity (Lee et al., 2020a; b), in hepatic drug metabolism and hepatoxicity (Corrado et al., 2019), in maintaining functional cell culture for long period (Ramaiahgari et al., 2017).

References

Achilli, T.-M., Meyer, J., & Morgan, J. R. (2012). Advances in the formation, use and understanding of multi-cellular spheroids. *Expert Opinion on Biological Therapy, 12*, 1347–1360. https://doi.org/10.1517/14712598.2012.707181.

Afshari, A., Shamdani, S., Uzan, G., Naserian, S., & Azarpira, N. (2020). Different approaches for transformation of mesenchymal stem cells into hepatocyte-like cells. *Stem Cell Research & Therapy, 11*, 54. https://doi.org/10.1186/s13287-020-1555-8.

Akhter, J., Johnson, L. A., Gunasegaram, A., Riordan, S. M., & Morris, D. L. (2007). Hepatocyte transplantation: A review of laboratory techniques and clinical experiences. *The Surgeon, 5*, 155–164. https://doi.org/10.1016/S1479-666X(07)80043-6.

Alqahtani, S. A. (2012). Update in liver transplantation. *Current Opinion in Gastroenterology, 28*, 230–238. https://doi.org/10.1097/MOG.0b013e3283527f16.

Andersson, T. B., Kanebratt, K. P., & Kenna, J. G. (2012). The HepaRG cell line: A unique in vitro tool for understanding drug metabolism and toxicology in human. *Expert Opinion on Drug Metabolism & Toxicology, 8*, 909–920. https://doi.org/10.1517/17425255.2012.685159.

Ang, L. T., Tan, A. K. Y., Autio, M. I., Goh, S. H., Choo, S. H., Lee, K. L., Tan, J., Pan, B., Lee, J. J. H., Lum, J. J., Lim, C. Y. Y., Yeo, I. K. X., Wong, C. J. Y., Liu, M., Oh, J. L. L., Chia, C. P. L., Loh, C. H., Chen, A., Chen, Q., ... Lim, B. (2018). A roadmap for human liver differentiation from pluripotent stem cells. *Cell Reports, 22*, 2190–2205. https://doi.org/10.1016/j.celrep.2018.01.087.

Aoki, T., Koizumi, T., Kobayashi, Y., Yasuda, D., Izumida, Y., Jin, Z., Nishino, N., Shimizu, Y., Kato, H., Murai, N., Niiya, T., Enami, Y., Mitamura, K., Yamamoto, T., & Kusano, M. (2005). A novel method of cryopreservation of rat and human hepatocytes by using encapsulation technique and possible use for cell transplantation. *Cell Transplantation, 14*, 609–620. https://doi.org/10.3727/000000005783982710.

Asgari, S., Moslem, M., Bagheri-lankarani, K., Pournasr, B., Miryounesi, M., & Baharvand, H. (2013). Differentiation and transplantation of human induced pluripotent stem cell-derived hepatocyte-like cells. *Stem Cell Reviews and Reports, 9*, 493–504. https://doi.org/10.1007/s12015-011-9330-y.

Asumda, F. Z., Hatzistergos, K. E., Dykxhoorn, D. M., Jakubski, S., Edwards, J., Thomas, E., & Schiff, E. R. (2018). Differentiation of hepatocyte-like cells from human pluripotent stem cells using small molecules. *Differentiation, 101*, 16–24. https://doi.org/10.1016/j.diff.2018.03.002.

Bachmann, A., Moll, M., Gottwald, E., Nies, C., Zantl, R., Wagner, H., Burkhardt, B., Sánchez, J. J. M., Ladurner, R., Thasler, W., Damm, G., & Nussler, A. K. (2015). 3D cultivation techniques for primary human hepatocytes. *Microarrays, 4*, 64–83. https://doi.org/10.3390/microarrays4010064.

Banas, A., Teratani, T., Yamamoto, Y., Tokuhara, M., Takeshita, F., Quinn, G., Okochi, H., & Ochiya, T. (2007). Adipose tissue-derived mesenchymal stem cells as a source of human hepatocytes. *Hepatology, 46*, 219–228. https://doi.org/10.1002/hep.21704.

Baudoin, R., Griscom, L., Prot, J. M., Legallais, C., & Leclerc, E. (2011). Behavior of HepG2/C3A cell cultures in a microfluidic bioreactor. *Biochemical Engineering Journal, 53*, 172–181. https://doi.org/10.1016/j.bej.2010.10.007.

Baust, J. M., Vogel, M. J., Van Buskirk, R., & Baust, J. G. (2001). A molecular basis of cryopreservation failure and its modulation to improve cell survival. *Cell Transplantation, 10*, 561–571. https://doi.org/10.3727/000000001783986413.

Beck, B. B., Habbig, S., Dittrich, K., Stippel, D., Kaul, I., Koerber, F., Goebel, H., Salido, E. C., Kemper, M., Meyburg, J., & Hoppe, B. (2012). Liver cell transplantation in severe infantile oxalosis—a potential bridging procedure to orthotopic liver transplantation? *Nephrology, Dialysis, Transplantation, 27*, 2984–2989. https://doi.org/10.1093/ndt/gfr776.

Bernasconi, C., Pelkonen, O., Andersson, T. B., Strickland, J., Wilk-Zasadna, I., Asturiol, D., Cole, T., Liska, R., Worth, A., Müller-Vieira, U., Richert, L., Chesne, C., & Coecke, S. (2019). Validation

of in vitro methods for human cytochrome P450 enzyme induction: Outcome of a multi-laboratory study. *Toxicology in Vitro, 60*, 212–228. https://doi.org/10.1016/j.tiv.2019.05.019.
Berry, M. N., & Friend, D. S. (1969). High-yield preparation of isolated rat liver parenchymal cells. *Journal of Cell Biology, 43*, 506–520.
Bharti, D., Shivakumar, S. B., Park, J.-K., Ullah, I., Subbarao, R. B., Park, J.-S., Lee, S.-L., Park, B.-W., & Rho, G.-J. (2018). Comparative analysis of human Wharton's jelly mesenchymal stem cells derived from different parts of the same umbilical cord. *Cell and Tissue Research, 372*, 51–65. https://doi.org/10.1007/s00441-017-2699-4.
Bilir, B. M., Guinette, D., Karrer, F., Kumpe, D. A., Krysl, J., Stephens, J., McGavran, L., Ostrowska, A., & Durham, J. (2000). Hepatocyte transplantation in acute liver failure. *Liver Transplantation, 6*, 32–40. https://doi.org/10.1016/S1527-6465(00)80030-1.
Bodzin, A. S., & Baker, T. B. (2018). Liver transplantation today: where we are now and where we are going. *Liver Transplantation, 24*, 1470–1475. https://doi.org/10.1002/lt.25320.
Bovard, D., Sandoz, A., Luettich, K., Frentzel, S., Iskandar, A., Marescotti, D., Trivedi, K., Guedj, E., Dutertre, Q., Peitsch, M. C., & Hoeng, J. (2018). A lung/liver-on-a-chip platform for acute and chronic toxicity studies. *Lab on a Chip, 18*, 3814–3829. https://doi.org/10.1039/C8LC01029C.
Branster, M. V., & Morton, R. K. (1957). Isolation of intact liver cells. *Nature, 180*, 1283–1284. https://doi.org/10.1038/1801283a0.
Bratt-Leal, A. M., Carpenedo, R. L., & McDevitt, T. C. (2009). Engineering the embryoid body microenvironment to direct embryonic stem cell differentiation. *Biotechnology Progress, 25*, 43–51. https://doi.org/10.1002/btpr.139.
Broutier, L., Andersson-Rolf, A., Hindley, C. J., Boj, S. F., Clevers, H., Koo, B.-K., & Huch, M. (2016). Culture and establishment of self-renewing human and mouse adult liver and pancreas 3D organoids and their genetic manipulation. *Nature Protocols, 11*, 1724–1743. https://doi.org/10.1038/nprot.2016.097.
Cameron, K., Tan, R., Schmidt-Heck, W., Campos, G., Lyall, M. J., Wang, Y., Lucendo-Villarin, B., Szkolnicka, D., Bates, N., Kimber, S. J., Hengstler, J. G., Godoy, P., Forbes, S. J., & Hay, D. C. (2015). Recombinant laminins drive the differentiation and self-organization of hESC-derived hepatocytes. *Stem Cell Reports, 5*, 1250–1262. https://doi.org/10.1016/j.stemcr.2015.10.016.
Campard, D., Lysy, P. A., Najimi, M., & Sokal, E. M. (2008). Native umbilical cord matrix stem cells express hepatic markers and differentiate into hepatocyte-like cells. *Gastroenterology, 134*, 833–848. https://doi.org/10.1053/j.gastro.2007.12.024.
Carpentier, A., Nimgaonkar, I., Chu, V., Xia, Y., Hu, Z., & Liang, T. J. (2016). Hepatic differentiation of human pluripotent stem cells in miniaturized format suitable for high-throughput screen. *Stem Cell Research, 16*, 640–650. https://doi.org/10.1016/j.scr.2016.03.009.
Chai, N.-L., Zhang, X.-B., Chen, S.-W., Fan, K.-X., & Linghu, E.-Q. (2016). Umbilical cord-derived mesenchymal stem cells alleviate liver fibrosis in rats. *World Journal of Gastroenterology, 22*, 6036–6048. https://doi.org/10.3748/wjg.v22.i26.6036.
Chan, H. F., Zhang, Y., & Leong, K. W. (2016). Efficient one-step production of microencapsulated hepatocyte spheroids with enhanced functions. *Small (weinheim an Der Bergstrasse, Germany), 12*, 2720–2730. https://doi.org/10.1002/smll.201502932.
Charni-Natan, M., & Goldstein, I. (2020). Protocol for primary mouse hepatocyte isolation. *STAR Protocols, 1*, 100086. https://doi.org/10.1016/j.xpro.2020.100086.
Chen, M.-Y., Lie, P.-C., Li, Z.-L., & Wei, X. (2009). Endothelial differentiation of Wharton's jelly–derived mesenchymal stem cells in comparison with bone marrow–derived mesenchymal stem cells. *Experimental Hematology, 37*, 629–640. https://doi.org/10.1016/j.exphem.2009.02.003.
Choi, S. M., Kim, Y., Shim, J. S., Park, J. T., Wang, R.-H., Leach, S. D., Liu, J. O., Deng, C.-X., Ye, Z., & Jang, Y.-Y. (2013). Efficient drug screening and gene correction for treating liver disease using patient-specific stem cells. *Hepatology, 57*, 2458–2468. https://doi.org/10.1002/hep.26237.
Chotinantakul, K., & Leeanansaksiri, W. (2012). Hematopoietic stem cell development, niches, and signaling pathways. *Bone Marrow Research, 2012*. https://doi.org/10.1155/2012/270425.
Clause, K. C., & Barker, T. H. (2013). Extracellular matrix signaling in morphogenesis and repair. *Current Opinion in Biotechnology, 24*, 830–833. https://doi.org/10.1016/j.copbio.2013.04.011.

Corbett, J. L., & Duncan, S. A. (2019). iPSC-derived hepatocytes as a platform for disease modeling and drug discovery. *Frontiers in Medicine, 6*, 265. https://doi.org/10.3389/fmed.2019.00265.

Corrado, B., Gregorio, V. D., Imparato, G., Attanasio, C., Urciuolo, F., & Netti, P. A. (2019). A three-dimensional microfluidized liver system to assess hepatic drug metabolism and hepatotoxicity. *Biotechnology and Bioengineering, 116*, 1152–1163. https://doi.org/10.1002/bit.26902.

Cui, H., Liu, Z., Wang, L., Bian, Y., Li, W., Zhou, H., Chu, X., & Zhao, Q. (2017). Icariin-treated human umbilical cord mesenchymal stem cells decrease chronic liver injury in mice. *Cytotechnology, 69*, 19–29. https://doi.org/10.1007/s10616-016-0034-7.

Cui, L., Shi, Y., Zhou, X., Wang, X., Wang, J., Lan, Y., Wang, M., Zheng, L., Li, H., Wu, Q., Zhang, J., Fan, D., & Han, Y. (2013). A set of microRNAs mediate direct conversion of human umbilical cord lining-derived mesenchymal stem cells into hepatocytes. *Cell Death & Disease, 4*, e918–e918. https://doi.org/10.1038/cddis.2013.429.

Czysz, K., Minger, S., & Thomas, N. (2015). DMSO efficiently down regulates pluripotency genes in human embryonic stem cells during definitive endoderm derivation and increases the proficiency of hepatic differentiation. *PLoS ONE, 10*, e0117689. https://doi.org/10.1371/journal.pone.0117689.

Dan, Y. Y., & Yeoh, G. C. (2008). Liver stem cells: a scientific and clinical perspective. *Journal of Gastroenterology and Hepatology, 23*, 687–698. https://doi.org/10.1111/j.1440-1746.2008.05383.x.

Danielyan, L., Beer-Hammer, S., Stolzing, A., Schäfer, R., Siegel, G., Fabian, C., Kahle, P., Biedermann, T., Lourhmati, A., Buadze, M., Novakovic, A., Proksch, B., Gleiter, C. H., Frey, W. H., & Schwab, M. (2014). Intranasal delivery of bone marrow-derived mesenchymal stem cells, macrophages, and microglia to the brain in mouse models of Alzheimer's and Parkinson's disease. *Cell Transplantation, 23*, 123–139. https://doi.org/10.3727/096368914X684970.

Danoy, M., Tauran, Y., Poulain, S., Arakawa, H., Mori, D., Araya, K., Kato, S., Kido, T., Kusuhara, H., Kato, Y., Miyajima, A., Plessy, C., Sakai, Y., & Leclerc, E. (2020). Analysis of hiPSCs differentiation toward hepatocyte-like cells upon extended exposition to oncostatin. *Differentiation, 114*, 36–48. https://doi.org/10.1016/j.diff.2020.05.006.

Demetriou, A. A., Brown, R. S., Busuttil, R. W., Fair, J., McGuire, B. M., Rosenthal, P., Am Esch, J. S., Lerut, J., Nyberg, S. L., Salizzoni, M., Fagan, E. A., de Hemptinne, B., Broelsch, C. E., Muraca, M., Salmeron, J. M., Rabkin, J. M., Metselaar, H. J., Pratt, D., De La Mata, M., ... Solomon, B. A. (2004). Prospective, randomized, multicenter, controlled trial of a bioartificial liver in treating acute liver failure. *Annals of Surgery, 239*, 660–670. https://doi.org/10.1097/01.sla.0000124298.74199.e5.

Deuse, T., Stubbendorff, M., Tang-Quan, K., Phillips, N., Kay, M. A., Eiermann, T., Phan, T. T., Volk, H.-D., Reichenspurner, H., Robbins, R. C., & Schrepfer, S. (2011). Immunogenicity and immunomodulatory properties of umbilical cord lining mesenchymal stem cells. *Cell Transplantation, 20*, 655–667. https://doi.org/10.3727/096368910X536473.

Dhawan, A., Mitry, R. R., Hughes, R. D., Lehec, S., Terry, C., Bansal, S., Arya, R., Wade, J. J., Verma, A., Heaton, N. D., Rela, M., & Mieli-Vergani, G. (2004). Hepatocyte transplantation for inherited factor VII deficiency. *Transplantation, 78*, 1812–1814. https://doi.org/10.1097/01.TP.0000146386.77076.47.

Dominici, M., Le Blanc, K., Mueller, I., Slaper-Cortenbach, I., Marini, F., Krause, D. S., Deans, R. J., Keating, A., Prockop, D. J., & Horwitz, E. M. (2006). Minimal criteria for defining multipotent mesenchymal stromal cells. The international society for cellular therapy position statement. *Cytotherapy, 8*, 315–317. https://doi.org/10.1080/14653240600855905.

Doulatov, S., Notta, F., Laurenti, E., & Dick, J. E. (2012). Hematopoiesis: a human perspective. *Cell Stem Cell, 10*, 120–136. https://doi.org/10.1016/j.stem.2012.01.006.

Duncan, S. A., Navas, M. A., Dufort, D., Rossant, J., & Stoffel, M. (1998). Regulation of a transcription factor network required for differentiation and metabolism. *Science, 281*, 692–695. https://doi.org/10.1126/science.281.5377.692.

Duval, K., Grover, H., Han, L.-H., Mou, Y., Pegoraro, A. F., Fredberg, J., & Chen, Z. (2017). Modeling physiological events in 2D vs. 3D cell culture. *Physiology, 32*, 266–277. https://doi.org/10.1152/physiol.00036.2016.

El Baz, H., Demerdash, Z., Kamel, M., Hammam, O., Abdelhady, D. S., Mahmoud, S., Hassan, S., Mahmoud, F., Atta, S., Riad, N. M., & Gaafar, T. (2020). Induction of hepatic regeneration in an experimental model using hepatocyte-differentiated mesenchymal stem cells. *Cellular Reprogramming, 22*, 134–146. https://doi.org/10.1089/cell.2019.0076.

Elkhafif, N., Baz, H. E., Hammam, O., Hassan, S., Salah, F., Mansour, W., Mansy, S., Yehia, H., Zaki, A., & Magdy, R. (2011). CD133+ human umbilical cord blood stem cells enhance angiogenesis in experimental chronic hepatic fibrosis. *APMIS, 119*, 66–75. https://doi.org/10.1111/j.1600-0463.2010.02693.x.

Fausto, N., Campbell, J. S., & Riehle, K. J. (2006). Liver regeneration. *Hepatology, 43*, S45–S53. https://doi.org/10.1002/hep.20969.

Feng, T., Zhang, J., Zeng, G., Zhou, R., Tang, X., Cui, C., Li, Y., Wang, H., Li, T., Zhu, W., & Yu, Z. (2015). Therapeutic potential of umbilical cord mesenchymal stem cells in mice with acute hepatic failure. *International Journal of Artificial Organs, 38*, 271–276. https://doi.org/10.5301/ijao.5000390.

Freyer, N., Greuel, S., Knöspel, F., Gerstmann, F., Storch, L., Damm, G., Seehofer, D., Foster Harris, J., Iyer, R., Schubert, F., & Zeilinger, K. (2018). Microscale 3D liver bioreactor for in vitro hepatotoxicity testing under perfusion conditions. *Bioengineering, 5*, 24. https://doi.org/10.3390/bioengineering5010024.

Friedenstein, A. J., Petrakova, K. V., Kurolesova, A. I., & Frolova, G. P. (1968). Heterotopic transplants of bone marrow. *Transplantation, 6*, 230–247.

Gabr, H., Zayed, R., ElBeshlawy, A., Hegazy, L., Fawzy, R., Samir, M., & Mosa, H. (2014). Differentiation of bone marrow hematopoietic stem cells into FVIII-producing hepatocytes: Approach to hemophilia treatment. *Comparative Clinical Pathology, 23*, 193–198. https://doi.org/10.1007/s00580-012-1595-2.

Gai, H., Nguyen, D. M., Joon Moon, Y., Aguila, J. R., Fink, L. M., Ward, D. C., & Ma, Y. (2010). Generation of murine hepatic lineage cells from induced pluripotent stem cells. *Differentiation, 79*, 171–181. https://doi.org/10.1016/j.diff.2010.01.002.

Gao, X., Li, R., Cahan, P., Zhao, Y., Yourick, J. J., & Sprando, R. L. (2020). Hepatocyte-like cells derived from human induced pluripotent stem cells using small molecules: Implications of a transcriptomic study. *Stem Cell Research & Therapy, 11*, 393. https://doi.org/10.1186/s13287-020-01914-1.

Gao, Y., Zhang, X., Zhang, L., Cen, J., Ni, X., Liao, X., Yang, C., Li, Y., Chen, X., Zhang, Z., Shu, Y., Cheng, X., Hay, D. C., Lai, D., Pan, G., Wei, G., & Hui, L. (2017). Distinct gene expression and epigenetic signatures in hepatocyte-like cells produced by different strategies from the same donor. *Stem Cell Reports, 9*, 1813–1824. https://doi.org/10.1016/j.stemcr.2017.10.019.

Ghosheh, N., Küppers-Munther, B., Asplund, A., Andersson, C. X., Björquist, P., Andersson, T. B., Carén, H., Simonsson, S., Sartipy, P., & Synnergren, J. (2020). Human pluripotent stem cell-derived hepatocytes show higher transcriptional correlation with adult liver tissue than with fetal liver tissue. *ACS Omega, 5*, 4816–4827. https://doi.org/10.1021/acsomega.9b03514.

Gilgenkrantz, H., de l'Hortet, A. C. (2018). Understanding liver regeneration: From mechanisms to regenerative medicine. *The American Journal of Pathology, 188*, 1316–1327. https://doi.org/10.1016/j.ajpath.2018.03.008.

Gómez-Lechón, M. J., Castell, J. V., & Donato, M. T. (2008). An update on metabolism studies using human hepatocytes in primary culture. *Expert Opinion on Drug Metabolism & Toxicology, 4*, 837–854. https://doi.org/10.1517/17425255.4.7.837.

Gómez-Lechón, M., Donato, M., & Jover, R. (2003). Human hepatocytes as a tool for studying toxicity and drug metabolism. *Current Drug Metabolism, 4*, 292–312. https://doi.org/10.2174/1389200033489424.

Goral, V. N., & Yuen, P. K. (2012). Microfluidic platforms for hepatocyte cell culture: New technologies and applications. *Annals of Biomedical Engineering, 40*, 1244–1254. https://doi.org/10.1007/s10439-011-0453-8.

Gordon, M. Y., Levičar, N., Pai, M., Bachellier, P., Dimarakis, I., Al-Allaf, F., M'Hamdi, H., Thalji, T., Welsh, J. P., Marley, S. B., Davies, J., Dazzi, F., Marelli-Berg, F., Tait, P., Playford, R., Jiao, L., Jensen, S., Nicholls, J. P., Ayav, A., … Habib, N. A. (2006). Characterization and clinical application of human CD34$^+$ stem/progenitor cell populations mobilized into the blood by granulocyte colony-stimulating factor. *Stem Cells, 24*, 1822–1830. https://doi.org/10.1634/stemcells.2005-0629.

Gramignoli, R. (2016). Therapeutic use of human amnion-derived products: Cell-based therapy for liver disease. *Current Pathobiology Reports, 4*, 157–167. https://doi.org/10.1007/s40139-016-0112-8.

Grix, T., Ruppelt, A., Thomas, A., Amler, A.-K., Noichl, B. P., Lauster, R., & Kloke, L. (2018). Bioprinting perfusion-enabled liver equivalents for advanced organ-on-a-chip applications. *Genes, 9*, 176. https://doi.org/10.3390/genes9040176.

Guo, L., Dial, S., Shi, L., Branham, W., Liu, J., Fang, J.-L., Green, B., Deng, H., Kaput, J., & Ning, B. (2011). Similarities and Differences in the expression of drug-metabolizing enzymes between human hepatic cell lines and primary human hepatocytes. *Drug Metabolism and Disposition, 39*, 528–538. https://doi.org/10.1124/dmd.110.035873.

Habibullah, C. M., Syed, I. H., Qamar, A., & Taher-Uz, Z. (1994). Human fetal hepatocyte transplantation in patients with fulminant hepatic failure. *Transplantation, 58*, 951–952.

Hallifax, D., Rawden, H. C., Hakooz, N., & Houston, J. B. (2005). Prediction of metabolic clearance using cryopreserved human hepatocytes: Kinetic characteristics for five benzodiazepines. *Drug Metabolism and Disposition, 33*, 1852–1858. https://doi.org/10.1124/dmd.105.005389.

Han, Y.-F., Tao, R., Sun, T.-J., Chai, J.-K., Xu, G., & Liu, J. (2013). Optimization of human umbilical cord mesenchymal stem cell isolation and culture methods. *Cytotechnology, 65*, 819–827. https://doi.org/10.1007/s10616-012-9528-0.

Hang, H., Shi, X., Gu, G. X., Wu, Y., Gu, J., & Ding, Y. (2010). In vitro analysis of cryopreserved alginate–poly-l-lysine–alginate-microencapsulated human hepatocytes. *Liver International, 30*, 611–622. https://doi.org/10.1111/j.1478-3231.2009.02197.x.

Hannan, N. R., Segeritz, C.-P., Touboul, T., & Vallier, L. (2013). Production of hepatocyte like cells from human pluripotent stem cells. *Nature Protocols, 8*, 430–437.

Hasan, M. H., Botros, K. G., El-Shahat, M. A., Abdallah, H. A., & Sobh, M. A. (2017). In vitro differentiation of human umbilical cord blood mesenchymal stem cells into functioning hepatocytes. *Alexandria Journal of Medicine, 53*, 167–173. https://doi.org/10.1016/j.ajme.2016.05.002.

Hass, R., Kasper, C., Böhm, S., & Jacobs, R. (2011). Different populations and sources of human mesenchymal stem cells (MSC): A comparison of adult and neonatal tissue-derived MSC. *Cell Communication and Signaling, 9*, 12. https://doi.org/10.1186/1478-811X-9-12.

Hay, D. C., Fletcher, J., Payne, C., Terrace, J. D., Gallagher, R. C. J., Snoeys, J., Black, J. R., Wojtacha, D., Samuel, K., Hannoun, Z., Pryde, A., Filippi, C., Currie, I. S., Forbes, S. J., Ross, J. A., Newsome, P. N., & Iredale, J. P. (2008). Highly efficient differentiation of hESCs to functional hepatic endoderm requires ActivinA and Wnt3a signaling. *Proc Natl Acad Sci U S A, 105*, 12301–12306. https://doi.org/10.1073/pnas.0806522105.

Heslop, J. A., Rowe, C., Walsh, J., Sison-Young, R., Jenkins, R., Kamalian, L., Kia, R., Hay, D., Jones, R. P., Malik, H. Z., Fenwick, S., Chadwick, A. E., Mills, J., Kitteringham, N. R., Goldring, C. E. P., & Park, B. K. (2017). Mechanistic evaluation of primary human hepatocyte culture using global proteomic analysis reveals a selective dedifferentiation profile. *Archives of Toxicology, 91*, 439–452. https://doi.org/10.1007/s00204-016-1694-y.

Hewitt, N. J., Lechón, M. J. G., Houston, J. B., Hallifax, D., Brown, H. S., Maurel, P., Kenna, J. G., Gustavsson, L., Lohmann, C., Skonberg, C., Guillouzo, A., Tuschl, G., Li, A. P., LeCluyse, E., Groothuis, G. M. M., & Hengstler, J. G. (2007). Primary hepatocytes: Current understanding of the regulation of metabolic enzymes and transporter proteins, and pharmaceutical practice for the use of hepatocytes in metabolism, enzyme induction, transporter, clearance, and hepatotoxicity studies. *Drug Metabolism Reviews, 39*, 159–234. https://doi.org/10.1080/03602530601093489.

Ho, C.-M., Dhawan, A., Hughes, R. D., Lehec, S. C., Puppi, J., Philippeos, C., Lee, P.-H., & Mitry, R. R. (2012). Use of indocyanine green for functional assessment of human hepatocytes for transplantation. *Asian Journal of Surgery, 35*, 9–15. https://doi.org/10.1016/j.asjsur.2012.04.017.

Holzman, M. D., Rozga, J. A., Neuzil, D. F., Griffin, D. O., Moscioni, A. D., & Demetriou, A. A. (1993). Selective intraportal hepatocyte transplantation in analbuminemic and Gunn rats [WWW Document]. *Transplantation*. https://doi.org/10.1097/00007890-199306000-00002.

Horslen, S. P., & Fox, I. J. (2004). Hepatocyte transplantation. *Transplantation, 77*, 1481–1486. https://doi.org/10.1097/01.TP.0000113809.53415.C2.

Hsieh, J.-Y., Fu, Y.-S., Chang, S.-J., Tsuang, Y.-H., & Wang, H.-W. (2010). Functional module analysis reveals differential osteogenic and stemness potentials in human mesenchymal stem cells from bone marrow and Wharton's Jelly of umbilical cord. *Stem Cells and Development, 19*, 1895–1910. https://doi.org/10.1089/scd.2009.0485.

Hu, J., Yu, X., Wang, Z., Wang, F., Wang, L., Gao, H., Chen, Y., Zhao, W., Jia, Z., Yan, S., & Wang, Y. (2013). Long term effects of the implantation of Wharton's jelly-derived mesenchymal stem cells from the umbilical cord for newly-onset type 1 diabetes mellitus. *Endocrine Journal, 60*, 347–357. https://doi.org/10.1507/endocrj.EJ12-0343.

Iansante, V., Mitry, R. R., Filippi, C., Fitzpatrick, E., & Dhawan, A. (2018). Human hepatocyte transplantation for liver disease: Current status and future perspectives. *Pediatric Research, 83*, 232–240. https://doi.org/10.1038/pr.2017.284.

Ibars, E. P., Cortes, M., Tolosa, L., Gómez-Lechón, M. J., López, S., Castell, J. V., & Mir, J. (2016). Hepatocyte transplantation program: Lessons learned and future strategies. *World Journal of Gastroenterology, 22*, 874–886. https://doi.org/10.3748/wjg.v22.i2.874.

Jaafarpour, Z., Soleimani, M., Hosseinkhani, S., Geramizadeh, B., Yaghmaei, P., Mobarra, N., & Karimi, M. H. (2020). Overexpression of microRNA-375 and microRNA-122 promotes the differentiation of human induced pluripotent stem cells into hepatocyte-like cells. *Biologicals, 63*, 24–32. https://doi.org/10.1016/j.biologicals.2019.12.005.

Jindal, R., Nahmias, Y., Tilles, A. W., Berthiaume, F., & Yarmush, M. L. (2009). Amino acid-mediated heterotypic interaction governs performance of a hepatic tissue model. *The FASEB Journal, 23*, 2288–2298. https://doi.org/10.1096/fj.08-114934.

Jing, R., Corbett, J. L., Cai, J., Beeson, G. C., Beeson, C. C., Chan, S. S., Dimmock, D. P., Lazcares, L., Geurts, A. M., Lemasters, J. J., & Duncan, S. A. (2018). A screen using iPSC-derived hepatocytes reveals NAD+ as a potential treatment for mtDNA depletion syndrome. *Cell Reports, 25*, 1469-1484.e5. https://doi.org/10.1016/j.celrep.2018.10.036.

Jitraruch, S., Dhawan, A., Hughes, R. D., Filippi, C., Lehec, S. C., Glover, L., & Mitry, R. R. (2017). Cryopreservation of hepatocyte microbeads for clinical transplantation. *Cell Transplantation, 26*, 1341–1354. https://doi.org/10.1177/0963689717720050.

Jung, J., Choi, J. H., Lee, Y., Park, J.-W., Oh, I.-H., Hwang, S.-G., Kim, K.-S., & Kim, G. J. (2013). Human placenta-derived mesenchymal stem cells promote hepatic regeneration in CCl4-injured rat liver model via increased autophagic mechanism. *Stem Cells, 31*, 1584–1596. https://doi.org/10.1002/stem.1396.

Jung, K. H., McCarthy, R. L., Zhou, C., Uprety, N., Barton, M. C., & Beretta, L. (2016). MicroRNA regulates hepatocytic differentiation of progenitor cells by targeting YAP1. *Stem Cells, 34*, 1284–1296. https://doi.org/10.1002/stem.2283.

Kao, S.-Y., Shyu, J.-F., Wang, H.-S., Hsiao, C.-Y., Su, C.-H., Chen, T.-H., Weng, Z.-C., & Tsai, P.-J. (2015). Transplantation of hepatocytelike cells derived from umbilical cord stromal mesenchymal stem cells to treat acute liver failure rat. *Journal of Biomedical Sciences, 4*. https://doi.org/10.4172/2254-609X.100002.

Khan, A. A., Habeeb, A., Parveen, N., Naseem, B., Babu, R. P., Capoor, A. K., & Habibullah, C. M.88 (2004). Peritoneal transplantation of human fetal hepatocytes for the treatment of acute fatty liver of pregnancy: a case report [WWW Document]. *Tropical Gastroenterology: Official*

Journal of the Digestive Diseases Foundation. URL http://pubmed.ncbi.nlm.nih.gov/15682663/?dopt=Abstract. Accessed August 12, 2020.

Kholodenko, I. V., Kurbatov, L. K., Kholodenko, R. V., Manukyan, G. V., & Yarygin, K. N. (2019). Mesenchymal stem cells in the adult human liver: Hype or hope? *Cells, 8*, 1127. https://doi.org/10.3390/cells8101127.

Khurana, S., & Mukhopadhyay, A. (2008). In vitro transdifferentiation of adult hematopoietic stem cells: An alternative source of engraftable hepatocytes. *Journal of Hepatology, 49*, 998–1007. https://doi.org/10.1016/j.jhep.2008.05.019.

Kidambi, S., Yarmush, R. S., Novik, E., Chao, P., Yarmush, M. L., & Nahmias, Y. (2009). Oxygen-mediated enhancement of primary hepatocyte metabolism, functional polarization, gene expression, and drug clearance. *Proceedings of the National Academy of Sciences of the United States of America, 106*, 15714. https://doi.org/10.1073/pnas.0906820106.

Kim, M. J., Shin, K. S., Jeon, J. H., Lee, D. R., Shim, S. H., Kim, J. K., Cha, D.-H., Yoon, T. K., & Kim, G. J. (2011). Human chorionic-plate-derived mesenchymal stem cells and Wharton's jelly-derived mesenchymal stem cells: A comparative analysis of their potential as placenta-derived stem cells. *Cell and Tissue Research, 346*, 53–64. https://doi.org/10.1007/s00441-011-1249-8.

Knobeloch, D., Ehnert, S., Schyschka, L., Büchler, P., Schoenberg, M., Kleeff, J., Thasler, W. E., Nussler, N. C., Godoy, P., Hengstler, J., Nussler, A. K. (2012). Human hepatocytes: Isolation, culture, and quality procedures. In *Human cell culture protocols, methods in molecular biology* (pp. 99–120). Humana Press. https://doi.org/10.1007/978-1-61779-367-7_8.

Kolanko, E., Kopaczka, K., Koryciak-Komarska, H., Czech, E., Szmytkowska, P., Gramignoli, R., & Czekaj, P. (2019). Increased immunomodulatory capacity of human amniotic cells after activation by pro-inflammatory chemokines. *European Journal of Pharmacology, 859*, 172545. https://doi.org/10.1016/j.ejphar.2019.172545.

Kusano, T., Aoki, T., Yasuda, D., Matsumoto, S., Jin, Z., Nishino, N., Hayashi, K., Odaira, M., Yamada, K., Koizumi, T., Izumida, Y., Mitamura, K., Enami, Y., Niiya, T., Murai, N., Kato, H., Shimizu, Y., Kou, K., Furukawa, Y., ... Kusano, M. (2008). Microencapsule technique protects hepatocytes from cryoinjury. *Hepatology Research, 38*, 593–600. https://doi.org/10.1111/j.1872-034X.2007.00311.x.

Lagasse, E., Connors, H., Al-Dhalimy, M., Reitsma, M., Dohse, M., Osborne, L., Wang, X., Finegold, M., Weissman, I. L., & Grompe, M. (2000). Purified hematopoietic stem cells can differentiate into hepatocytes in vivo. *Nature Medicine, 6*, 1229–1234. https://doi.org/10.1038/81326.

Langer, R., & Vacanti, J. (2016). Advances in tissue engineering. *Journal of Pediatric Surgery, 51*, 8–12. https://doi.org/10.1016/j.jpedsurg.2015.10.022.

Lázaro-Diéguez, F., & Müsch, A. (2014). The special case of hepatocytes. *Bioarchitecture, 4*, 47–52. https://doi.org/10.4161/bioa.29012.

Le, T. V., Nguyen, N. H., Do, H. Q., Le, H. M., & Truong, N. H. (2017). Transplantation of umbilical cord blood-derived mesenchymal stem cells to treat liver cirrhosis in mice: a comparison of tail and portal vein injection. *Progress Stem Cell, 4*, 201–216. https://doi.org/10.15419/psc.v4i2.365.

Lee, H., Chae, S., Kim, J. Y., Han, W., Kim, J., Choi, Y., & Cho, D.-W. (2019). Cell-printed 3D liver-on-a-chip possessing a liver microenvironment and biliary system. *Biofabrication, 11*, 025001. https://doi.org/10.1088/1758-5090/aaf9fa.

Lee, J.-Y., Han, H.-J., Lee, S.-J., Cho, E.-H., Lee, H.-B., Seok, J.-H., Lim, H. S., & Son, W.-C. (2020). Use of 3D Human Liver Organoids To Predict Drug-Induced Phospholipidosis. *International Journal of Molecular Sciences, 21*, 2982. https://doi.org/10.3390/ijms21082982.

Lee, S. M. L., Bertinetti-Lapatki, C., Schiergens, T. S., Jauch, K.-W., Roth, A. B., & Thasler, W. E. (2020). Concurrent isolation of hepatic stem cells and hepatocytes from the human liver. *Vitro Cellular Developmental Biology-Animal, 56*, 253–260. https://doi.org/10.1007/s11626-020-00433-w.

Lee, S.-G., Moon, S.-H., Kim, H.-J., Lee, J. Y., Park, S.-J., Chung, H.-M., Ha, T. Y., Song, G.-W., Jung, D.-H., Park, H., Kwon, T.-W., & Cho, Y.-P. (2015). Bone marrow-derived progenitor cells

in de novo liver regeneration in liver transplant. *Liver Transplantation, 21*, 1186–1194. https://doi.org/10.1002/lt.24099.
Leist, M., Ghallab, A., Graepel, R., Marchan, R., Hassan, R., Bennekou, S. H., Limonciel, A., Vinken, M., Schildknecht, S., Waldmann, T., Danen, E., van Ravenzwaay, B., Kamp, H., Gardner, I., Godoy, P., Bois, F. Y., Braeuning, A., Reif, R., Oesch, F., … Hengstler, J. G. (2017). Adverse outcome pathways: Opportunities, limitations and open questions. *Archives of Toxicology, 91*, 3477–3505. https://doi.org/10.1007/s00204-017-2045-3.
Li, L., Miu, K.-K., Gu, S., Cheung, H.-H., & Chan, W.-Y. (2018). Comparison of multi-lineage differentiation of hiPSCs reveals novel miRNAs that regulate lineage specification. *Scientific Reports, 8*, 1–15. https://doi.org/10.1038/s41598-018-27719-0.
Li, Y. C., Wang, D. P., & Chiang, J. Y. (1990). Regulation of cholesterol 7 alpha-hydroxylase in the liver. Cloning, sequencing, and regulation of cholesterol 7 alpha-hydroxylase mRNA. *Journal of Biological Chemistry, 265*, 12012–12019.
Liang, J., & Sun, L. (2015). Mesenchymal stem cells transplantation for systemic lupus erythematosus. *International Journal of Rheumatic Diseases, 18*, 164–171. https://doi.org/10.1111/1756-185X.12531.
Lillegard, J. B., Fisher, J. E., Nedredal, G., Luebke-Wheeler, J., Bao, J., Wang, W., Amoit, B., & Nyberg, S. L. (2011). Normal atmospheric oxygen tension and the use of antioxidants improve hepatocyte spheroid viability and function. *Journal of Cellular Physiology, 226*, 2987–2996. https://doi.org/10.1002/jcp.22651.
Lim, R., Hodge, A., Moore, G., Wallace, E. M., & Sievert, W. (2017). A pilot study evaluating the safety of intravenously administered human amnion epithelial cells for the treatment of hepatic fibrosis. *Frontiers in Pharmacology, 8*, 549. https://doi.org/10.3389/fphar.2017.00549.
Lin, R.-Z., & Chang, H.-Y. (2008). Recent advances in three-dimensional multicellular spheroid culture for biomedical research. *Biotechnology Journal, 3*, 1172–1184. https://doi.org/10.1002/biot.200700228.
Loh, K. M., Palaria, A., & Ang, L. T. (2019). Efficient differentiation of human pluripotent stem cells into liver cells. *JoVE (Journal of Visualized Experiments)* e58975. https://doi.org/10.3791/58975.
Louisse, J., Alewijn, M., Peijnenburg, A. A. C. M., Cnubben, N. H. P., Heringa, M. B., Coecke, S., & Punt, A. (2020) Towards harmonization of test methods for in vitro hepatic clearance studies. *Toxicology in Vitro, 63*, 104722. https://doi.org/10.1016/j.tiv.2019.104722.
Maleki, M., Ghanbarvand, F., Reza Behvarz, M., Ejtemaei, M., & Ghadirkhomi, E. (2014). Comparison of mesenchymal stem cell markers in multiple human adult stem cells. *International Journal of Stem Cells, 7*, 118–126. https://doi.org/10.15283/ijsc.2014.7.2.118.
Matsumura, M., Imura, T., Inagaki, A., Ogasawara, H., Fukuoka, K., Fathi, I., Miyagi, S., Ohashi, K., Unno, M., Kamei, T., Satomi, S., & Goto, M. (2019). A simple and useful predictive assay for evaluating the quality of isolated hepatocytes for hepatocyte transplantation. *Science and Reports, 9*, 1. https://doi.org/10.1038/s41598-019-42720-x.
McKenzie, T. J., Lillegard, J. B., & Nyberg, S. L. (2008). Artificial and bioartificial liver support. *Seminars in Liver Disease, 28*, 210–217. https://doi.org/10.1055/s-2008-1073120.
McNiven, M. A., Wolkoff, A. W., & Hubbard, A. (2009). A stimulus needed for the study of membrane traffic in hepatocytes. *Hepatology, 50*, 345–348. https://doi.org/10.1002/hep.23004.
Medine, C. N., Lucendo-Villarin, B., Zhou, W., West, C. C., & Hay, D. C. (2011). Robust generation of hepatocyte-like cells from human embryonic stem cell populations. *Journal of Visualized Experiments,* e2969. https://doi.org/10.3791/2969.
Mehrotra, M., Rosol, M., Ogawa, M., & LaRue, A. C. (2010). Amelioration of a mouse model of osteogenesis imperfecta with hematopoietic stem cell transplantation: Micro-computed tomography studies. *Experimental Hematology, 38*, 593–602. https://doi.org/10.1016/j.exphem.2010.04.008.
Meyburg, J., Das, A. M., Hoerster, F., Lindner, M., Kriegbaum, H., Engelmann, G., Schmidt, J., Ott, M., Pettenazzo, A., Luecke, T., Bertram, H., Hoffmann, G. F., & Burlina, A. (2009). One

liver for four children: First clinical series of liver cell transplantation for severe neonatal urea cycle defects. *Transplantation, 87*, 636–641. https://doi.org/10.1097/TP.0b013e318199936a.
Michalopoulos, G. K. (2007). Liver regeneration. *Journal of Cellular Physiology, 213*, 286–300. https://doi.org/10.1002/jcp.21172.
Michalopoulos, G. K., Bowen, W. C., Mulè, K., & Stolz, D. B. (2001). Histological organization in hepatocyte organoid cultures. *American Journal of Pathology, 159*, 1877–1887.
Minamiguchi, H., Ishikawa, F., Fleming, P. A., Yang, S., Drake, C. J., Wingard, J. R., & Ogawa, M. (2008). Transplanted human cord blood cells generate amylase-producing pancreatic acinar cells in engrafted mice. *Pancreas, 36*, e30. https://doi.org/10.1097/MPA.0b013e3181584656.
Mita, S., Suzuki, H., Akita, H., Hayashi, H., Onuki, R., Hofmann, A. F., & Sugiyama, Y. (2006). Inhibition of bile acid transport across Na^+/taurocholate cotransporting polypeptide (SLC10A1) and bile salt export pump (ABCB 11)-coexpressing LLC-PK1 cells by cholestasis-inducing drugs. *Drug Metabolism and Disposition, 34*, 1575–1581. https://doi.org/10.1124/dmd.105.008748.
Mitry, R. R., Dhawan, A., Hughes, R. D., Bansal, S., Lehec, S., Terry, C., Heaton, N. D., Karani, J. B., Mieli-Vergani, G., & Rela, M. (2004). one liver, three recipients: Segment IV from split-liver procedures as a source of hepatocytes for cell transplantation. *Transplantation, 77*, 1614–1616. https://doi.org/10.1097/01.TP.0000122224.98318.19.
Mitry, R. R., Hughes, R. D., Aw, M. M., Terry, C., Mieli-Vergani, G., Girlanda, R., Muiesan, P., Rela, M., Heaton, N. D., & Dhawan, A. (2003). Human hepatocyte isolation and relationship of cell viability to early graft function. *Cell Transplantation, 12*, 69–74. https://doi.org/10.3727/000000003783985197.
Moradi, E., Jalili-Firoozinezhad, S., & Solati-Hashjin, M. (2020). Microfluidic organ-on-a-chip models of human liver tissue. *Acta Biomaterialia, 116*, 67–83. https://doi.org/10.1016/j.actbio.2020.08.041.
Muraca, M., Gerunda, G., Neri, D., Vilei, M. T., Granato, A., Feltracco, P., Giron, G., & Burlina, A. B. (2002). Hepatocyte transplantation as a treatment for glycogen storage disease type IA. *Journal of Hepatology, 36*, 41. https://doi.org/10.1016/S0168-8278(02)80123-4.
Nagamura-Inoue, T., & He, H. (2014). Umbilical cord-derived mesenchymal stem cells: Their advantages and potential clinical utility. *World J Stem Cells, 6*, 195–202. https://doi.org/10.4252/wjsc.v6.i2.195.
Navarro-Alvarez, N., Soto-Gutierrez, A., Rivas-Carrillo, J. D., Chen, Y., Yamamoto, T., Yuasa, T., Misawa, H., Takei, J., Tanaka, N., & Kobayashi, N. (2006). Self-assembling peptide nanofiber as a novel culture system for isolated porcine hepatocytes. *Cell Transplantation, 15*, 921–927. https://doi.org/10.3727/000000006783981387.
Nazarov, I., Lee, J. W., Soupene, E., Etemad, S., Knapik, D., Green, W., Bashkirova, E., Fang, X., Matthay, M. A., Kuypers, F. A., & Serikov, V. B. (2012). Multipotent stromal stem cells from human placenta demonstrate high therapeutic potential. *Stem Cells Translational Medicine, 1*, 359–372. https://doi.org/10.5966/sctm.2011-0021.
Pediaditakis, P., Lopez-Talavera, J. C., Petersen, B., Monga, S. P. S., & Michalopoulos, G. K. (2001). The processing and utilization of hepatocyte growth factor/scatter factor following partial hepatectomy in the rat. *Hepatology, 34*, 688–693. https://doi.org/10.1053/jhep.2001.27811.
Phan, T. G., Ma, H., Lim, R., Sobey, C. G., & Wallace, E. M. (2018). Phase 1 trial of amnion cell therapy for ischemic stroke. *Frontiers in Neurology, 9*, 198. https://doi.org/10.3389/fneur.2018.00198.
Pilat, N., Unger, L., & Berlakovich, G. A. (2013). Implication for Bone marrow derived stem cells in hepatocyte regeneration after orthotopic liver transplantation [WWW Document]. *International Journal of Hepatology*. https://doi.org/10.1155/2013/310612.
Rajan, S. A. P., Aleman, J., Wan, M., Pourhabibi Zarandi, N., Nzou, G., Murphy, S., Bishop, C. E., Sadri-Ardekani, H., Shupe, T., Atala, A., Hall, A. R., & Skardal, A. (2020). Probing prodrug metabolism and reciprocal toxicity with an integrated and humanized multi-tissue organ-on-a-chip platform. *Acta Biomaterialia, 106*, 124–135. https://doi.org/10.1016/j.actbio.2020.02.015.
Ramaiahgari, S. C., den Braver, M. W., Herpers, B., Terpstra, V., Commandeur, J. N. M., van de Water, B., & Price, L. S. (2014). A 3D in vitro model of differentiated HepG2 cell spheroids

with improved liver-like properties for repeated dose high-throughput toxicity studies. *Archives of Toxicology, 88*, 1083–1095. https://doi.org/10.1007/s00204-014-1215-9.
Ramaiahgari, S. C., Waidyanatha, S., Dixon, D., DeVito, M. J., Paules, R. S., & Ferguson, S. S. (2017). From the cover: three-dimensional (3D) HepaRG spheroid model with physiologically relevant xenobiotic metabolism competence and hepatocyte functionality for liver toxicity screening. *Toxicological Sciences, 159*, 124–136. https://doi.org/10.1093/toxsci/kfx122.
Raoufil, A., Aminil, A., Azadbakht, M., Farhadifar, F., Rahram Nikhn Frrfin Fthi, N. F. (2015). Production of hepatocyte-like cells from human umbilical vein mesenchymal stem cells. *Italian Journal of Anatomy and Embryology = Archivio italiano di anatomia ed embriologia, 120*, 150–161.
Raut, A., & Khanna, A. (2016). Enhanced expression of hepatocyte-specific microRNAs in valproic acid mediated hepatic trans-differentiation of human umbilical cord derived mesenchymal stem cells. *Experimental Cell Research, 343*, 237–247. https://doi.org/10.1016/j.yexcr.2016.03.015.
Resca, E., Zavatti, M., Maraldi, T., Bertoni, L., Beretti, F., Guida, M., La Sala, G. B., Guillot, P. V., David, A. L., Sebire, N. J., De Pol, A., & De Coppi, P. (2015). Enrichment in c-Kit improved differentiation potential of amniotic membrane progenitor/stem cells. *Placenta, 36*, 18–26. https://doi.org/10.1016/j.placenta.2014.11.002.
Riley, R. J., & Grime, K. (2004). Metabolic screening in vitro: Metabolic stability, CYP inhibition and induction. *Drug Discovery Today: Technologies, 1*, 365–372. https://doi.org/10.1016/j.ddtec.2004.10.008.
Roelandt, P., Pauwelyn, K. A., Sancho-Bru, P., Subramanian, K., Bose, B., Ordovas, L., Vanuytsel, K., Geraerts, M., Firpo, M., De Vos, R., Fevery, J., Nevens, F., Hu, W.-S., & Verfaillie, C. M. (2010). Human embryonic and rat adult stem cells with primitive endoderm-like phenotype can be fated to definitive endoderm, and finally hepatocyte-like cells. *PLoS ONE, 5*, e12101. https://doi.org/10.1371/journal.pone.0012101.
Ryu, N.-E., Lee, S.-H., & Park, H. (2019). Spheroid culture system methods and applications for mesenchymal stem cells. *Cells, 8*, 1620. https://doi.org/10.3390/cells8121620.
Salama, H., Zekri, A.-R., Zern, M., Bahnassy, A., Loutfy, S., Shalaby, S., Vigen, C., Burke, W., Mostafa, M., Medhat, E., Alfi, O., & Huttinger, E. (2010). Autologous hematopoietic stem cell transplantation in 48 patients with end-stage chronic liver diseases. *Cell Transplantation, 19*, 1475–1486. https://doi.org/10.3727/096368910X514314.
Schneider, A., Attaran, M., Meier, P. N., Strassburg, C., Manns, M. P., Ott, M., Barthold, M., Arseniev, L., Becker, T., & Panning, B. (2006). Hepatocyte transplantation in an acute liver failure due to mushroom poisoning. *Transplantation, 82*, 1115–1116. https://doi.org/10.1097/01.tp.0000232451.93703.ab.
Schulze, R. J., Schott, M. B., Casey, C. A., Tuma, P. L., & McNiven, M. A. (2019). The cell biology of the hepatocyte: A membrane trafficking machine. *Journal of Cell Biology, 218*, 2096–2112. https://doi.org/10.1083/jcb.201903090.
Seglen, P. O. (1976). Preparation of isolated rat liver cells. In Prescott, D. M. (Ed.), *Methods in cell biology* (pp. 29–83). Academic Press. https://doi.org/10.1016/S0091-679X(08)61797-5.
Sellamuthu, S., Manikandan, R., Thiagarajan, R., Babu, G., Dinesh, D., Prabhu, D., & Arulvasu, C. (2011). In vitro trans-differentiation of human umbilical cord derived hematopoietic stem cells into hepatocyte like cells using combination of growth factors for cell based therapy. *Cytotechnology, 63*, 259–268. https://doi.org/10.1007/s10616-011-9337-x.
Sera, Y., LaRue, A. C., Moussa, O., Mehrotra, M., Duncan, J. D., Williams, C. R., Nishimoto, E., Schulte, B. A., Watson, P. M., Watson, D. K., & Ogawa, M. (2009). Hematopoietic stem cell origin of adipocytes. *Experimental Hematology, 37*, 1108-1120.e4. https://doi.org/10.1016/j.exphem.2009.06.008.
Sharma, M., Rao, P. N., Sasikala, M., Kuncharam, M. R., Reddy, C., Gokak, V., Raju, B., Singh, J. R., Nag, P., & Reddy, D. N. (2015). Autologous mobilized peripheral blood CD34+ cell infusion in non-viral decompensated liver cirrhosis. *World Journal of Gastroenterology, 21*, 7264–7271. https://doi.org/10.3748/wjg.v21.i23.7264.

Shi, M., Liu, Z., Wang, Y., Xu, R., Sun, Y., Zhang, M., Yu, X., Wang, H., Meng, L., Su, H., Jin, L., & Wang, F. (2017). A pilot study of mesenchymal stem cell therapy for acute liver allograft rejection. *Stem Cells Translational Medicine, 6*, 2053–2061. https://doi.org/10.1002/sctm.17-0134.

Siller, R., Greenhough, S., Naumovska, E., & Sullivan, G. J. (2015). Small-molecule-driven hepatocyte differentiation of human pluripotent stem cells. *Stem Cell Reports, 4*, 939–952. https://doi.org/10.1016/j.stemcr.2015.04.001.

Sirenko, O., Hancock, M. K., Hesley, J., Hong, D., Cohen, A., Gentry, J., Carlson, C. B., & Mann, D. A. (2016). Phenotypic characterization of toxic compound effects on liver spheroids derived from iPSC using confocal imaging and three-dimensional image analysis. *ASSAY and Drug Development Technologies, 14*, 381–394. https://doi.org/10.1089/adt.2016.729.

Si-Tayeb, K., Noto, F. K., Nagaoka, M., Li, J., Battle, M. A., Duris, C., North, P. E., Dalton, S., & Duncan, S. A. (2010). Highly efficient generation of human hepatocyte–like cells from induced pluripotent stem cells. *Hepatology, 51*, 297–305. https://doi.org/10.1002/hep.23354.

Snykers, S., Vanhaecke, T., De Becker, A., Papeleu, P., Vinken, M., Van Riet, I., & Rogiers, V. (2007). Chromatin remodeling agent trichostatin A: A key-factor in the hepatic differentiation of human mesenchymal stem cells derived of adult bone marrow. *BMC Developmental Biology, 7*, 24. https://doi.org/10.1186/1471-213X-7-24.

Sokal, E. M., Smets, F., Bourgois, A., Van Maldergem, L., Buts, J.-P., Reding, R., Bernard Otte, J., Evrard, V., Latinne, D., Vincent, M. F., Moser, A., & Soriano, H. E. (2003). Hepatocyte transplantation in a 4-year-old girl with peroxisomal biogenesis disease: Technique, safety, and metabolic follow-up1. *Transplantation, 76*, 735–738. https://doi.org/10.1097/01.TP.0000077420.81365.53.

Soldatow, V. Y., LeCluyse, E. L., Griffith, L. G., & Rusyn, I. (2013). In vitro models for liver toxicity testing. *Toxicol Res (camb), 2*, 23–39. https://doi.org/10.1039/C2TX20051A.

Soltys, K. A., Setoyama, K., Tafaleng, E. N., Soto Gutiérrez, A., Fong, J., Fukumitsu, K., Nishikawa, T., Nagaya, M., Sada, R., Haberman, K., Gramignoli, R., Dorko, K., Tahan, V., Dreyzin, A., Baskin, K., Crowley, J. J., Quader, M. A., Deutsch, M., Ashokkumar, C., ... Fox, I. J. (2017). Host conditioning and rejection monitoring in hepatocyte transplantation in humans. *Journal of Hepatology, 66*, 987–1000. https://doi.org/10.1016/j.jhep.2016.12.017.

Spada, M., Porta, F., Righi, D., Gazzera, C., Tandoi, F., Ferrero, I., Fagioli, F., Sanchez, M. B. H., Calvo, P. L., Biamino, E., Bruno, S., Gunetti, M., Contursi, C., Lauritano, C., Conio, A., Amoroso, A., Salizzoni, M., Silengo, L., Camussi, G., & Romagnoli, R. (2020). Intrahepatic administration of human liver stem cells in infants with inherited neonatal-onset hyperammonemia: A phase I study. *Stem Cell Reviews and Reports, 16*, 186–197. https://doi.org/10.1007/s12015-019-09925-z.

Stéphenne, X., Debray, F. G., Smets, F., Jazouli, N., Sana, G., Tondreau, T., Menten, R., Goffette, P., Boemer, F., Schoos, R., Gersting, S. W., Najimi, M., Muntau, A. C., Goyens, P., & Sokal, E. M. (2012). Hepatocyte transplantation using the domino concept in a child with tetrabiopterin nonresponsive phenylketonuria. *Cell Transplantation, 21*, 2765–2770. https://doi.org/10.3727/096368912X653255.

Stéphenne, X., Najimi, M., Sibille, C., Nassogne, M., Smets, F., & Sokal, E. M. (2006). Sustained engraftment and tissue enzyme activity after liver cell transplantation for argininosuccinate lyase deficiency. *Gastroenterology, 130*, 1317–1323. https://doi.org/10.1053/j.gastro.2006.01.008.

Strom, S. C., Jirtle, R. L., Jones, R. S., Novicki, D. L., Rosenberg, M. R., Novotny, A., Irons, G., McLain, J. R., & Michalopoulos, G. (1982). Isolation, culture, and transplantation of human hepatocytes. *Journal of the National Cancer Institute, 68*(5), 771–778.

Tanaka, K., Soto-Gutierrez, A., Navarro-Alvarez, N., Rivas-Carrillo, J. D., Jun, H.-S., & Kobayashi, N. (2006). Functional hepatocyte culture and its application to cell therapies. *Cell Transplantation, 15*, 855–864. https://doi.org/10.3727/000000006783981332.

Tasnim, F., Phan, D., Toh, Y.-C., & Yu, H. (2015). Cost-effective differentiation of hepatocyte-like cells from human pluripotent stem cells using small molecules. *Biomaterials, 70*, 115–125. https://doi.org/10.1016/j.biomaterials.2015.08.002.

Thi, V. L. D., Wu, X., Belote, R. L., Andreo, U., Takacs, C. N., Fernandez, J. P., Vale-Silva, L. A., Prallet, S., Decker, C. C., Fu, R. M., Qu, B., Uryu, K., Molina, H., Saeed, M., Steinmann, E., Urban, S., Singaraja, R. R., Schneider, W. M., Simon, S. M., & Rice, C. M. (2020). Stem cell-derived polarized hepatocytes. *Nature Communications, 11*, 1–13. https://doi.org/10.1038/s41467-020-15337-2.

Tolosa, L., Caron, J., Hannoun, Z., Antoni, M., López, S., Burks, D., Castell, J. V., Weber, A., Gomez-Lechon, M.-J., & Dubart-Kupperschmitt, A. (2015). Transplantation of hESC-derived hepatocytes protects mice from liver injury. *Stem Cell Research & Therapy, 6*, 1–17. https://doi.org/10.1186/s13287-015-0227-6.

Tostões, R. M., Leite, S. B., Serra, M., Jensen, J., Björquist, P., Carrondo, M. J. T., Brito, C., & Alves, P. M. (2012). Human liver cell spheroids in extended perfusion bioreactor culture for repeated-dose drug testing. *Hepatology, 55*, 1227–1236. https://doi.org/10.1002/hep.24760.

Treyer, A., & Müsch, A. (2013). Hepatocyte Polarity. *Comprehensive Physiology, 3*, 243–287. https://doi.org/10.1002/cphy.c120009.

Tümpel, S., & Rudolph, K. L. (2019). Quiescence: Good and bad of stem cell aging. *Trends in Cell Biology, 29*, 672–685. https://doi.org/10.1016/j.tcb.2019.05.002.

Ullah, I., Subbarao, R. B., & Rho, G. J. (2015). Human mesenchymal stem cells—current trends and future prospective. *Bioscience Reports, 35*, e00191. https://doi.org/10.1042/BSR20150025.

Utech, S., Prodanovic, R., Mao, A. S., Ostafe, R., Mooney, D. J., & Weitz, D. A. (2015). Microfluidic generation of monodisperse, structurally homogeneous alginate microgels for cell encapsulation and 3D cell culture. *Advanced Healthcare Materials, 4*, 1628–1633. https://doi.org/10.1002/adhm.201500021.

Varghese, D. S., Alawathugoda, T. T., & Ansari, S. A. (2019) Fine tuning of hepatocyte differentiation from human embryonic stem cells: Growth factor vs. small molecule-based approaches. *Stem Cells International, 2019*. https://doi.org/10.1155/2019/5968236.

Vassilopoulos, G., Wang, P.-R., & Russell, D. W. (2003). Transplanted bone marrow regenerates liver by cell fusion. *Nature, 422*, 901–904. https://doi.org/10.1038/nature01539.

Viiri, L. E., Rantapero, T., Kiamehr, M., Alexanova, A., Oittinen, M., Viiri, K., Niskanen, H., Nykter, M., Kaikkonen, M. U., & Aalto-Setälä, K. (2019). Extensive reprogramming of the nascent transcriptome during iPSC to hepatocyte differentiation. *Scientific Reports, 9*, 1–12. https://doi.org/10.1038/s41598-019-39215-0.

Vinken, M., & Hengstler, J. G. (2018). Characterization of hepatocyte-based in vitro systems for reliable toxicity testing. *Archives of Toxicology, 92*, 2981–2986. https://doi.org/10.1007/s00204-018-2297-6.

Wang, L., Li, J., Liu, H., Li, Y., Fu, J., Sun, Y., Xu, R., Lin, H., Wang, S., Lv, S., Chen, L., Zou, Z., Li, B., Shi, M., Zhang, Z., & Wang, F.-S. (2013). A pilot study of umbilical cord-derived mesenchymal stem cell transfusion in patients with primary biliary cirrhosis. *Journal of Gastroenterology and Hepatology, 28*, 85–92. https://doi.org/10.1111/jgh.12029.

Wang, Q., Sun, D., Liang, Z., Wang, J., Zhong, X., Lyu, Y., Cao, J., Lin, Z., Du, Y., Miao, Z., Lu, S., Li, C., Xu, J., Shi, Y., & Deng, H. (2020). Generation of human hepatocytes from extended pluripotent stem cells. *Cell Research, 30*, 810–813. https://doi.org/10.1038/s41422-020-0293-x.

Wang, X., Willenbring, H., Akkari, Y., Torimaru, Y., Foster, M., Al-Dhalimy, M., Lagasse, E., Finegold, M., Olson, S., & Grompe, M. (2003). Cell fusion is the principal source of bone-marrow-derived hepatocytes. *Nature, 422*, 897–901. https://doi.org/10.1038/nature01531.

Wei, X., Wang, C., Liu, Q., Li, J., Li, D., Zhao, F., Lian, J., Xie, Y., Wang, P., Bai, X., & Jia, Z. (2008). In vitro hepatic differentiation of mesenchymal stem cells from human fetal bone marrow. *Journal of International Medical Research, 36*, 721–727. https://doi.org/10.1177/147323000803600414.

Woods, E. J., Perry, B. C., Hockema, J. J., Larson, L., Zhou, D., & Goebel, W. S. (2009). Optimized cryopreservation method for human dental pulp-derived stem cells and their tissues of origin for banking and clinical use. *Cryobiology, 59*, 150–157. https://doi.org/10.1016/j.cryobiol.2009.06.005.

Wu, X., Xia, Y., Zhou, O., Song, Y., Zhang, X., Tian, D., Li, Q., Shu, C., Liu, E., Yuan, X., He, L., Liu, C., Li, J., Liang, X., Yang, K., Fu, Z., Zou, L., Bao, L., & Dai, J. (2020). Allogeneic human umbilical cord-derived mesenchymal stem cells for severe bronchopulmonary dysplasia in children: Study protocol for a randomized controlled trial (MSC-BPD trial). *Trials, 21*, 125. https://doi.org/10.1186/s13063-019-3935-x.

Wu, Y.-H., Hu, S.-Q., Liu, J., Cao, H.-C., Xu, W., Li, Y.-J., & Li, L.-J. (2014). Nature and mechanisms of hepatocyte apoptosis induced by D-galactosamine/lipopolysaccharide challenge in mice. *International Journal of Molecular Medicine, 33*, 1498–1506. https://doi.org/10.3892/ijmm.2014.1730.

Xu, H., Jiao, Y., Qin, S., Zhao, W., Chu, Q., & Wu, K. (2018). Organoid technology in disease modelling, drug development, personalized treatment and regeneration medicine. *Experimental Hematology & Oncology, 7*, 1–12. https://doi.org/10.1186/s40164-018-0122-9.

Xu, W., He, H., Pan, S., Chen, Y., Zhang, M., Zhu, S., Gao, Z., Peng, L., & Li, J. (2019). Combination treatments of plasma exchange and umbilical cord-derived mesenchymal stem cell transplantation for patients with hepatitis B virus-related acute-on-chronic liver failure: A clinical trial in China [WWW Document]. *Stem Cells International*. https://doi.org/10.1155/2019/4130757.

Xue, G., Han, X., Ma, X., Wu, H., Qin, Y., Liu, J., Hu, Y., Hong, Y., & Hou, Y. (2016). Effect of microenvironment on differentiation of human umbilical cord mesenchymal stem cells into hepatocytes in vitro and in vivo. *BioMed Research International, 2016*, 8916534. https://doi.org/10.1155/2016/8916534.

Yagi, T., Hardin, J. A., Valenzuela, Y. M., Miyoshi, H., Gores, G. J., & Nyberg, S. L. (2001). Caspase inhibition reduces apoptotic death of cryopreserved porcine hepatocytes. *Hepatology, 33*, 1432–1440. https://doi.org/10.1053/jhep.2001.24560.

Yang, J.-F., Cao, H.-C., Pan, Q.-L., Yu, J., Li, J., & Li, L.-J. (2015). Mesenchymal stem cells from the human umbilical cord ameliorate fulminant hepatic failure and increase survival in mice. *Hepatobiliary & Pancreatic Diseases International, 14*, 186–193. https://doi.org/10.1016/S1499-3872(15)60354-X.

Yin, L., Zhu, Y., Yang, J., Ni, Y., Zhou, Z., Chen, Y., & Wen, L. (2015). Adipose tissue-derived mesenchymal stem cells differentiated into hepatocyte-like cells in vivo and in vitro. *Molecular Medicine Reports, 11*, 1722–1732. https://doi.org/10.3892/mmr.2014.2935.

Yoon, H.-H., Jung, B.-Y., Seo, Y.-K., Song, K.-Y., & Park, J.-K. (2010). In vitro hepatic differentiation of umbilical cord-derived mesenchymal stem cell. *Process Biochemistry, 45*, 1857–1864. https://doi.org/10.1016/j.procbio.2010.06.009.

Yu, Y.-D., Kim, K.-H., Lee, S.-G., Choi, S.-Y., Kim, Y.-C., Byun, K.-S., Cha, I.-H., Park, K.-Y., Cho, C.-H., & Choi, D.-H. (2011). Hepatic differentiation from human embryonic stem cells using stromal cells. *Journal of Surgical Research, 170*, e253–e261. https://doi.org/10.1016/j.jss.2011.06.032.

Yu, Y.-B., Song, Y., Chen, Y., Zhang, F., & Qi, F.-Z. (2018). Differentiation of umbilical cord mesenchymal stem cells into hepatocytes in comparison with bone marrow mesenchymal stem cells. *Molecular Medicine Reports, 18*, 2009–2016. https://doi.org/10.3892/mmr.2018.9181.

Zhang, G.-Z., Sun, H.-C., Zheng, L.-B., Guo, J.-B., & Zhang, X.-L. (2017). In vivo hepatic differentiation potential of human umbilical cord-derived mesenchymal stem cells: Therapeutic effect on liver fibrosis/cirrhosis. *World Journal of Gastroenterology, 23*, 8152–8168. https://doi.org/10.3748/wjg.v23.i46.8152.

Zhang, Y., Li, Y., Li, W., Cai, J., Yue, M., Jiang, L., Xu, R., Zhang, L., Li, J., & Zhu, C. (2018). Therapeutic effect of human umbilical cord mesenchymal stem cells at various passages on acute liver failure in rats. *Stem Cells International, 2018*, 7159465. https://doi.org/10.1155/2018/7159465.

Zhang, Y.-N., Lie, P.-C., & Wei, X. (2009). Differentiation of mesenchymal stromal cells derived from umbilical cord Wharton's jelly into hepatocyte-like cells. *Cytotherapy, 11*, 548–558. https://doi.org/10.1080/14653240903051533.

Zhang, Z., Lin, H., Shi, M., Xu, R., Fu, J., Lv, J., Chen, L., Lv, S., Li, Y., Yu, S., Geng, H., Jin, L., Lau, G. K. K., & Wang, F.-S. (2012). Human umbilical cord mesenchymal stem cells improve liver function and ascites in decompensated liver cirrhosis patients. *Journal of Gastroenterology and Hepatology, 27*, 112–120. https://doi.org/10.1111/j.1440-1746.2011.07024.x.

Zhao, X., Zhu, Y., Laslett, A. L., & Chan, H. F. (2020). Hepatic differentiation of stem cells in 2D and 3D biomaterial systems. *Bioengineering, 7*, 47. https://doi.org/10.3390/bioengineering7020047.

Zheng, G., Liu, Y., Jing, Q., & Zhang, L. (2015). Differentiation of human umbilical cord-derived mesenchymal stem cells into hepatocytes in vitro. *Bio-Medical Materials and Engineering, 25*, 145–157. https://doi.org/10.3233/BME-141249.

Zheng, J., Li, H., He, L., Huang, Y., Cai, J., Chen, L., Zhou, C., Fu, H., Lu, T., Zhang, Y., Yao, J., & Yang, Y. (2019). Preconditioning of umbilical cord-derived mesenchymal stem cells by rapamycin increases cell migration and ameliorates liver ischaemia/reperfusion injury in mice via the CXCR4/CXCL12 axis. *Cell Proliferation, 52*, e12546. https://doi.org/10.1111/cpr.12546.

Ziolkowska, K., Jedrych, E., Kwapiszewski, R., Lopacinska, J. M., Skolimowski, M., & Chudy, M. (2010). PDMS/glass microfluidic cell culture system for cytotoxicity tests and cells passage. *Sensor Actuator B-CHEM, Sensors and Actuators B, Sensors and Actuators B: Chemical : International Journal Devoted to Research and Development of Physical and Chemical Transducers, Sensors and Actuators b: Chemicalinterna, 145*, 533–542. https://doi.org/10.1016/j.snb.2009.11.010.

Chapter 7
Differentiation of Stem Cells into Pancreatic Lineage: In vitro Cell Culture, in vivo Transplantation in Animal Models

Reham M. Balahmar

Abstract The pancreas is an abdominal glandular organ which is involved in the maintenance of the nutritional balance, through the synthesis and secretion of hormones and enzymes. It consists of two main parts: endocrine and exocrine glands. Endocrine part of the pancreas is composed of clusters of cells which are collectively called as islets of Langerhans, containing five types of cells; these cells produce different hormones that are responsible for maintaining the balance of glucose in blood. Exocrine part of the pancreas consists of acinar cells and ductal cells. They are involved in the secretion of enzymes that assist digestion. The absence of proper functioning β-cells causes diabetes. Diabetes mellitus is a chronic metabolic disorder characterized by deficiency or loss of the insulin-producing β-cells of the pancreas. Stem cells are a revolution in modern medicine and have become the most promising therapeutic approach for treating diabetes that can offer an alternative source of insulin-producing cells. This chapter reviews some attempts that have been used depicting different differentiation methodologies of transforming stem cells into β-cells in vitro and in vivo. It also sheds light on some of the human clinical trials, and the results used for stem cells for diabetes treatment that have been achieved.

Keywords Exocrine · Endocrine · Islets of Langerhans · Ductal cells · Differentiation · Stem cells · Insulin

Pancreas

The pancreas is a glandular organ, located in the upper abdomen behind the stomach (Githens, 1994). The pancreas anatomy is an irregular shape and classified into four main parts: head, neck, body, and tail (Frantz et al., 2012). It has a significant function in controlling blood glucose homeostasis by producing digestive enzymes and hormones. It consists of two major compartments: exocrine and endocrine parts of pancreas. The exocrine tissue forms up to 90% of the pancreas which contain acinar

---

R. M. Balahmar (✉)
Nottingham Trent University, Nottingham, UK

F. A. Khan (ed.), *Advances in Application of Stem Cells: From Bench to Clinics*, Stem Cell Biology and Regenerative Medicine 69,
https://doi.org/10.1007/978-3-030-78101-9_7

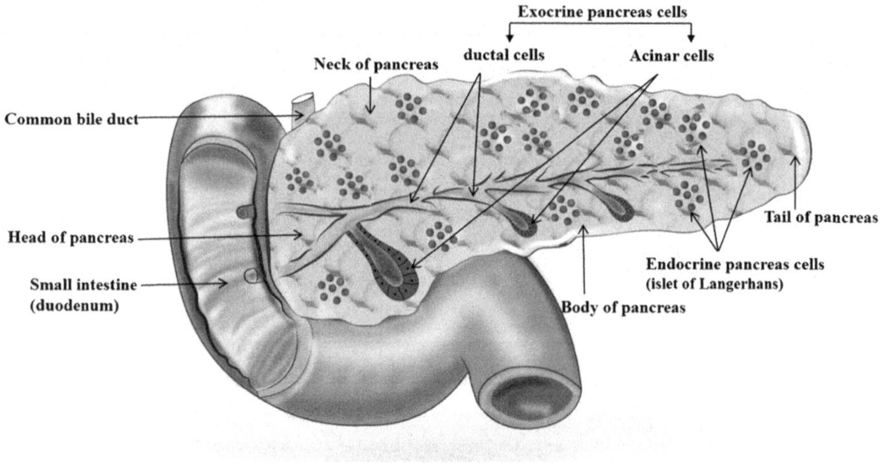

Fig. 7.1 Structure of the pancreas. The anatomy of pancreas is divided into four parts: head, neck, body, and tail. Pancreas consist of two main sections: exocrine and endocrine cells, each performs different functions. Exocrine cells consist of acinar cells and ductal cells. Endocrine cells are clustered into the islets of Langerhans

cells and ductal cells that produce various digestive enzymes into the digestive tract. The endocrine tissue forms up to 10% of the entire pancreas. It synthesizes and produces hormones that participate in regulating the metabolism of glucose, carbohydrates, lipid, and protein. The endocrine cells are located in groups of cells called the islets of Langerhans. The islets of Langerhans consist of different cell types that produce different hormones including α-(glucagon), β-(insulin), δ-(somatostatin), ε-(ghrelin), and PP-cells (pancreatic polypeptide) (Bastidas-Ponce et al., 2017; Cabrera et al., 2006). A simple illustration of the pancreas structure is given in Fig. 7.1.

Pancreas Development

Development Pancreas in Embryonic Stage and Endoderm Specification

During the early embryonic development stage of gastrulation, three primary germ layers, i.e., ectoderm, mesoderm, and endoderm are formed. These layers were developed from intensive cell migration of the inner cell mass. These three germ layers differentiate into specific tissues and organs during embryonic development. The gastrointestinal organs and the pancreas are originated from the endoderm. An epithelial sheet of cells of definitive endoderm (DE) formation begins between embryonic day (E) 6.5–7.5 in mice. The (DE) is defined as a group of multi-functional stem cells that are allocated as a single germ layer occurred during gastrulation. Endoderm patterning process is regulated interactions with surrounding mesoderm tissues (Guo & Hebrok, 2009; Zorn & Wells, 2009). Development of pancreas is a dynamic activity which is controlled by a multifaceted regulatory network consisting of various transcription factors and signaling molecules, critical for growth and development of the pancreas in early stages. The wingless-type MMTV integration site (Wnt) signaling pathway and the transforming growth factor beta (TGF-β) are important for the generation of definitive endoderm during gastrulation (D'Amour et al., 2005). Studies show that effect the deleting β-catenin can be decreased the numbering of β-cell in pancreas. This is attributed to the defect in multipotent progenitor cell expansion rather than the subsequent differentiation steps (Bastidas-Ponce et al., 2017). Wnt signaling regulates the pancreatic progenitor cells proliferation, β-cell, and acinar cell replication (Guo & Hebrok, 2009). Another important signaling in this stage is a member of the TGF-B family called nodal. Nodal signaling is essential for determining the anterior/posterior axis (D'Amour et al., 2005). It is also necessary for the starting of gastrulation through regulation of Wnt signaling, fibroblast growth factors (FGFs). Nodal signaling induces the endoderm transcription factors expression such as Sox17, GATA4, and GATA6 (Zorn & Wells, 2009).

Specification of the Dorsal and Ventral Pancreas

The first pancreatic specification begins at embryonic day E8.5 in mice and third week after the fertilization in humans. After gastrulation, the pancreas originates as two distinct buds in the dorsal and ventral pancreatic buds foregut (Watt et al., 2007). In mice, the first appearance of the dorsal pancreatic bud is at E9.0, while the ventral bud develops at E9.5. This distinction is caused by different signals from nearby tissues derived from the mesoderm. The dorsal bud produces the gastric and splenic lobes, and the ventral bud develops the lobe that runs along the proximal duodenum in the mouse (Dassaye et al., 2016; Watt et al., 2007). The main part of the head, body, and tail of the developed pancreas was generated by the dorsal bud, whereas the ventral bud participates in the inferior part of the head organ in humans (Piper et al., 2004). The dorsal bud develops adjacent to the notochord, composed of cells originating from the mesoderm. At E11.5 in mice, the ductal epithelium produces and branches into the nearby mesenchyme; giving rise to highly branched structures (Zhou et al., 2007). There are several important signaling pathways involved in pancreas development and formation of the dorsal bud including FGFs, TGF-B, retinoic acid, Sonic hedgehog (Shh), Wnt, and Notch.

The Primary Transition

The pancreatic development is divided into three main stages: The primary transition is characterized by active proliferation of multipotent pancreatic progenitor cells that generate a stratified epithelium, with microlumen formation that are fused at the end to develop (Pictet et al., 1972). The first wave of insulin, glucagon, and double positive cells show in the dorsal bud of mice at this phase (Herrera, 2000). The primary transition in humans forms a proto-differentiated epithelium (Sarkar et al., 2008). In different transition stages, transcription factors (extrinsic and intrinsic signals) play essential roles during pancreatic development that includes the initiation of transcription, regulating pancreatic morphogenesis, controlling cell differentiation, and maintenance of the cellular phenotypes and function (Dassaye et al., 2016). Transcription factors involved in the primary transition stage are summarized below.

Pancreatic Duodenal Homeobox Gene 1 (Pdx1)

Pdx1 is a homeodomain transcription factor that has a crucial regulator function in normal pancreatic development, β-cell survival and function. It is called as insulin promoter factor 1 (IPF1). The expression of Pdx1 appears early in the dorsal and ventral buds development at E8.5 (Ahlgren et al., 1996). It is considered as an essential entry point to the dissected complex signaling and transcriptional regulatory networks which plays critical roles in proliferation and differentiation of the pancreas (Pan & Wright, 2011). Pdx1 is expressed in all cell development from endoderm-endocrine, exocrine and ductal (Dassaye et al., 2016). By lineage, tracing methods show that early Pdx1 positive cells produce both exocrine and endocrine components (Gannon et al., 2000). In mice, the expression of Pdx1 is limited to β-cells by E15.5, as it controls the insulin gene expression in the β-cells (Stoffers et al., 1997). In humans, the expression of Pdx1 occurs approximately on the 7th week of gestation and continues until the mature β-cell is formed (Kaneto & Matsuoka, 2015). Both humans and rodents, the whole deficiency of Pdx1 expression leads to pancreatic agenesis, while lack Pdx1expression causes dysfunction of β-cell, exponential death of β-cell, and eventually diabetes (Fujimoto & Polonsky, 2009).

Pancreas Specific Transcription Factor 1a (Ptf1a)

Ptf1a is a heterotrimeric of the basic helix-loop-helix (bHLH) transcription factor composing of p48 subunit (Beres et al., 2006). It plays a serious role for the dorsal pancreatic endoderm development and exocrine gene transcription (Yoshitomi & Zaret, 2004). During development of pancreas, the Ptf1a expression level determines the cell fate. Low expression determines the endocrine cell fate while the high level of Ptf1a promotes the exocrine and inhibits the endocrine cell fate (Dassaye et al.,

2016). Lineage tracing study showed that Ptf1a is expressed in pancreatic progenitors, parallel with Pdx1, and its expression becomes gradually limited to the exocrine cells by E13.5, whereas it regulated enzyme release in exocrine tissue (Kawaguchi et al., 2002). In addition, Ptf1a is essential for Notch signaling mediated control in early development of pancreas by regulating the delta-like ligand (DII1) expression (Ahnfelt-Rønne et al., 2012). Recessive Ptf1a mutations in humans generate isolated neonatal diabetes mellitus (NDM), and this type of diabetes is associated with cerebral agenesis (Sellick et al., 2004).

Hepatocyte Nuclear Factor 1 Beta (Hnf1b)

Hnf1b is a nuclear transcription factor of the homeodomain family. It has a serious role in the specification of endocrine and exocrine cell fate and the endocrine precursors generation (De Vas et al., 2015). In the mice model, Hnf1b was first detected in the primitive endoderm on E4.5 and which is required for specification of the primitive endoderm lineage. Hnf1b is expressed in the early pancreatic development. It has a role in the precursor neurogenin 3 (Ngn3) positive cells which are destined to become islet cells (Maestro et al., 2007). Moreover, Hnf1b acts as a key player in early pancreas development. The expression of Hnf1b showed in the pre-pancreatic foregut endoderm and in initial pancreatic progenitor cells (Lau et al., 2018). Knockout mouse models for Hnf1b have both exocrine and endocrine defects, which serves as evidence for the importance of the Hnf1b factor in early development and specification (Haumaitre et al., 2005).

GATA Binding Protein 4 and 6 (GATA 4) and (GATA 6)

GATA 4 and 6 are zinc finger transcription factor family members; both are associated with the early stages of pancreatic development (Decker et al., 2006). Both have been co-expressed during initial foregut endoderm, later the dorsal and ventral pancreatic buds epithelial. GATA4 expression in acinar differentiated cells (Ketola et al., 2004). GATA6 expression is limited to endocrine and ductal cells, and it is required for the mature acinar cell maintenance (Decker et al., 2006). The ventral pancreatic is not formed in null GATA4 mice, while the dorsal pancreatic is formed normally (Watt et al., 2007). Moreover, GATA4 mutations caused congenital heart defects in humans. GATA6 haploinsufficiency mutations in humans cause pancreatic agenesis and cardiac abnormalities (Chao et al., 2008).

SRY Sex Determining Region Y Box 17 (Sox17)

Sox17 is a high mobility group (HMG) box transcription factor, which controls endoderm organ formation. It is required to maintain boundaries of the biliary system between the liver and the ventral pancreas (Spence et al., 2009). It acts as a transcriptional regulator for other essential transcription factors during endoderm development, including Hnf1b, which is known to regulate postnatal β-cell function. Additionally, Sox17 is involved in insulin production and traffic regulation in β-cells in normal and diabetic conditions (Jonatan et al., 2014).

Sex Determining Region Y Box 9 (Sox9)

Sox9 is another HMG box transcription factor that is co-expressed with Pdx1 in multipotent progenitor cells (MPCs) at E9.5–12.5. Sox9 has an essential function in the formation of the pancreatic lineages that include ductal, islet and acinar lineages by stimulating their proliferation, persistence, and survival under an undifferentiated condition (Seymour et al., 2007). The inactivation of Sox9 expression caused hypoplasia of the dorsal and ventral buds in mice (Akiyama et al., 2005). The deficiency of Sox9 in mice leads to a failure in development of pancreas and this resulting in death. Positive cells of Sox9 in humans produce mature endocrine cells; however, Sox9 expression is limited to ductal cells in the mature pancreas (Seymour et al., 2007). Also, Sox9 is important for maintaining the identity of MPCs through a process that can be related to FGF signals of mesenchymal cells (Seymour et al., 2012).

It has a role in maintenance of multipotent progenitors by regulator Hnf1b and other factors (Lynn et al., 2007).

Insulin Gene Enhancer Binding Protein (Isl-1)

Isl1 is a LIM-homeodomain (LIM-HD) transcription factor belonging to the most important subfamilies of homeobox genes (Wang et al., 2014). It has a regulating role in pancreas primary and secondary transitions which is involved in dorsal pancreatic mesenchyme development in primary transitions, while secondary transition is participated in all endocrine cell and dorsal exocrine and formation (Dassaye et al., 2016). In the dorsal bud, and the mesenchyme surrounded the dorsal bud Isl1 was expressed. Isl1 expression is essential for the formation of the dorsal bud and differentiation of insulin-producing cells (Ahlgren et al., 1997). It is also expressed in all pancreatic islet cells. Its deficiency not only decreases islet cell proliferation but also leads to the continued loss of islet cells in mice pancreas (Dassaye et al., 2016).

The Secondary Transition

This stage is characterized by the major wave of growth and differentiation toward the three lineages of pancreatic, acinar, ductal, and islet cells (Pictet et al., 1972). In this stage, the epithelium grows outward and forms MPCs around the periphery of the epithelium. This branching happens for several days till the population of the MPCs dwindles completely. After E14.5, these progenitors differentiate into endocrine, acinar, and duct cells that are committed to exocrine fate (Zhou et al., 2007). Complete differentiation of β- and α-cells takes place from the epithelial tissue between E13–15. During the secondary transition, several transcription factors and signaling molecules have been identified as pancreatic markers and define pancreatic lineages, as summarized below.

Neurogenin 3 (Ngn3)

Ngn3 is a basic helix-loop-helix (bHLH) transcription factor, expressed at secondary transition from E9.5–16.5. Ngn3 plays an important role as a master regulator and activator of gene transcription in endocrine progenitor cells during pancreatic development. Moreover, Ngn3 expression stimulates NeuroD1, Pax4, Nkx2.2, and Rfx6 that are involved in additional differentiation and subtype specification of different endocrine hormones produced by pancreas. Ngn3 is important for differentiation of endocrine cells, and its expression considered as signs of an endocrine pancreas (Dassaye et al., 2016). Loss of Ngn3 expression leads to a failure in the development of all pancreatic islet cells including α-, β-, δ-, ε-, and PP-cells, and thus, the hormones produced by them (Heller et al., 2005). Therefore, targeted disruption of Ngn3 in humans causes the failure of developed islet growth, neonatal diabetes mellitus, and early death (Schwitzgebel, 2014). High level of Ngn3 expression is essential to direct progenitor pancreatic cells into the endocrine cell fate and to initiate endocrine differentiation. While, low Ngn3 expression improves both acinar and duct cells development (Wang et al., 2010).

Neurogenic Differentiation Factor 1 (NeuroD1)

NeuroD1 is a bHLH transcription factor, expressed at E9.5 in a subset of pancreatic epithelial cells and later expressed in α-, β-, and δ-cells. It is essential for β-cell maturation and maintenance of glucose response (Dassaye et al., 2016). Targeted disruption of Neurod1 in mice leads to reduction of insulin-producing cells and a decrease in glucagon-producing cells resulting in diabetes and early death of newborn mice (Naya et al., 1997). Homozygous mutation of NeuroD1 in humans results in the development of neonatal diabetes along with cerebellar hypoplasia (Rubio-Cabezas et al., 2011).

V-Maf Musculoaponeurotic Fibrosarcoma Oncogene Family Protein A and B (MafA) and (MafB)

The Maf family of proteins is a subgroup of the basic region-leucine zipper (bZIP) transcription factors. The Maf protein family has two main groups according to their molecular size as large or small Maf proteins (Motohashi et al., 2002). The large Maf proteins include MafA and MafB. Both are regulators for tissue-specific gene expression and cell differentiation in pancreas and other organs (Hang & Stein, 2011). MafA is activated by transcription factors such as Pdx1 and Nkx2.2 (Raum et al., 2006). MafA plays a critical role in activation genes significant for β-cell role, such as insulin (Aramata et al., 2005). Interestingly, the islet-enriched transcription factors, such as Pdx1 and Ngn3 together with MafA, have the ability to reprogram adult pancreatic acinar cells to β-like cells in mice. MafA with Pdx1 induces the development of β-cell from Ngn3-positive endocrine precursors and also permits Pdx1 to produce β-cells from α-cells and adult acinar cells (Dassaye et al., 2016; Zhou et al., 2007). MafB plays important roles in a variety of cell development, which is essential for differentiation α- and β-cell (Artner et al., 2010). It is expressed in α-cells of adult pancreas and is important for their function. During pancreas development, MafB is generated in glucagon and insulin cells. MafB is expressed earlier than MafA, detected pancreatic epithelium around E10.5, and MafA production was detected at E13.5 only in insulin cells (Hang & Stein, 2011).

Paired Box Genes 4 and 6 (Pax4) and (Pax6)

The paired box (Pax) gene family consists of nine developmental control genes (Walther et al., 1991). Pax4 was expressed in dorsal and ventral buds at E9.5 but the expression was limited to β-cells and not found in adult islets while the loss of Pax4 resulted in the loss of β-cells it also resulted in the increase of glucagon cells (Sosa-Pineda et al., 1997). Pax6 is expressed between E9.5–E10.5 throughout the ventral and dorsal pancreatic buds, but later its expression is restricted to endocrine cell lineage (Turque et al., 1994). Moreover, it has a main function in the formation and cell differentiation of endocrine pancreas, brain, and various organs. In addition, Pax6 developmental regulator is critical for adult mouse maintenance of glucose regulation and of the endocrine role (Hart et al., 2013). A homozygous deletion of the Pax6 in mice caused diabetes and a decreased number of islet cell types (Sander et al., 1997). The mutations in Pax6 in humans lead to neonatal diabetes (Solomon et al., 2009).

Aristaless-Related Homeobox (Arx)

Arx encodes a transcription factor belonging to the Aristaless-related paired-class homeobox proteins (Bienvenu et al., 2002). Both transcription factors, Arx and Pax4, have roles in differential endocrine cell subtype specifications. Moreover, both Arx and Pax4 factors exhibit antagonistic functions in endocrine specification (Collombat et al., 2003). Arx mutant mice showed upregulated Pax4 mRNA expression, while Pax4 mutant mice showed high levels of Arx mRNA. Arx is confined to α- and PP-cells fates and represses the β-/δ-cell lineage, whereas Pax4 promotes the β-/δ-cell lineages (Collombat et al., 2005). Arx-deficient mice showed hypoglycemia, symptoms of dehydration and weakness associated with early total lack of adult α-cells, and increase in β- and δ-cell types (Collombat et al., 2003). In addition, double deficiency of Arx/Pax4 showed severe hyperglycemia with the loss of early developed α- and β-cell total and a significant rise in δ-cells numbers. Both cases of deficiency of Arx or double deficiency of Arx/Pax4 resulted in death of the animal in two days after birth (Collombat et al., 2005).

Homeobox Genes (Nkx 2.2) and (Nkx 6.1)

Nkx2.2 and Nkx6.1 are the NK class of homeodomain transcription factors encoding genes 2.2 and 6.1. Nkx2.2 expressed at E9.5 in the dorsal pancreatic epithelium which plays a main role in differentiation of β-cell. Nkx2.2 is also involved in the late differentiation of β-, α-, and PP-cells formation. It has a regulated role in endocrine cell differentiation by interacting with other transcription factors (Sussel et al., 1998). Nkx6.1 expressed at E9.5 in both pancreatic buds, whereas it is limited to developing β-cells at E14. It plays an important role in pancreas development. Nkx6.1 is also required for β-cells differentiation (Dassaye et al., 2016).

Regulatory Factor X 6 (Rfx6)

Rfx6 is a member of the regulatory factor X family of winged-helix transcription factors. Rfx6 is important for islet cell differentiation and insulin secretion in mice and humans (Aftab et al., 2008). It is expressed in the definitive endoderm at E7.5, then co-expressed with Nkx2.2. and Ngn3. At E9.0, it is limited to the pancreatic buds and later is limited to all adult endocrine cells (Dassaye et al., 2016). In humans, the mutations in Rfx6 lead to Mitchell-Riley syndrome, an autosomal-recessive disease of neonatal diabetes (Concepcion et al., 2014). At embryonic stages, Rfx6 knockout mice failed to develop any type of islet cell and died shortly after birth (Smith et al., 2010).

The Final Transition

During final transition, the pancreatic differentiation cells undergo further remodeling, including other processes such as replication and neogenesis leading to the formation of adult pancreas. In this stage, the expansion of the ductal, acinar, and islet cells occurs. The islet cells are not fully developed till E19 up to 2 weeks after birth (Pictet et al., 1972). After differentiation, the endocrine cells delaminate, migrate into adjacent exocrine tissues, and aggregate into clusters of cells to form mature islets (Dassaye et al., 2016). Mouse islets are composed of 75% β-cells and 20% α-cells (Brissova et al., 2005). Mouse islets cells are organized as a central core of β-cells, which are enclosed by α-, δ-, ε-, and PP-cells (Prado et al., 2004). However, in

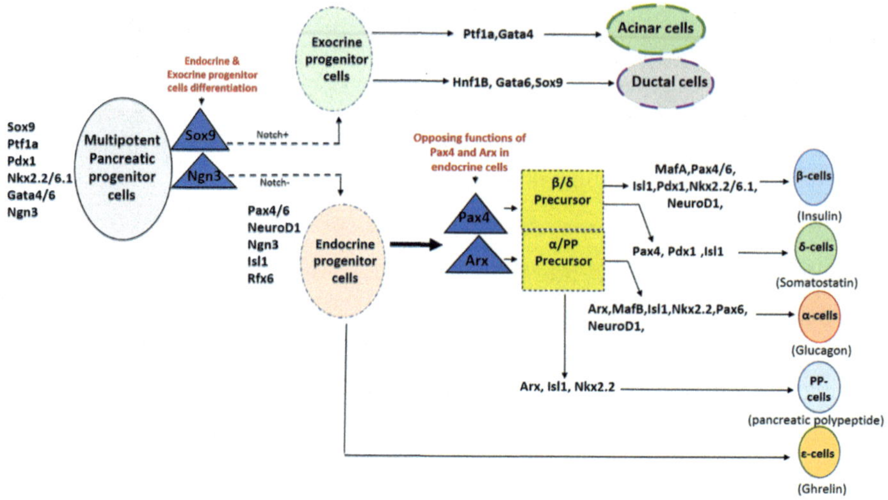

Fig. 7.2 Transcription factors network regulating the pancreas development (Figure modified from Dassaye et al., (2016))

human, islets cells contain 50% of β-cells and 40% of α-cells (Brissova et al., 2005), and human β-cells are intermingled with α- and δ- cells through the islet (Cabrera et al., 2006). The schematic diagram in Fig. 7.2 summarizes some important keys of transcription factors during different transitions regulating pancreas development.

Pancreatic Exocrine Cells

The exocrine pancreas part consists of two major cell types: acinar and duct cells. It accounts for at least 80% of pancreatic mass. Acini cells are organized into lobules with networks split into tubules. Each acinus is collected by pyramidal acinar cells (Frantz et al., 2012). The acinus (from the Latin term meaning "berry in a cluster") is specialized in synthesis, storage active enzymes, and inactive precursors (zymogens) and secret digestive enzymes into the duodenum (Williams, 2001). The acinar cells produce enzymes such as trypsinogen, chymotrypsinogen, carboxypeptidases, and nucleases. The acinar cell basal region has the nucleus and abundant rough endoplasmic reticulum for synthesis of digestive enzymes, whereas the apical region contains zymogen granules (storage for the enzymes). The digestive enzymes are secreted by the lumen of the acinar (Muallem et al., 1995). However, duct cells play essential roles for pancreatic exocrine activity by (**a**) producing both bicarbonate-rich fluid and mucins (that helps in neutralizing the stomach acid), in addition (**b**) forming extensive networks of tubules which transports enzymes for the digestive tract (Githens, 1988). The duct cells are ciliated, and polarized epithelial cells consist of cells that are cuboidal to pyramidal and have abundant mitochondria, that is important for energy products and required for its ion transport functions (Benitez et al., 2012). Centroacinar cells located at the junction of the acinus and ductile, and this has a ductal cell characteristic. The duct cell as well as the centroacinar cells have carbonic anhydrase, and this is significant for their capability in secretion of bicarbonate (Steward et al., 2005).

Pancreatic Endocrine Cells

The endocrine part represents only 2% of total volume of the pancreas. It is an aggregate of cells called the islets of Langerhans. The islet contains five main types of cells producing different hormones controlling and maintaining the homeostasis of glucose in the bloodstream include; **α-**, **β**-, **δ-**, **ε**-, and **PP-cells** (Frantz et al., 2012).

α-cells make up about 30–45% of the islet cells. Glucagon is released from α-cells in order to stimulate glycogenolysis in the liver when blood glucose levels are low (Freychet et al., 1988). Glucagon is a catabolic hormone that has an essential role in regulating glucose homeostasis in blood by stimulating hepatic glucose production (Röder et al., 2016). High glucose concentrations stimulate secretion of insulin from β-cells and inhibit secretion of glucagon, while low glucose concentrations (hypoglycemia) stimulate glucagon secretion (Brereton et al., 2015). Glucagon is released in response to hypoglycemic condition, prolonged fasting, exercise, and meals rich in protein. During long fasting hours, glucagon drives the liver and renal gluconeogenesis to increase endogenous blood glucose (Freychet et al., 1988).

β-cells produce the insulin hormone and make up 50–60% of each islet. The high level of blood glucose is a stimulator of insulin release (Komatsu et al., 2013). Insulin is an anabolic hormone; it is synthesized from β-cells as preproinsulin. A signal peptide from preproinsulin is cleaved into proinsulin in endoplasmic reticulum, where proinsulin is further cleaved by an endopeptidase identified as prohormone convertase to insulin and C-peptide. The resulting insulin is then stored in secretory granules by Golgi-apparatus waiting upon suitable stimuli for releasing from the cells. β-cell releases the insulin in response to various nutrients in the blood (Fu et al., 2013). Initially, for the release of insulin, glucose primarily enters into the β-cell through glucose transporter 2 (GLUT2) by means of a facilitated diffusion mechanism. GLUT2 is transmembrane located on the surface of the β-cells carrier proteins which permit passive transport of glucose into the cells (Marshall et al., 1993). It was expressed in β-cells, liver, and the expression was reduced in renal and absorptive cells of the intestinal. After entering β-cells, glucose is phosphorylated by glucokinase, which is a key enzyme that converts the glucose to glucose-6 phosphate (Efrat et al., 1994). The endpoint of glycolysis results in the surge of ATP/ADP ratio. This for a short period leads the cells into oxidative stress. A subsequent increase in the intracellular calcium levels due to oxidative stress results in the closure of ATP-sensitive K^+ channels (KATP). This results in the depolarization of membrane and dumping of insulin from the vesicles. Some products resulting from these mechanisms acting as insulin secretion signals, such as NADPH, malonyl-CoA, and glutamate (Henquin, 2000).

δ-cells produce the hormone somatostatin and comprise only 5% of the islet cells (Cabrera et al., 2006). The δ–cells existing in the gastrointestinal tract, the central peripheral nervous system, and hypothalamus (Arimura et al., 1975). Somatostatin commonly known as somatotropin release-inhibiting factor or growth hormone inhibiting hormone (Brazeau et al., 1973). It is an inhibitor hormone released from the pituitary gland as well as an inhibitor of insulin and glucagon secretion (Youos, 2011).

PP-cells produce and secrete pancreatic polypeptide hormone (PP). These pancreatic cells are known as F-cells, which form the lowest amount of the total islet cells <5% (Cabrera et al., 2006). PP is a polypeptide hormone of 36 amino acids belonging to the peptide YY (PYY) and neuropeptide Y (NPY) family peptide (Kimmel et al., 1975). PP-hormone regulates the exocrine and endocrine production function of the pancreas. It is released rapidly into the circulation after food ingestion, cholinergic stimulation, and hypoglycemia. However, the glucose presence inhibits PP-hormone production. The physiological function of PP-hormone has not been established, but known to play a role in acid secretion and gallbladder relaxation, it effected upon digestive secretion and motility have been described (Frantz et al., 2012).

ε-cells comprise only <1% of the islet cells. These pancreatic cells produce the hormone ghrelin which is known as a hunger hormone (Pradhan et al., 2013). Ghrelin is a peptide containing 28 amino acids with n-octanoylation at serine 3 (Kojima et al., 1999). Mainly, it is produced in the stomach, increases secretion of growth hormones from the pituitary gland by growth hormone secretagogue receptor (GSH-R), and stimulates food intake and energy balance. Ghrelin has an essential role in secretory functions and development of pancreas. In addition, the majority of studies indicate that it has a functional in the glucose regulation that occurs by modulating insulin release (Pradhan et al., 2013). It increases blood glucose levels by suppressing insulin releasing from β-cells. Also, it has a role in β-cells growth and proliferation and prevents β-cell apoptosis (Sakata et al., 2019). Ghrelin signaling plays an important regulator role in obesity, insulin resistance, and diabetes. Interestingly, it has many regulatory physiological functions, most of these functions seem to be different from its effect on stimulating food intake, including cardiac functions, gastric motility stimulating, development of cancer, immunity, and inflammation system (Pradhan et al., 2013).

β-cell Regeneration

Studying biological development of β-cell regeneration is essential for developing new treatments for diabetes. In both animals and humans, β-cell mass expansion slows significantly in adulthood due to very low rates of β-cell replication and neogenesis process (Teta et al., 2007). However, the regeneration of β-cells occurs during different physiological and pathophysiological conditions such as pregnancy, obesity, and partial pancreatectomy.

β-cells Regeneration During Pregnancy

Pancreatic β-cells regeneration occurs during pregnancy in humans and other animals. The maternal pancreas in pregnancy adapts to increased insulin resistance and metabolic demand by upregulating the proliferation β-cell mass which eventually returns to normal levels after delivery. This happens by decreasing β-cell proliferation and size and increasing its death (Sorenson & Brelje, 1997). Changes that affect β-cell mass increase in pregnancy time can lead to unregulated glucose homeostasis and gestational diabetes (Zhang et al., 2010). The proliferation of β-cells during pregnancy in rodents and humans is induced by placental lactogen (PL) and prolactin (PRL) hormones (Nielsen et al., 1999). It has proved this by experiment in which a short infusion of prolactin is enough to decrease menin (gene name multiple endocrine neoplasia type 1 MEN1) expression, which acts as a tumor suppressor blocked β-cell replication (Karnik et al., 2007). In pregnant rodents, β-cell proliferation is improved (two to five-fold) at gestational days 13–15 returning to normal levels at day 18–19 of delivery. Proliferation and hypertrophy of pre-existing β-cells are two of key developments of cells involved in this increase in β-cell mass (Ernst et al., 2011). In human pregnancy, there was an increase in the relative capacity of maternal islets and hyperplasia of β-cells (Van Assche et al., 1978). This the adaptive development in β-cell numbers is achieved by β-cell neogenesis rather than duplication of β-cells in existing islets (Butler et al., 2010). But, the clear proof of the proliferation process in these islet cells remains inconclusive and needs to be further investigated.

β-cells Regeneration During Obesity

Another physiological condition of β-cells regeneration is the insulin resistance or obesity, where β-cell mass can multiply by several fold in obese mice. Using stained pancreatic sections from obese mice and humans, it has been detected as insulin producing cells that express Ki67 (a marker associated with cell proliferation). The regeneration β-cells ability is much higher in mice than in humans; however, the basic process is not completely clear (Butler et al., 2003).

β-cells Regeneration After Partial Pancreatectomy

After pancreatectomy procedure, islet cells regeneration occurred in response to this injury. In rats, eight weeks after a 90% pancreatectomy, there was a regeneration to 27% and 45% of pancreatic weight and β-cell mass of sham-operated pancreas, respectively, (Bonner-Weir et al., 1993). The examination of potential of β-cell regeneration showed that in adult pigs 6 weeks after 60% pancreatectomy, there was a 19% increase in β-cell mass. Likewise, an 80% pancreatectomy caused a 56% increase in β-cell mass. Moreover, there was no improvement of insulin secretion or β-cell mass in the pancreas remnant in adult dogs after 80% pancreatectomy (Löhr et al., 1989). In humans, a 50% pancreatectomy does not prompt increased β-cell mass or regeneration. However, it causes impairments in secretion of insulin, and increases diabetes risk. Unlike humans, in mice diabetes does not develop spontaneously without changes in specific diet regimen, genetic predisposition, or chemical interventions. Therefore, differences between humans and rodents in β-cell turnover must be studied when estimating new treatment choices that aim to restore β-cell mass in diabetics (Menge et al., 2008).

Pancreatic Diseases

Pancreatic disorders are divided into two categories depending on which part was affected, whether it is pancreatic exocrine or endocrine. The most common disorders that affect the exocrine pancreas are pancreatic cancer and pancreatitis. Pancreatic ductal adenocarcinoma is the most common type of pancreatic cancer. Diabetes and rare pancreatic neuroendocrine tumors affect the endocrine islets of Langerhans (Zhou & Melton, 2018). Pancreatitis is an inflammation of the pancreas that happens due to injury produced by enzymes action in pancreatic tissue. These enzymes are activated normally once they exit the pancreas; however, blockage due to infections or gallbladder stones can cause accumulation of these enzymes and activation within the pancreas. Pancreatitis can be acute or chronic. The most common causes of pancreatitis are alcohol consumption, cystic fibrosis, and hypercalcemia (Banks et al., 2010). Pancreatic ductal adenocarcinoma is poor differentiation of ductal/glandular structures which is believed to develop from progressive changes in the tissue. The prognosis of this disease is very poor with a low survival rate. Although the success of treatment is limited, there are many types of treatment. However, detection happens at the late stage of disease. Therefore, most of the current research is focused on early detection methods of pancreatic cancer (Castellanos et al., 2011).

Diabetes Mellitus

Diabetes is the most common health challenges facing the modern world that creates a striking impact on health, society, and economy (Zhou & Melton, 2018). According to the latest data published in the International Diabetes Federation (IDF) Atlas, 9th edition illustrates that 463 million people worldwide are currently living with diabetes. DM is a metabolic disorder caused by an increase in blood sugar levels resulting from insulin resistance or a decrease in insulin production by the β-cells of pancreas or both. This disease leads to kidney failure, heart diseases, stroke, neuropathy, and retinopathy (Pagliuca & Melton, 2013). Diabetes is divided into two groups: type 1 and type 2 diabetes. Type 1 diabetes (T1D) produced by T cell-mediated autoimmune destruction of insulin-producing β-cells in the pancreatic causes insulin deficiency that leads to hyperglycemia (Ashcroft & Rorsman, 2012). The main susceptibility genes associated with diabetes are genes which regulate the human leukocyte antigen immune system (Singal & Blajchman, 1973). Type 2 diabetes (T2D) is impaired insulin production and insulin resistance, which is most often associated with different conditions such as age, obesity, and genetic factors. The most prevalent affecting approximately 85% of diabetic patients. Patients can be treated through diet, good nutrition, and exercise during the early stages of the disease (Ashcroft & Rorsman, 2012).

Alternative Sources of Pancreatic β-cells

The main treatment for patients with type 1 diabetes is based on daily injection of exogenous insulin and combined with blood glucose monitoring. Although traditional insulin treatment supports blood glucose control levels, it is ineffective in the long term. Another alternative treatment is replacement of β-cells by transplantation of pancreas or pancreatic islets. However, it is currently challenging because there are many challenges facing these methods, including the shortage of human donor tissue, ethical conflict, and rejection of the organ by the immune system, use of immunosuppressive drugs, and other complications following these methods. On the other hand, type 2 diabetes is regulated by small-molecule drugs (such as phenformin, metformin) to stimulate the function of β-cell, to promote secretion sensitivity of insulin. In general, for patient type 1 diabetes (T1D) and advanced patient type 2 diabetes (T2D), it is difficult to control blood glucose perfectly by insulin therapy. Therefore, most of the researchers are trying to develop and find new strategies to generate pancreatic β-cells by focusing on the stem cell research, which received much consideration in this view and showed promising possibility as an alternate source of β-cells and other cells. Due to the regeneration and differentiation potential of stem cells, they are best candidates for diabetes treatment (Guney & Gannon, 2009; Peng et al., 2018). Although, there was difficulty in producing and developing the adult derived pancreatic β-cells in vitro, and there are many studies that have used different stem cells models in order to obtain successful positive differentiation of β-cell in vitro. In order to provide diverse sources of β-cells for transplantation in diabetes patients, including embryonic stem cells (ESCs), induced pluripotent stem cells (iPSCs), mesenchymal stem cells (MSCs), and adult pancreatic stem cells and other cell sources. It will be mentioned in some of these sources in this chapter.

Differentiation of Embryonic Stem Cells (ESCs) into Pancreatic β-cells

ESCs are pluripotent stem cells obtained from the undifferentiated inner cell mass of a blastocyst during embryonic development, and they have the capacity to differentiate into all cell types, including germ cells (Bradley et al., 1984). Over the past decade, various protocols have been described with the aim to mimic normal pancreatic development to produce pancreatic β-cells (insulin-secreting cells) in vitro. Some of these protocols were cultured mouse/human ESCs and iPSCs either in two-dimensional culture 2D methods, while other groups have used three-dimensional culture 3D methods. This is accomplished through the inclusion of several signaling molecules in the culture media that simulates those growth factors secreted from adjacent tissues such as sonic hedgehog, retinoic acid, and FGFs. Briefly, the ESCs differentiation development includes definitive endoderm formation and induction expression of Neurod1 and Ngn3, which they are critical in pancreatic growth, such as and finally the development of β-cell lineage by the induction of insulin and Nkx6.1 (Tse et al., 2015).

Random Differentiation of Stem Cells into Pancreatic β-cells

The first positive protocols to promote differentiation of insulin-producing cells from ESCs used agents such as dimethyl sulfoxide (DMSO), nicotinamide, and antibiotic selection for isolating insulin-positive clones. The insulin-producing cells were able to restore hypoglycaemia in mice with diabetes; however, out of 784 clones, only 8 clones of geneticin-resistant cells were identified (Soria et al., 2000). Although this experiment provided evidence in principle that ESCs can produce insulin, the low success rate represented limitations to its clinical application. A parallel study confirmed that human ESCs (H9 cell line) were able to differentiate spontaneously into insulin-positive cells after cultured as embryoid bodies in suspension. These cells expressed transcription factors which are necessary for pancreas development as Pdx1, Ngn3, β-cell markers insulin, glucokinase, and Glut2 (Assady et al., 2001). These studies showed clear evidence that ESCs differentiate into endocrine cells spontaneously; however, the generated cell populations were very low, and they had a very low insulin content. This leads to insufficient differentiation development. Another attempt to improve the efficiency and yield of insulin-producing cells by using signaling molecules to direct the ESCs toward differentiation in vitro (by means of a five-step protocol) (Lumelsky et al., 2001). This protocol depends on selected nestin-positive cells, which is an intermediate filament protein initially in neural stem cells (NSCs) of embryo and adult brains. It was used as a marker for stem/progenitor cell populations in different tissues (Lendahl et al., 1990). Nestin-positive cells were selected by plated embryoid bodies in medium containing insulin, selenium, transferrin, and fibronectin (ITSFn), expanded in medium containing FGF2, N2, and B27 supplements, and further differentiated by withdrawal of FGF2 in the presence nicotinamide. Even though the cells secreted insulin in response to glucose, these cells were also unsuccessful to normalize levels of blood glucose when subcutaneous transplantation into diabetic mice (Lumelsky et al., 2001). Several other studies found similar results by using mouse or human ES cell lines with a slight change in growth factors supplementation and found that there was an increase (30-fold) in insulin content when differentiated cells cultured in suspension (Baharvand et al., 2006; Kania et al., 2003). Other modifications applied according to the original protocol of Lumelsky et al. (2001) by replacing the B27 supplements with a phosphatidylinositol-3 kinase (PI3K) inhibitor at the last stage of differentiation. As a result, there was a 30-fold increase in insulin content. These cells prolonged survival, but they also failed to normalize the blood glucose level in transplanted animals. Another finding established that when using exendin-4 (its analog glucagon-like peptide 1 GLP1) or glucose-dependent insulinotropic polypeptide to differentiation culture. Cells significantly increased Pdx1 expression, insulin content and insulin secretion, resulting in reversal of hyperglycemia in diabetic mice (Lester et al., 2004).

Direct Differentiation into Pancreatic β-cells

The production of β-cells directly occurred after differentiated cells, by a procedure named direct reprogramming, by passing the pluripotent condition (Takahashi et al., 2016). Generally, one should follow the normal endodermal pathway to produce β-cells from ESCs, due to the limitation of previous protocols in establishing a definitive endoderm progenitor population. This may be due to a lack of information regarding factors that stimulate the formation of the endoderm. Therefore, some studies used different culture conditions in order to mimic the properties of molecular signals which are known to initiate and/or control the development of pancreas from the endoderm *in vivo* (Zhou & Melton, 2008). The first differentiation of hESCs into definitive endoderm which was confirmed by expression of endodermal markers Sox17 and GATA4 (D'Amour et al., 2005). Using a viral system to express certain genes, such as Ngn3, Pdx1, and MafA, have successfully promoted insulin-producing cells from acinar cells in adult mice (Zhou & Melton, 2008), when these cells were transplanted in diabetic animals it resulted in normalized blood glucose in these animals. Retinoic acid is another molecule required for pancreas development in mouse embryos. It significantly induces Pdx1 expression by ESCs progenies. Some protocols used it alone or mixtures with other factors such as activin-A, sodium butyrate, FGF2, and nicotinamide (McKiernan et al., 2007). Different studies improved earlier protocols of differentiation and generation of pancreatic β-cells by using novel small molecules. A seven-stage protocol was described to produce β-like cells from hESCs which maintained the expression of Pdx1, Nkx6.1, and NeuroD1, while also expressing MafA, which is a key β-cell maturation transcription factor. These cells were generated in serum-free conditions and addition of small molecules improved pancreatic specification and improved the generation of Pdx1 and Nkx6.1. These molecules include growth differentiation factor 8 (GDF8), GSK3β inhibitor (Inhibition of Glycogen Synthase Kinase 3β), FGF7, vitamin C, Retinoic acid, TPB ((Trifluoromethyl)phenyl)-2,4-pentadienoylamino)benzolactam), LDN (signaling inhibitor), and Sonic hedgehog agonist-1 (SANT-1) (Rezania et al., 2014). Moreover, a group of small molecules were used, such as R428, ALK5 inhibitor II, and N-acetyl cysteine (N-Cys), which might promote pancreatic β-cells differentiation with MafA expression at stage (S) 7 cells named (S7). The S7 cells are able to ameliorate hyperglycemia when transplanted into diabetic mice. However, these cells performed similarly, yet not identically to human β-cells. Additional studies carried out a modified protocol that showed improved glucose-responsive cells, which displayed properties similar to β-cells and an enrichment of Nkx6.1 and C-peptide expression. They tested several combinations of compounds and growth factors to generate stem cell-derived β (SC-β). They also found that the SC-β cells were able to generate insulin-secreting cells and with important expression markers of β-cells maturation such as MafA. Moreover, the SC-β cells have the ability to maintain euglycemia after transplanting into recipient mice (Pagliuca et al., 2014).

Induced Pluripotent Stem Cells (iPSCs)

The iPSCs are resulting from reprogramming of human skin cells and other cells (Takahashi et al., 2016). Generation pluripotent stem cells (iPSCs) generated from somatic cells by the transduction of the main four stem cell transcription factors, namely Oct3/4, Sox2, Klf4c, and c-Myc (Takahashi & Yamanaka, 2006). The iPSCs have ESCs properties, and they can proliferate and self-renew in vitro and differentiate into different germ layers. These cells are similar to ESCs in morphological, surface antigens, expression genes and epigenetic status of pluripotent cell-specific genes (Takahashi et al., 2007). However, some differences have been discovered between hESCs and hiPSCs regarding the profiles of gene expression, genetic stability, and epigenetic modifications, for instance, DNA methylation profiles, stability of genomic imprinting, potential epigenetic modifications, and ability to model disease. The iPSCs were not identical to ESCs due to their leftover memory of their somatic origin. This memory may affect their protection, but there is no confirmed evidence if this memory can be fatal in cellular therapies (Shahjalal et al., 2018). Furthermore, the recent development in induced iPSCs has led to the avoidance of ethical debate of using hESCs. Studies have been established that induced insulin-secreting cells from iPSCs generated from fibroblasts, and these cells have properties similar to differentiated ESCs (Tateishi et al., 2008). Pancreatic β-cells generated from iPSCs are a useful technique for analyzing pathological type 1 and 2 diabetes if the cells are generated from iPSCs established from diabetic patients (Pagliuca et al., 2014). The original protocol for iPSCs generation is used the retroviral or lentiviral-mediated expression of Oct3/4, Sox2, c-Myc, and Klf4. This was not an appropriate method to generate iPSCs that can be used in therapeutic applications due to the risks caused during insertion mutations and use the c-Myc oncogene, which produced tumorigenesis in chimeric mice obtained from these cells (Takahashi et al., 2007). Some of the studies have established iPSCs generation using non-integrating methods of gene delivery with potentially reduced risks such as plasmid transfection, episomal plasmid vectors, the PiggyBac transposon, and adenoviral transduction (Mayhew & Wells, 2010).

Differentiation of Mesenchymal Stem Cells (MSCs) into Pancreatic β-cells

Mesenchymal stromal cells (MSCs) are heterogeneous population of stromal stem cells, which have capacities to differentiate into different cell types of all germ layers producing osteoblasts, adipocytes, myoblasts, and endocrine cells (Pittenger et al., 1999). MSCs isolated and cultured from several tissues such as bone marrow, skin, fat, umbilical cord blood, and placenta (Hass et al., 2011). Furthermore, MSCs are able to adhere to plastic in culture, easy to maintain under standard culture conditions, expansion in vitro and not only have the highest capacity to proliferate but also they can retain their pluripotent features even after different number of passages (Ueyama et al., 2012). Moreover, MSCs secrete factors, such as chemokines, cytokines, which improve the tissue microenvironment under injury conditions (Tögel et al., 2007). MSCs regulated the adaptive and innate immune systems by inhibiting both cells T- and B-activation and proliferation, inhibiting the dendritic cells differentiation, inhibiting proliferation and cytotoxicity of natural killer (NK) cells (Wang et al., 2018). They have low antigenicity, thereby reducing the toxicity (Solis et al., 2019). All these properties make MSCs a good alternative source for producing differentiated cells for therapy compared to other stem cell types. Interestingly, MSCs have potential to differentiate into insulin-producing cells (IPCs) in vitro by adopting specific methods (Chen et al., 2004). The microenvironment has an essential functional in the stem cells differentiation and survival, where it was found that conditioned medium prepared from redeveloped pancreatic tissue after partial pancreatectomy might promote rat bone marrow (BM)-MSCs to differentiate into IPCs (Choi et al., 2005). It was used rat BM-MSCs to differentiate islet cells, and transplantation of these cells reduces glucose level in non-obese (NOD) diabetic rats (Wu et al., 2007). It was also confirmed that human MSCs produced from the Wharton's jelly of the umbilical cord which has the potential to differentiate into islet cell clusters. These islet-like clusters can produce insulin in vitro and in vivo (Chao et al.,

Table 7.1 Summary of some advantages and disadvantages of different stem cell types used in diabetes treatment (Lilly et al., 2016; Shahjalal et al., 2018)

Cells type	Advantages	Disadvantages
ESCs	• Highly pluripotent differentiation capacity • Unlimited self-renewal capacity	• Limited source with ethical issues • Risk of tumor development after transplantation • Immune rejection problems
iPSCs	• iPSCs have ESCs properties • Easily obtainable as a source of stem cells without ethical issues	• Risk of tumor formation after transplantation • Mutagenic potential in reprogramming procedures
MSCs	• Easy to isolate and expand without ethical issues • High immunomodulatory properties	• Replicative lifespan is limited • Contamination risk during differentiation and manufacturing in large amounts

ESCs: embryonic stem cells; **iPSCs**: induced pluripotent stem cells; **MSCs**: mesenchymal stem cells

2008). MSCs obtained from Wharton's jelly can be used for xenotransplantation, as they do not show any stimulation of immune rejection responses (Weiss et al., 2008). Two methods used in vitro to promote MSC differentiation into IPCs. The first method is using genetic engineering to modulate gene expression via introducing key transcriptional regulatory Pdx-1 and Beta2 (Wu et al., 2007). The second method is using culture medium with specific soluble inducers or small-molecule compounds for inducing and promoting β-cell differentiation (Parnaud et al., 2008). An extract injured pancreatic tissue of rat was used for MSCs differentiation into IPCs using traditional two-step induction protocol. In stage 1 of protocol, BM-MSCs were induced with EGF, B27, and bFGF. In stage 2, serum-free high-glucose culture medium with activin-A, betacellulin, hepatocyte growth factor, nicotinamide, and other cytokines was used. Nicotinamide promotes fetal pancreatic cell differentiation, increases the amount of β-cell, and helps to synthesize insulin, while activin-A and betacellulin induce the differentiation of MSCs into β-cell. It indicated that the derived IPCs were effective in vivo and able to reverse hyperglycemia in diabetic rats (Xie et al., 2013). Some of the advantages and disadvantages of stem cells used for diabetes treatment are summarized in Table 7.1

Differentiation of Adult Pancreatic Stem/Progenitor Cells into Pancreatic β-cells

Adult pancreatic stem cells are considered as potential sources of β-cells as they have the characteristics of stem cells including clonogenicity, multi-potency, and self-renewal. It has been suggested that all pancreatic exocrine cells, pancreatic ducts cells, and the islets of Langerhans are potential sources of a pancreatic stem/progenitor cells. The cells of pancreas, such as ductal cells and acinar cells, share the same embryological origin with β-cells, which can be differentiated and re-programmed for insulin production (Kim & Lee, 2016; Pan et al., 2019). It is observed that almost all ductal cells express Pdx-1, which is important for pancreas development, especially in islet neogenesis of β-cells (Heimberg et al., 2000). Therefore, it was speculated that the ductal cells are the main source of new islet cells in the formation of new islet cells (Liao et al., 2007). Insulin-producing human islet-like clusters may be developed from human ductal tissue (Bonner-Weir et al., 2000). Using ductal tissue from a mouse or human reported the potential of generating islet-like clusters with identical or replicated protocols (Gao et al., 2003). The ductal cells probably switch to a less differentiated stage of expressing Pdx-1, as result the ductal cells acted as progenitor cells in the mature pancreas. Several investigations have tried to identify stem/progenitor cells in the pancreas (Bonner-Weir et al., 2004). Nestin is an important neural stem cell marker. Stem cells expressing nestin were isolated from human and rat pancreatic islets, and these cells can be cultured in vitro for a long period and have the ability to form insulin producing cells (Zulewski et al., 2001). The spherical neural stem cell clusters which were previously isolated from islet and ductal cells of pancreas show a precursor phenotype of both pancreatic and neural lineage. The progenitor cells have the ability to induce different populations of neurons and glial cells and also differentiate into pancreatic endocrine α-, β-, and δ-cells. Moreover, generated β-like cells derived from these progenitors demonstrate glucose-dependent $Ca2^{+}$ responsiveness and insulin secretion (Seaberg et al., 2004). The adult pancreatic stem cells successfully differentiate into islet-like cells. The human pancreatic ductal cells proliferate and differentiate into IPCs in vitro using combinations of growth factors, extracellular matrix proteins, and transcription factors (Corritore et al., 2016). In addition, ductal epithelial cells are considered as a source of pancreatic progenitors that can be generated in adult pancreas in diabetic mice after the partial pancreatectomy (Bhartiya, 2016). These results indicate that the existence of stem/progenitor cells in the pancreas might be a hopeful source for in vitro generation of islet cells, which can be useful in diabetes treatment. However, the specific marker identification is required to improve isolation populations of these cells (Pan et al., 2019).

Generation of Pancreatic β-cells Using the Three-Dimensional 3D Cell Culture Method

Why Use 3-D Cell Culture Systems?

3D cell culture is a more suitable technique used to study stem cell differentiation. 3D cultures can be produced in three different methods, such as cultures in matrigel, cultures on scaffolds, or suspension cultures on non-adherent plates and using floating culture. Different methods and materials are used in 3D cultures to simulate and mimic the environment growth in vivo, and this allows cells to grow and migrate in 3-D full space (Wang et al., 2019). The 3D cultures provide different physiological features for testing the delivery and toxicity of drug. Also, it offers many advantages including the interaction between cells-cells and cells-extracellular matrix. The cellular heterogeneity of spheroid generated from 3D culture can closely mimic the morphological cells in vivo as well as their functions such as proliferation and induce the differentiation of cells which is helpful to its function, gene expression, etc., (Mehta et al., 2012). Interestingly, the morphology and polarity of the cells are maintained in 3D cultures, and they can be returned to cells before cultured in 2D (Kapałczyńska et al., 2018). All these features made 3D culture a powerful tool to increase stem cell differentiation or to support the cells reaching to the last stage of differentiation. However, 3D culture involves high cost and consumes more time (Wang et al., 2019). In contrast, the 2D monolayer cultures have low-cost culture maintenance and are simple to use. The 2D culture provides unlimited access to the components of the user medium as oxygen, nutrients, and molecules. However, the 2D cultures have many limitations, such as cells growing in 2D cannot mimic a growth environment in vivo. It cannot provide cell-cell and cell-extracellular environment interactions (Kapałczyńska et al., 2018)

Generation of Pancreatic β-cells Using the 3D Cell Culture Method

The 3D cell cultures offer several advantages and have increased the interest of many research groups to generate pancreatic progenitors and insulin-producing cells from ESCs/iPSCs in vitro. The iPSCs derived from type 1 diabetes (T1D) patients were used to generate glucose-responsive and IPCs by using 3D culture methods. T1D iPSCs were originally shown resistant to differentiation, but the demethylation treatment effects showed a major improvement in IPCs yield. These cells release insulin in response to high-glucose stimulation in vitro. Moreover, these cells showed similar shape, size, and number of their granules that originated in cadaveric β-cells. The IPCs were transplanted into immunodeficient mice with streptozotocin (STZ)-induced diabetes, and hyperglycemia was gradually reduced. It can be considered that T1D iPSCs-derived β-cells are a suitable candidate for diabetes treatment (Manzar et al., 2017). However, it still needs a more efficient culture system that can be useful for future research and clinical applications. Organoid is once such a promising alternative, which is a 3D cellular cluster in vitro consisting of a group of primary cells, ESCs, or iPSCs. It has the capacity to regenerate to new cell types, self-regulate, and show functions similar to the original tissue in vivo (Fatehullah et al., 2016). The islet-like organoids clusters obtained from human pluripotent stem cells (hPSC) were capable of glucose-responsive insulin secretion and have therapeutic effects which could be used as alternative sources for diabetes treatment. It was demonstrated that the pancreatic endocrine cells (ECs) differentiated from hESCs allowed the formation of cell clusters with 3D structures (100–150 μm in diameter). Moreover, the hESC-derived clustered endocrine cells secreted insulin and other pancreatic endocrine hormones. These EC clusters (ECCs) improved the secretion of insulin response to glucose (Kim et al., 2016). The generation of islet organoids would be valuable for research in diabetes pathophysiology, treatment, and screening of drugs (Wang et al., 2017). Generally, in the absence of capillary vessels in the 3D islet-like structure, the physiological oxygen circulation is insufficient. Both oxygen circulation and extracellular matrix (ECM) are essential for the reconstruction of the pancreatic β-like cells in differentiation. Therefore, components of the ECM such as laminin, collagen, and fibronectin membranes were used to control the tension of oxygen (Thakur et al., 2020). The development of islet organoids from hESCs in biomimetic 3D scaffolds. Matrigel and collagen type I used to form biocompatible scaffold, a porous, for supporting pancreatic islet differentiation. The porous plays a main function by supplying the cells with suitable energy, nutrients, and oxygen. Organoid biomimetic scaffolds could mimic the in vivo environments and also support (ECM)–cell and cell–cell interactions, which is an important regulator of cellular developments that help several functions. The organoids resulted from this study consist of α-, β-, δ-, and PP-cells. Remarkably, the generation of insulin-secreting cells did not co-express glucagon, somatostatin, or pancreatic polypeptide. The expression of Pdx1, MafA, Ngn3, and Glut2 was noticed in cell clusters in 3D culture. The cells grown in the scaffolds showed an increase in insulin expression compared to those grown in 2D cultures (Wang et al., 2017). A mixture of polycaprolactone (PCL) and polyvinyl alcohol-based (PVA) scaffold has been used to differentiate hPSCs into pancreatic lineage cells. The PCL/PVA has an important function in maintenance of the microenvironment, metabolic activation, and the expression of transcription factors needed for pancreatic cell differentiation. A study was carried out to investigate hiPSCs differentiation ability to insulin-secreting cells in which 3D culture was compared with 2D culture. The expressions of Insulin, Pdx1, Glut2, and Ngn3 in PCL/PVA scaffold were significantly higher than those expressed in 2D cultures. These results showed that the improved differentiation of IPCs from hiPSCs might be a result of PCL/PVA nanofibrous scaffolds used (Abazari et al., 2018). Additionally, Amikagel system permitted the coaggregation of hESC-pancreatic progenitor cells and endothelial cells by which pancreatic organoids, were closer to natural islet physiology, are formed in vitro. Amikagel encouraged spontaneous pancreatic progenitor spheroids differentiation into β-like cells, showing C-peptide protein expression and the capacity of glucose stimulation in vitro (Huang et al., 2020). Decellularization is the procedure used in biomedical engineering, in which ECM is separated from its native cells of a tissue or organ and retaining the real structure, biochemical, and biomechanical signals for producing a natural 3D scaffold, which might allow the integration of features like vasculature. This process could be accomplished using different methods including physical, chemical, and biological with each method having both advantages and disadvantages (Gilpin & Yang, 2017). The decellularized rat pancreatic ECM (dpECM) can induce self-assembly of human islet organoids during induced iPSCs differentiation. The iPSC-derived islet organoids of dpECM were secreted main hormones including

Table 7.2 Summary of some attempts that used different protocols to differentiate stem cells into pancreatic β-cells in vitro and in vivo

Cell line	Culture technique	GSIS	References
Mouse ESCs	Bacterial Petri dish culture; suspension culture (8–10 days) and the results of embryoid bodies were plated onto plastic cell culture dishes	Yes	Soria et al. (2000)
Mouse ESCs	2D cell culture	Yes	Lumelsky et al. (2001)
Human ESCs	2D cell culture	No	D'Amour et al. (2005)
Human MSCs	2D cell culture	Yes	Chao et al. (2008)
Human ESCs Human iPSCs	2D cell culture (stages 1–4); 3D suspension culture (stages 5–7)	Yes	Rezania et al. (2014)
Human ESCs Human iPSCs	Suspension-based culture system	Yes	Pagliuca et al. (2014)
Human ESCs	Matrigel-coated 4-well plates and suspension culture	Yes	Kim et al. (2016)
Human ESCs	3D collagen scaffolds method	Yes	Wang et al. (2017)
T1D human iPSCs line	Matrigel /3D cell culture	Yes	Manzar et al., (2017)
Human iPSCs	2D cell culture (18 days); 3D suspension culture (10 days)	Yes	Bi et al. (2020)

ESCs: embryonic stem cells; **iPSCs**: induced pluripotent stem cells; **MSCs**: mesenchymal stem cells; **GSIS**: Glucose-stimulated insulin secretion; **T1D**: Type 1 diabetes

glucagon and insulin. These organoids contained α-, β-, δ-, and PP-cells. The exposure of iPSCs to the dpECM at differentiation stage showed higher expression of Pdx1, MafA, and Nkx6.1 (Bi et al., 2020). Table 7.2 summarizes some methods used to differentiate different types of stem cells into β-cells.

Transdifferentiation of Pancreatic Cells

Transdifferentiation is a term that refers to changes in the cellular phenotype, such as conversion of differentiated cell type to another (Tosh et al., 2002). It is a process of phenotypic plasticity in a mature cell. Phenotypic change occurs in chronically damaged tissues and in tissue regeneration (Shen et al., 2000). It is considered as the most attractive method of developing β-cell sources which can be used for cell therapy. This procedure is based on cell reprogramming including neogenesis and regeneration of β-cell from progenitor cells (Kim et al., 2019). The transdifferentiation in the pancreas is acinar to ductal metaplasia (ADM) and is the process where acinar cells differentiate into duct cells, which plays a role in regeneration injured pancreas. Moreover, under certain microenvironment conditions, the acinar cells can differentiate into hepatocyte-like cells and adipocytes (Lardon et al., 2004). The over-expression of polymorphisms of T cell factor 7-like 2 induced ductal epithelial cell proliferation and differentiation into islet-like clusters (Shu et al., 2012). The AR42J cell line derived from a pancreatic tumor, which has features of pancreatic acinar cells can transdifferentiate; by reprogramming these cells toward β-cells phenotype using of Pdx1, Ngn3, and MafA, induction of endocrine markers was observed (Akinci et al., 2012).

Clinical Trials of Stem Cell Therapy for Diabetes

There is a growing global interest in stem cells research and the possibility to use it for treating various diseases such as diabetes. Stem cells have great therapeutic potential in this field. They have the potential for self-renewing, repairing damaged tissues cells, immunomodulatory properties, and their ability to provide an unlimited source of insulin-producing β-cells (Pathak et al., 2019; Peng et al., 2018). There have been several attempts of human clinical research studies using different types of stem cells in diabetic treatment, and some of these applications are summarized in this chapter. The first human clinical trial used autologous nonmyeloablative hematopoietic stem cells transplantation (AHST) to treat recent type 1 diabetic (T1D) patients. Also, it was evaluated the safety and metabolic effects of immunosuppression therapy. The results showed that most newly T1D patients accomplished different times of insulin independence and treatment-related toxicity was acceptable, no mortality reported. Moreover, with AHST, β-cell function was improved promisingly. However, this study needs further follow-up to confirm the time of insulin independence, randomness sample, and a control group (Voltarelli et al., 2007). A clinical study was carried out to estimate the effects of AHST in clinical and molecular processes in 9 recent T1D patients. The results showed that AHST increased the islet cell function due to removal of the islet specific autoreactive T cells; the difference in T1D patient reactions to AHST could be referred to these different transcriptional actions in the peripheral blood mononuclear cell (Zhang et al., 2012). Although less clinical trials have been performed in developing stem cell therapy for T2D, some encouraging results have been reported. It was studied the combination of intrapancreatic autologous stem cell (ASC) infusion with hyperbaric oxygen treatment (HBOT) in 25 T2D patients. In the follow-up period, hemoglobin A1c (glycosylated hemoglobin) levels were decreased, the insulin dose requirements reduced and increased C-peptide levels. These results suggest that ASC infusion and HBOT have positive therapeutic effects for T2D patients by improving metabolic control and reducing insulin requirements (Estrada et al., 2008). However, this study requires randomized controlled samples to confirm it. Another study that evaluated the combination of autologous bone marrow stem cell transplantation (ABMSCT) and (HBOT) on 31 T2D patients. Significant reductions in the dose of oral hypoglycemic drugs and decreased exogenous insulin dose have been demonstrated in all patients who used this therapy, but the functional development of pancreatic β-cell may be transient (Wang et al., 2011).

The intra-arterial injection of stem cells derived from bone marrow to T2D patients showed positive results, which confirm the efficacy and safety of this treatment for diabetics (Bhansali et al., 2014). An example of current human clinical trial which used stem cells in diabetes treatment was a pilot study of the therapeutic possibility of educator stem cells treatment in T1D phase 1 for both genders (18 years and older) in Hackensack, US. The aim of this study was to achieve patient's apheresis and then have their own blood returned to them with the "educated" lymphocytes (ClinicalTrials.gov NCT02624804). It was found that the stem cell educator treatment can develop the clinical treatment of diabetes and other diseases by cord blood-derived multipotent stem cells (CB-SCs) immune education and immune balance induction without the ethical and safety issues associated with traditional stem cell methods (Cheng et al., 2016).

The Challenges of Stem Cell Therapy

Stem cells therapy is a promising potential therapeutic method for treating diabetes. Nevertheless, the results of stem cell clinical trials for diabetes treatments need further improvements to make them readily available for treatments. There are many challenges and obstacles that remain to be resolved in adopting this technology. The major challenges include how to (**a**) generate more developed functional β-like cells *in vitro* from hPSCs; (**b**) improve the efficiency differentiation of IPCs from hPSCs; (**c**) protect transplanted IPCs from autoimmune system; (**d**) generate enough numbers of cell types that required for clinical transplantation trial; (**e**) establish overall of insulin independence; and (**f**) avoid the carcinogenic properties that stem cells form, and maintaining the function and integrity of their stem cell-like characteristics in their production development (Chen et al., 2020). One of the most challenging goals that must be faced and overcome when using stem cell therapy is the immune rejection of the host. ViaCyte Inc. has successfully developed an encapsulation system named "Encaptra," in which microencapsulated pancreatic progenitors derived from stem cells are implanted subcutaneously to T1D patients in a phase 1 and 2 trial to evaluate efficacy and safety (ClinicalTrials.gov NCT02239354). The encapsulation device can provide a physical barrier protecting transplanted cells from the immune system while allowing oxygen and nutrients to pass through the membrane. Also, this system allows protecting patients from the risk of stem cell-derived β-cell oncogenic transformation (Chen et al., 2020; Sneddon et al., 2018). Sernova a clinical company, is developing a treatment for T1D (ClinicalTrials.gov NCT01652911) using implantable therapy device named "Cell Pouch," containing a scaffold with chambers that allows islet cells to vascularize, mimicking an environment similar to a natural organ. This device is inserted under the skin for a month to allow integration of vascular with the surrounding tissues (Sneddon et al., 2018). Systems like these can be potential solutions for future researchers to develop protocols using stem cells.

Summary

The rate of diabetic disease is frighteningly increased around the world. The absence or loss of insulin-producing β-cells causes diabetes. The traditional treatment of diabetes has many limitations and cannot mimic natural pancreatic insulin production. As an alternative treatment for diabetes, islet transplantation maintains glucose homeostasis, and it is limited due to the lack of islet donation and other complications. Therefore, it is significant to determine advanced approaches to gain functional β-cells. Stem cell therapy offers a powerful promising potential for treating diabetes. Several types of stem cells have been proven effective in treating diabetes with clear limitations such as (ESCs), (iPSCs), (MSCs), and adult pancreatic stem cells. In conclusion, further human stem cell clinical trials are needed to overcome the challenges associated with stem cell and to make stem cell therapy a viable option for treating diabetes in the future.

References

Abazari, M. F., Soleimanifar, F., Nouri Aleagha, M., Torabinejad, S., Nasiri, N., Khamisipour, G., Amini Mahabadi, J., Mahboudi, H., Enderami, S. E., Saburi, E., Hashemi, J., & Kehtari, M. (2018). PCL/PVA nanofibrous scaffold improve insulin-producing cells generation from human induced pluripotent stem cells. *Gene, 671*, 50–57. https://doi.org/10.1016/j.gene.2018.05.115

Aftab, S., Semenec, L., Chu, J. S., & Chen, N. (2008). Identification and characterization of novel human tissue-specific RFX transcription factors. *BMC Evolutionary Biology, 8*, 226. https://doi.org/10.1186/1471-2148-8-226

Ahlgren, U., Jonsson, J., & Edlund, H. (1996). The morphogenesis of the pancreatic mesenchyme is uncoupled from that of the pancreatic epithelium in IPF1/PDX1-deficient mice. *Development, 122*(5), 1409–1416.

Ahlgren, U., Pfaff, S. L., Jessell, T. M., Edlund, T., & Edlund, H. (1997). Independent requirement for ISL1 in formation of pancreatic mesenchyme and islet cells. *Nature, 385*(6613), 257–260. https://doi.org/10.1038/385257a0

Ahnfelt-Rønne, J., Jørgensen, M. C., Klinck, R., Jensen, J. N., Füchtbauer, E. M., Deering, T., MacDonald, R. J., Wright, C. V., Madsen, O. D., & Serup, P. (2012). Ptf1a-mediated control of Dll1 reveals an alternative to the lateral inhibition mechanism. *Development, 139*(1), 33–45. https://doi.org/10.1242/dev.071761

Akinci, E., Banga, A., Greder, L. V., Dutton, J. R., & Slack, J. M. (2012). Reprogramming of pancreatic exocrine cells towards a beta (β) cell character using Pdx1, Ngn3 and MafA. *The Biochemical Journal, 442*(3), 539–550. https://doi.org/10.1042/BJ20111678

Akiyama, H., Kim, J. E., Nakashima, K., Balmes, G., Iwai, N., Deng, J. M., Zhang, Z., Martin, J. F., Behringer, R. R., Nakamura, T., & de Crombrugghe, B. (2005). Osteo-chondroprogenitor cells are derived from Sox9 expressing precursors. *Proceedings of the National Academy of Sciences of the United States of America, 102*(41), 14665–14670. https://doi.org/10.1073/pnas.0504750102

Aramata, S., Han, S. I., Yasuda, K., & Kataoka, K. (2005). Synergistic activation of the insulin gene promoter by the beta-cell enriched transcription factors MafA, Beta2, and Pdx1. *Biochimica Et Biophysica Acta, 1730*(1), 41–46. https://doi.org/10.1016/j.bbaexp.2005.05.009

Arimura, A., Sato, H., Dupont, A., Nishi, N., & Schally, A. V. (1975). Somatostatin: Abundance of immunoreactive hormone in rat stomach and pancreas. *Science, 189*(4207), 1007–1009. https://doi.org/10.1126/science.56779

Artner, I., Hang, Y., Mazur, M., Yamamoto, T., Guo, M., Lindner, J., Magnuson, M. A., & Stein, R. (2010). MafA and MafB regulate genes critical to β-cells in a unique temporal manner. *Diabetes, 59*(10), 2530–2539. https://doi.org/10.2337/db10-0190

Ashcroft, F. M., & Rorsman, P. (2012). Diabetes mellitus and the β-cell: The last ten years. *Cell, 148*(6), 1160–1171. https://doi.org/10.1016/j.cell.2012.02.010

Assady, S., Maor, G., Amit, M., Itskovitz-Eldor, J., Skorecki, K. L., & Tzukerman, M. (2001). Insulin production by human embryonic stem cells. *Diabetes, 50*(8), 1691–1697. https://doi.org/10.2337/diabetes.50.8.1691

Baharvand, H., Jafary, H., Massumi, M., & Ashtiani, S. K. (2006). Generation of insulin-secreting cells from human embryonic stem cells. *Development, Growth and Differentiation, 48*(5), 323–332. https://doi.org/10.1111/j.1440-169X.2006.00867.x

Banks, P. A., Conwell, D. L., & Toskes, P. P. (2010). The management of acute and chronic pancreatitis. *Gastroenterol Hepatol (NY), 6*(2 Suppl 3), 1–16.

Bastidas-Ponce, A., Scheibner, K., Lickert, H., & Bakhti, M. (2017). Cellular and molecular mechanisms coordinating pancreas development. *Development, 144*(16), 2873–2888. https://doi.org/10.1242/dev.140756

Benitez, C. M., Goodyer, W. R., & Kim, S. K. (2012). Deconstructing pancreas developmental biology. *Cold Spring Harb Perspect Biol, 4*(6). https://doi.org/10.1101/cshperspect.a012401

Beres, T. M., Masui, T., Swift, G. H., Shi, L., Henke, R. M., & MacDonald, R. J. (2006). PTF1 is an organ-specific and Notch-independent basic helix-loop-helix complex containing the mammalian suppressor of hairless (RBP-J) or its paralogue, RBP-I. *Molecular and Cellular Biology, 26*(1), 117–130. https://doi.org/10.1128/MCB.26.1.117-130.2006

Bhansali, A., Upreti, V., Walia, R., Gupta, V., Bhansali, S., Sharma, R. R., Grover, S., Marwaha, N., & Khandelwal, N. (2014). Efficacy and safety of autologous bone marrow derived hematopoietic stem cell transplantation in patients with type 2 DM: A 15 months follow-up study. *Indian Journal Endocrinologica Metabolism, 18*(6), 838–845. https://doi.org/10.4103/2230-8210.140257

Bhartiya, D. (2016). Stem cells to replace or regenerate the diabetic pancreas: Huge potential and existing hurdles. *Indian Journal of Medical Research, 143*(3), 267–274. https://doi.org/10.4103/0971-5916.182615

Bi, H., Karanth, S. S., Ye, K., Stein, R., & Jin, S. (2020). Decellularized tissue matrix enhances self-assembly of islet organoids from pluripotent stem cell differentiation. *ACS Biomaterials Science and Engineering, 6*(7), 4155–4165. https://doi.org/10.1021/acsbiomaterials.0c00088

Bienvenu, T., Poirier, K., Friocourt, G., Bahi, N., Beaumont, D., Fauchereau, F., Ben Jeema, L., Zemni, R., Vinet, M. C., Francis, F., Couvert, P., Gomot, M., Moraine, C., van Bokhoven, H., Kalscheuer, V., Frints, S., Gecz, J., Ohzaki, K., Chaabouni, H., … Chelly, J. (2002). ARX, a novel Prd-class-homeobox gene highly expressed in the telencephalon, is mutated in X-linked mental retardation. *Human Molecular Genetics, 11*(8), 981–991. https://doi.org/10.1093/hmg/11.8.981

Bonner-Weir, S., Baxter, L. A., Schuppin, G. T., & Smith, F. E. (1993). A second pathway for regeneration of adult exocrine and endocrine pancreas. A possible recapitulation of embryonic development. *Diabetes, 42*(12), 1715–1720. https://doi.org/10.2337/diab.42.12.1715

Bonner-Weir, S., Taneja, M., Weir, G. C., Tatarkiewicz, K., Song, K. H., Sharma, A., & O'Neil, J. J. (2000). In vitro cultivation of human islets from expanded ductal tissue. *Proceedings of the National Academy Od Sciences USA, 97*(14), 7999–8004. https://doi.org/10.1073/pnas.97.14.7999

Bonner-Weir, S., Toschi, E., Inada, A., Reitz, P., Fonseca, S. Y., Aye, T., & Sharma, A. (2004). The pancreatic ductal epithelium serves as a potential pool of progenitor cells. *Pediatric Diabetes, 5*(Suppl 2), 16–22. https://doi.org/10.1111/j.1399-543X.2004.00075.x

Bradley, A., Evans, M., Kaufman, M. H., & Robertson, E. (1984). Formation of germ-line chimaeras from embryo-derived teratocarcinoma cell lines. *Nature, 309*(5965), 255–256. https://doi.org/10.1038/309255a0

Brazeau, P., Vale, W., Burgus, R., Ling, N., Butcher, M., Rivier, J., & Guillemin, R. (1973). Hypothalamic polypeptide that inhibits the secretion of immunoreactive pituitary growth hormone. *Science, 179*(4068), 77–79. https://doi.org/10.1126/science.179.4068.77

Brereton, M., Vergari, E., Zhang, Q., & Clark, A. (2015). Alpha-, Delta- and PP-cells: Are they the architectural cornerstones of islet structure and co-ordination? *Journal of Histochemistry and Cytochemistry, 63*, 575–591. https://doi.org/10.1369/0022155415583535

Brissova, M., Fowler, M. J., Nicholson, W. E., Chu, A., Hirshberg, B., Harlan, D. M., & Powers, A. C. (2005). Assessment of human pancreatic islet architecture and composition by laser scanning confocal microscopy. *Journal of Histochemistry and Cytochemistry, 53*(9), 1087–1097. https://doi.org/10.1369/jhc.5C6684.2005

Butler, A. E., Janson, J., Bonner-Weir, S., Ritzel, R., Rizza, R. A., & Butler, P. C. (2003). β-cell deficit and increased β-cell apoptosis in humans with type 2 diabetes. *Diabetes, 52*(1), 102–110. https://doi.org/10.2337/diabetes.52.1.102

Butler, A. E., Cao-Minh, L., Galasso, R., Rizza, R. A., Corradin, A., Cobelli, C., & Butler, P. C. (2010). Adaptive changes in pancreatic β-cell fractional area and beta cell turnover in human pregnancy. *Diabetologia, 53*(10), 2167–2176. https://doi.org/10.1007/s00125-010-1809-6

Cabrera, O., Berman, D. M., Kenyon, N. S., Ricordi, C., Berggren, P. O., & Caicedo, A. (2006). The unique cytoarchitecture of human pancreatic islets has implications for islet cell function. *Proceedings of the National Academy of Science USA, 103*(7), 2334–2339. https://doi.org/10.1073/pnas.0510790103

Castellanos, E., Berlin, J., & Cardin, D. B. (2011). Current treatment options for pancreatic carcinoma. *Current Oncology Reports, 13*(3), 195–205. https://doi.org/10.1007/s11912-011-0164-1

Chao, K. C., Chao, K. F., Chen, C. F., & Liu, S. H. (2008). A novel human stem cell coculture system that maintains the survival and function of culture islet-like cell clusters. *Cell Transplantation, 17*(6), 657–664. https://doi.org/10.3727/096368908786092801

Chen, L. B., Jiang, X. B., & Yang, L. (2004). Differentiation of rat marrow mesenchymal stem cells into pancreatic islet beta-cells. *World Journal of Gastroenterology, 10*(20), 3016–3020. https://doi.org/10.3748/wjg.v10.i20.3016

Chen, S., Du, K., & Zou, C. (2020). Current progress in stem cell therapy for type 1 diabetes mellitus. *Stem Cell Research and Therapy, 11*(1), 275. https://doi.org/10.1186/s13287-020-017 93-6

Cheng, S. K., Park, E. Y., Pehar, A., Rooney, A. C., & Gallicano, G. I. (2016). Current progress of human trials using stem cell therapy as a treatment for diabetes mellitus. *American Journal Stem Cells, 5*(3), 74–86.

Choi, K. S., Shin, J. S., Lee, J. J., Kim, Y. S., Kim, S. B., & Kim, C. W. (2005). In vitro trans-differentiation of rat mesenchymal cells into insulin-producing cells by rat pancreatic extract. *Biochemical and Biophysical Research Communications, 330*(4), 1299–1305. https://doi.org/10.1016/j.bbrc.2005.03.111

Collombat, P., Mansouri, A., Hecksher-Sorensen, J., Serup, P., Krull, J., Gradwohl, G., & Gruss, P. (2003). Opposing actions of Arx and Pax4 in endocrine pancreas development. *Genes and Development, 17*(20), 2591–2603. https://doi.org/10.1101/gad.269003

Collombat, P., Hecksher-Sørensen, J., Broccoli, V., Krull, J., Ponte, I., Mundiger, T., Smith, J., Gruss, P., Serup, P., & Mansouri, A. (2005). The simultaneous loss of Arx and Pax4 genes promotes a somatostatin-producing cell fate specification at the expense of the alpha- and beta-cell lineages in the mouse endocrine pancreas. *Development, 132*(13), 2969–2980. https://doi.org/10.1242/dev.01870

Concepcion, J. P., Reh, C. S., Daniels, M., Liu, X., Paz, V. P., Ye, H., Highland, H. M., Hanis, C. L., & Greeley, S. A. (2014). Neonatal diabetes, gallbladder agenesis, duodenal atresia, and intestinal malrotation caused by a novel homozygous mutation in RFX6. *Pediatric Diabetes, 15*(1), 67–72. https://doi.org/10.1111/pedi.12063

Corritore, E., Lee, Y. S., Sokal, E. M., & Lysy, P. A. (2016). β-cell replacement sources for type 1 diabetes: A focus on pancreatic ductal cells. *Therapeutic Advances in Endocrinology and Metabolism, 7*(4), 182–199. https://doi.org/10.1177/2042018816652059

D'Amour, K. A., Agulnick, A. D., Eliazer, S., Kelly, O. G., Kroon, E., & Baetge, E. E. (2005). Efficient differentiation of human embryonic stem cells to definitive endoderm. *Nature Biotechnology, 23*(12), 1534–1541. https://doi.org/10.1038/nbt1163

Dassaye, R., Naidoo, S., & Cerf, M. E. (2016). Transcription factor regulation of pancreatic organogenesis, differentiation and maturation. *Islets, 8*(1), 13–34. https://doi.org/10.1080/19382014.2015.1075687

De Vas, M. G., Kopp, J. L., Heliot, C., Sander, M., Cereghini, S., & Haumaitre, C. (2015). Hnf1b controls pancreas morphogenesis and the generation of Ngn3+ endocrine progenitors. *Development, 142*(5), 871–882. https://doi.org/10.1242/dev.110759

Decker, K., Goldman, D. C., Grasch, C. L., & Sussel, L. (2006). Gata6 is an important regulator of mouse pancreas development. *Developmental Biology, 298*(2), 415–429. https://doi.org/10.1016/j.ydbio.2006.06.046

Efrat, S., Tal, M., & Lodish, H. F. (1994). The pancreatic beta-cell glucose sensor. *Trends in Biochemical Sciences, 19*(12), 535–538. https://doi.org/10.1016/0968-0004(94)90056-6

Ernst, S., Demirci, C., Valle, S., Velazquez-Garcia, S., & Garcia-Ocaña, A. (2011). Mechanisms in the adaptation of maternal β-cells during pregnancy. *Diabetes Management (lond), 1*(2), 239–248. https://doi.org/10.2217/dmt.10.24

Estrada, E. J., Valacchi, F., Nicora, E., Brieva, S., Esteve, C., Echevarria, L., Froud, T., Bernetti, K., Cayetano, S. M., Velazquez, O., Alejandro, R., & Ricordi, C. (2008). Combined treatment of intrapancreatic autologous bone marrow stem cells and hyperbaric oxygen in type 2 diabetes mellitus. *Cell Transplantation, 17*(12), 1295–1304. https://doi.org/10.3727/096368908787648119

Fatehullah, A., Tan, S. H., & Barker, N. (2016). Organoids as an in vitro model of human development and disease. *Nature Cell Biology, 18*(3), 246–254. https://doi.org/10.1038/ncb3312

Frantz, E., Souza-Mello, V., & Mandarim-de-Lacerda, C. (2012). *Pancreas: Anatomy, diseases and health implications.*

Freychet, L., Rizkalla, S. W., Desplanque, N., Basdevant, A., Zirinis, P., Tchobroutsky, G., & Slama, G. (1988). Effect of intranasal glucagon on blood glucose levels in healthy subjects and hypoglycaemic patients with insulin-dependent diabetes. *Lancet, 1*(8599), 1364–1366. https://doi.org/10.1016/s0140-6736(88)92181-2

Fu, Z., Gilbert, E. R., & Liu, D. (2013). Regulation of insulin synthesis and secretion and pancreatic Beta-cell dysfunction in diabetes. *Current Diabetes Review, 9*(1), 25–53.

Fujimoto, K., & Polonsky, K. S. (2009). Pdx1 and other factors that regulate pancreatic beta-cell survival. *Diabetes, Obesity and Metabolism, 11*(Suppl 4), 30–37. https://doi.org/10.1111/j.1463-1326.2009.01121.x

Gannon, M., Ray, M. K., Van Zee, K., Rausa, F., Costa, R. H., & Wright, C. V. (2000). Persistent expression of HNF6 in islet endocrine cells causes disrupted islet architecture and loss of beta cell function. *Development, 127*(13), 2883–2895.

Gao, R., Ustinov, J., Pulkkinen, M. A., Lundin, K., Korsgren, O., & Otonkoski, T. (2003). Characterization of endocrine progenitor cells and critical factors for their differentiation in human adult pancreatic cell culture. *Diabetes, 52*(8), 2007–2015. https://doi.org/10.2337/diabetes.52.8.2007

Gilpin, A., & Yang, Y. (2017). Decellularization strategies for regenerative medicine: From processing techniques to applications. *BioMed Research International, 2017*, 9831534. https://doi.org/10.1155/2017/9831534

Githens, S. (1988). The pancreatic duct cell: Proliferative capabilities, specific characteristics, metaplasia, isolation, and culture. *Journal of Pediatric Gastroenterology and Nutrition, 7*(4), 486–506.

Githens, S. (1994). Pancreatic duct cell cultures. *Annual Review of Physiology, 56*, 419–443. https://doi.org/10.1146/annurev.ph.56.030194.002223

Guney, M. A., & Gannon, M. (2009). Pancreas cell fate. *Birth Defects Research. Part c, Embryo Today, 87*(3), 232–248. https://doi.org/10.1002/bdrc.20156

Guo, T., & Hebrok, M. (2009). Stem cells to pancreatic beta-cells: New sources for diabetes cell therapy. *Endocrine Reviews, 30*(3), 214–227. https://doi.org/10.1210/er.2009-0004

Hang, Y., & Stein, R. (2011). MafA and MafB activity in pancreatic β-cells. *Trends in Endocrinology and Metabolism, 22*(9), 364–373. https://doi.org/10.1016/j.tem.2011.05.003

Hart, A. W., Mella, S., Mendrychowski, J., van Heyningen, V., & Kleinjan, D. A. (2013). The developmental regulator Pax6 is essential for maintenance of islet cell function in the adult mouse pancreas. *PLoS ONE, 8*(1), e54173. https://doi.org/10.1371/journal.pone.0054173

Hass, R., Kasper, C., Böhm, S., & Jacobs, R. (2011). Different populations and sources of human mesenchymal stem cells (MSC): A comparison of adult and neonatal tissue-derived MSC. *Cell Communication and Signaling: CCS, 9*, 12. https://doi.org/10.1186/1478-811X-9-12

Haumaitre, C., Barbacci, E., Jenny, M., Ott, M. O., Gradwohl, G., & Cereghini, S. (2005). Lack of TCF2/vHNF1 in mice leads to pancreas agenesis. *Proceedings of the National Academy of Science USA, 102*(5), 1490–1495. https://doi.org/10.1073/pnas.0405776102

Heimberg, H., Bouwens, L., Heremans, Y., Van De Casteele, M., Lefebvre, V., & Pipeleers, D. (2000). Adult human pancreatic duct and islet cells exhibit similarities in expression and differences in phosphorylation and complex formation of the homeodomain protein Ipf-1. *Diabetes, 49*(4), 571–579. https://doi.org/10.2337/diabetes.49.4.571

Heller, R. S., Jenny, M., Collombat, P., Mansouri, A., Tomasetto, C., Madsen, O. D., Mellitzer, G., Gradwohl, G., & Serup, P. (2005). Genetic determinants of pancreatic epsilon-cell development. *Developmental Biology, 286*(1), 217–224. https://doi.org/10.1016/j.ydbio.2005.06.041

Henquin, J. C. (2000). Triggering and amplifying pathways of regulation of insulin secretion by glucose. *Diabetes, 49*(11), 1751–1760. https://doi.org/10.2337/diabetes.49.11.1751

Herrera, P. L. (2000). Adult insulin- and glucagon-producing cells differentiate from two independent cell lineages. *Development, 127*(11), 2317–2322.

Huang, H., Bader, T. N., & Jin, S. (2020). Signaling Molecules Regulating pancreatic endocrine development from pluripotent stem cell differentiation. *International Journal of Molecular of Science, 21*(16). https://doi.org/10.3390/ijms21165867

Jonatan, D., Spence, J. R., Method, A. M., Kofron, M., Sinagoga, K., Haataja, L., Arvan, P., Deutsch, G. H., & Wells, J. M. (2014). Sox17 regulates insulin secretion in the normal and pathologic mouse β-cell. *PLoS ONE, 9*(8), e104675. https://doi.org/10.1371/journal.pone.0104675

Kaneto, H., & Matsuoka, T. A. (2015). Role of pancreatic transcription factors in maintenance of mature β-cell function. *International Journal of Molecular Sciences, 16*(3), 6281–6297. https://doi.org/10.3390/ijms16036281

Kania, G., Blyszczuk, P., Czyz, J., Navarrete-Santos, A., & Wobus, A. M. (2003). Differentiation of mouse embryonic stem cells into pancreatic and hepatic cells. *Methods in Enzymology, 365*, 287–303. https://doi.org/10.1016/s0076-6879(03)65021-4

Kapałczyńska, M., Kolenda, T., Przybyła, W., Zajączkowska, M., Teresiak, A., Filas, V., Ibbs, M., Bliźniak, R., Łuczewski, Ł., & Lamperska, K. (2018). 2D and 3D cell cultures—a comparison of different types of cancer cell cultures. *Archives of Medical Science, 14*(4), 910–919. https://doi.org/10.5114/aoms.2016.63743

Karnik, S. K., Chen, H., McLean, G. W., Heit, J. J., Gu, X., Zhang, A. Y., Fontaine, M., Yen, M. H., & Kim, S. K. (2007). Menin controls growth of pancreatic beta-cells in pregnant mice and promotes gestational diabetes mellitus. *Science, 318*(5851), 806–809. https://doi.org/10.1126/science.1146812

Kawaguchi, Y., Cooper, B., Gannon, M., Ray, M., MacDonald, R. J., & Wright, C. V. (2002). The role of the transcriptional regulator Ptf1a in converting intestinal to pancreatic progenitors. *Nature Genetics, 32*(1), 128–134. https://doi.org/10.1038/ng959

Ketola, I., Otonkoski, T., Pulkkinen, M. A., Niemi, H., Palgi, J., Jacobsen, C. M., Wilson, D. B., & Heikinheimo, M. (2004). Transcription factor GATA-6 is expressed in the endocrine and GATA-4 in the exocrine pancreas. *Molecular and Cellular Endocrinology, 226*(1–2), 51–57. https://doi.org/10.1016/j.mce.2004.06.007

Kim, H. S., & Lee, M. K. (2016). β-Cell regeneration through the transdifferentiation of pancreatic cells: Pancreatic progenitor cells in the pancreas. *Journal of Diabetes Investigation, 7*(3), 286–296. https://doi.org/10.1111/jdi.12475

Kim, Y., Kim, H., Ko, U. H., Oh, Y., Lim, A., Sohn, J. W., Shin, J. H., & Han, Y. M. (2016). Islet-like organoids derived from human pluripotent stem cells efficiently function in the glucose responsiveness in vitro and in vivo. *Science and Reports, 6*, 35145. https://doi.org/10.1038/srep35145

Kim, J., Shim, I. K., Hwang, D. G., Lee, Y. N., Kim, M., Kim, H., Kim, S. W., Lee, S., Kim, S. C., Cho, D. W., & Jang, J. (2019). 3D cell printing of islet-laden pancreatic tissue-derived extracellular matrix bioink constructs for enhancing pancreatic functions. *Journal of Materials Chemistry B, 7*(10), 1773–1781. https://doi.org/10.1039/c8tb02787k

Kimmel, J. R., Hayden, L. J., & Pollock, H. G. (1975). Isolation and characterization of a new pancreatic polypeptide hormone. *Journal of Biological Chemistry, 250*(24), 9369–9376.

Kojima, M., Hosoda, H., Date, Y., Nakazato, M., Matsuo, H., & Kangawa, K. (1999). Ghrelin is a growth-hormone-releasing acylated peptide from stomach. *Nature, 402*(6762), 656–660. https://doi.org/10.1038/45230

Komatsu, M., Takei, M., Ishii, H., & Sato, Y. (2013). Glucose-stimulated insulin secretion: A newer perspective. *Journal of Diabetes Investigation, 4*(6), 511–516. https://doi.org/10.1111/jdi.12094

Lardon, J., De Breuck, S., Rooman, I., Van Lommel, L., Kruhøffer, M., Orntoft, T., Schuit, F., & Bouwens, L. (2004). Plasticity in the adult rat pancreas: Transdifferentiation of exocrine to hepatocyte-like cells in primary culture. *Hepatology, 39*(6), 1499–1507. https://doi.org/10.1002/hep.20213

Lau, H. H., Ng, N. H. J., Loo, L. S. W., Jasmen, J. B., & Teo, A. K. K. (2018). The molecular functions of hepatocyte nuclear factors—In and beyond the liver. *Journal of Hepatology, 68*(5), 1033–1048. https://doi.org/10.1016/j.jhep.2017.11.026

Lendahl, U., Zimmerman, L. B., & McKay, R. D. (1990). CNS stem cells express a new class of intermediate filament protein. *Cell, 60*(4), 585–595. https://doi.org/10.1016/0092-8674(90)90662-x

Lester, L. B., Kuo, H. C., Andrews, L., Nauert, B., & Wolf, D. P. (2004). Directed differentiation of rhesus monkey ES cells into pancreatic cell phenotypes. *Reproductive Biology and Endocrinology, 2*, 42. https://doi.org/10.1186/1477-7827-2-42

Liao, Y. H., Verchere, C. B., & Warnock, G. L. (2007). Adult stem or progenitor cells in treatment for type 1 diabetes: Current progress. *Canadian Journal of Surgery, 50*(2), 137–142.

Lilly, M. A., Davis, M. F., Fabie, J. E., Terhune, E. B., & Gallicano, G. I. (2016). Current stem cell based therapies in diabetes. *American Journal Stem Cells, 5*(3), 87–98.

Löhr, M., Lübbersmeyer, J., Otremba, B., Klapdor, R., Grossner, D., & Klöppel, G. (1989). Increase in β-cells in the pancreatic remnant after partial pancreatectomy in pigs. An immunocytochemical and functional study. *Virchows Archiv B Cell Pathology Including Molecular Pathology, 56*(4), 277–286. https://doi.org/10.1007/BF02890027

Lumelsky, N., Blondel, O., Laeng, P., Velasco, I., Ravin, R., & McKay, R. (2001). Differentiation of embryonic stem cells to insulin-secreting structures similar to pancreatic islets. *Science, 292*(5520), 1389–1394. https://doi.org/10.1126/science.1058866

Lynn, F. C., Smith, S. B., Wilson, M. E., Yang, K. Y., Nekrep, N., & German, M. S. (2007). Sox9 coordinates a transcriptional network in pancreatic progenitor cells. *Proceedings of the National Academy of Science USA, 104*(25), 10500–10505. https://doi.org/10.1073/pnas.0704054104

Maestro, M. A., Cardalda, C., Boj, S. F., Luco, R. F., Servitja, J. M., & Ferrer, J. (2007). Distinct roles of HNF1 β, HNF1 α, and HNF4 α in regulating pancreas development, β-cell function and growth. *Endocrine Development, 12*, 33–45. https://doi.org/10.1159/000109603

Manzar, G. S., Kim, E. M., & Zavazava, N. (2017). Demethylation of induced pluripotent stem cells from type 1 diabetic patients enhances differentiation into functional pancreatic β-cells. *Journal of Biological Chemistry, 292*(34), 14066–14079. https://doi.org/10.1074/jbc.M117.784280

Marshall, M. O., Thomas, H. M., Seatter, M. J., Greer, K. R., Wood, P. J., & Gould, G. W. (1993). Pancreatic β-cells express a low affinity glucose transporter: Functional consequences in normal and diabetic states. *Biochemical Society Transactions, 21*(1), 164–168. https://doi.org/10.1042/bst0210164

Mayhew, C. N., & Wells, J. M. (2010). Converting human pluripotent stem cells into β-cells: Recent advances and future challenges. *Current Opinion in Organ Transplantation, 15*(1), 54–60. https://doi.org/10.1097/MOT.0b013e3283337e1c

McKiernan, E., O'Driscoll, L., Kasper, M., Barron, N., O'Sullivan, F., & Clynes, M. (2007). Directed differentiation of mouse embryonic stem cells into pancreatic-like or neuronal- and glial-like phenotypes. *Tissue Engineering, 13*(10), 2419–2430. https://doi.org/10.1089/ten.2006.0373

Mehta, G., Hsiao, A. Y., Ingram, M., Luker, G. D., & Takayama, S. (2012). Opportunities and challenges for use of tumor spheroids as models to test drug delivery and efficacy. *Journal of Controlled Release, 164*(2), 192–204. https://doi.org/10.1016/j.jconrel.2012.04.045

Menge, B. A., Tannapfel, A., Belyaev, O., Drescher, R., Müller, C., Uhl, W., Schmidt, W. E., & Meier, J. J. (2008). Partial pancreatectomy in adult humans does not provoke β-cell regeneration. *Diabetes, 57*(1), 142–149. https://doi.org/10.2337/db07-1294

Motohashi, H., O'Connor, T., Katsuoka, F., Engel, J. D., & Yamamoto, M. (2002). Integration and diversity of the regulatory network composed of Maf and CNC families of transcription factors. *Gene, 294*(1–2), 1–12. https://doi.org/10.1016/s0378-1119(02)00788-6

Muallem, S., Kwiatkowska, K., Xu, X., & Yin, H. L. (1995). Actin filament disassembly is a sufficient final trigger for exocytosis in nonexcitable cells. *Journal of Cell Biology, 128*(4), 589–598. https://doi.org/10.1083/jcb.128.4.589

Naya, F. J., Huang, H. P., Qiu, Y., Mutoh, H., DeMayo, F. J., Leiter, A. B., & Tsai, M. J. (1997). Diabetes, defective pancreatic morphogenesis, and abnormal enteroendocrine differentiation in BETA2/neuroD-deficient mice. *Genes and Development, 11*(18), 2323–2334. https://doi.org/10.1101/gad.11.18.2323

Nielsen, J. H., Svensson, C., Galsgaard, E. D., Møldrup, A., & Billestrup, N. (1999). β-cell proliferation and growth factors. *Journal of Molecular Medicine (berlin, Germany), 77*(1), 62–66. https://doi.org/10.1007/s001090050302

Pagliuca, F. W., & Melton, D. A. (2013). How to make a functional β-cell. *Development, 140*(12), 2472–2483. https://doi.org/10.1242/dev.093187

Pagliuca, F. W., Millman, J. R., Gürtler, M., Segel, M., Van Dervort, A., Ryu, J. H., Peterson, Q. P., Greiner, D., & Melton, D. A. (2014). Generation of functional human pancreatic β-cells in vitro. *Cell, 159*(2), 428–439. https://doi.org/10.1016/j.cell.2014.09.040

Pan, F. C., & Wright, C. (2011). Pancreas organogenesis: From bud to plexus to gland. *Developmental Dynamics, 240*(3), 530–565. https://doi.org/10.1002/dvdy.22584

Pan, G., Mu, Y., Hou, L., & Liu, J. (2019). Examining the therapeutic potential of various stem cell sources for differentiation into insulin-producing cells to treat diabetes. *Annales D'endocrinologie, 80*(1), 47–53. https://doi.org/10.1016/j.ando.2018.06.1084

Parnaud, G., Bosco, D., Berney, T., Pattou, F., Kerr-Conte, J., Donath, M. Y., Bruun, C., Mandrup-Poulsen, T., Billestrup, N., & Halban, P. A. (2008). Proliferation of sorted human and rat β-cells. *Diabetologia, 51*(1), 91–100. https://doi.org/10.1007/s00125-007-0855-1

Pathak, V., Pathak, N. M., O'Neill, C. L., Guduric-Fuchs, J., & Medina, R. J. (2019). Therapies for Type 1 diabetes: Current scenario and future perspectives. *Clinical Medicine Insights: Endocrinology and Diabetes, 12*, 1179551419844521. https://doi.org/10.1177/1179551419844521

Peng, B. Y., Dubey, N. K., Mishra, V. K., Tsai, F. C., Dubey, R., Deng, W. P., & Wei, H. J. (2018). Addressing stem cell therapeutic approaches in pathobiology of diabetes and its complications. *Journal of Diabetes Research, 2018*, 7806435. https://doi.org/10.1155/2018/7806435

Pictet, R. L., Clark, W. R., Williams, R. H., & Rutter, W. J. (1972). An ultrastructural analysis of the developing embryonic pancreas. *Developmental Biology, 29*(4), 436–467. https://doi.org/10.1016/0012-1606(72)90083-8

Piper, K., Brickwood, S., Turnpenny, L. W., Cameron, I. T., Ball, S. G., Wilson, D. I., & Hanley, N. A. (2004). Beta cell differentiation during early human pancreas development. *Journal of Endocrinology, 181*(1), 11–23. https://doi.org/10.1677/joe.0.1810011

Pittenger, M. F., Mackay, A. M., Beck, S. C., Jaiswal, R. K., Douglas, R., Mosca, J. D., Moorman, M. A., Simonetti, D. W., Craig, S., & Marshak, D. R. (1999). Multilineage potential of adult human mesenchymal stem cells. *Science, 284*(5411), 143–147. https://doi.org/10.1126/science.284.5411.143

Pradhan, G., Samson, S. L., & Sun, Y. (2013). Ghrelin: Much more than a hunger hormone. *Current Opinion in Clinical Nutrition and Metabolic Care, 16*(6), 619–624. https://doi.org/10.1097/MCO.0b013e328365b9be

Prado, C. L., Pugh-Bernard, A. E., Elghazi, L., Sosa-Pineda, B., & Sussel, L. (2004). Ghrelin cells replace insulin-producing beta cells in two mouse models of pancreas development. *Proceedings of the National Academy of Sciences USA, 101*(9), 2924–2929. https://doi.org/10.1073/pnas.0308604100

Raum, J. C., Gerrish, K., Artner, I., Henderson, E., Guo, M., Sussel, L., Schisler, J. C., Newgard, C. B., & Stein, R. (2006). FoxA2, Nkx2.2, and PDX-1 regulate islet β-cell-specific mafA expression through conserved sequences located between base pairs −8118 and −7750 upstream from the transcription start site. *Molecular and Cell Biology, 26*(15), 5735–5743. https://doi.org/10.1128/MCB.00249-06

Rezania, A., Bruin, J. E., Arora, P., Rubin, A., Batushansky, I., Asadi, A., O'Dwyer, S., Quiskamp, N., Mojibian, M., Albrecht, T., Yang, Y. H., Johnson, J. D., & Kieffer, T. J. (2014). Reversal of diabetes with insulin-producing cells derived in vitro from human pluripotent stem cells. *Nature Biotechnology, 32*(11), 1121–1133. https://doi.org/10.1038/nbt.3033

Röder, P. V., Wu, B., Liu, Y., & Han, W. (2016). Pancreatic regulation of glucose homeostasis. *Experimental and Molecular Medicine, 48*, e219. https://doi.org/10.1038/emm.2016.6

Rubio-Cabezas, O., Jensen, J. N., Hodgson, M. I., Codner, E., Ellard, S., Serup, P., & Hattersley, A. T. (2011). Permanent neonatal diabetes and enteric anendocrinosis associated with biallelic mutations in NEUROG3. *Diabetes, 60*(4), 1349–1353. https://doi.org/10.2337/db10-1008

Sakata, N., Yoshimatsu, G., & Kodama, S. (2019). Development and Characteristics of Pancreatic Epsilon Cells. *International Journal of Molecular of Science, 20*(8). https://doi.org/10.3390/ijms20081867

Sander, M., Neubüser, A., Kalamaras, J., Ee, H. C., Martin, G. R., & German, M. S. (1997). Genetic analysis reveals that PAX6 is required for normal transcription of pancreatic hormone genes and islet development. *Genes and Development, 11*(13), 1662–1673. https://doi.org/10.1101/gad.11.13.1662

Sarkar, S. A., Kobberup, S., Wong, R., Lopez, A. D., Quayum, N., Still, T., Kutchma, A., Jensen, J. N., Gianani, R., Beattie, G. M., Jensen, J., Hayek, A., & Hutton, J. C. (2008). Global gene expression profiling and histochemical analysis of the developing human fetal pancreas. *Diabetologia, 51*(2), 285–297. https://doi.org/10.1007/s00125-007-0880-0

Schwitzgebel, V. M. (2014). Many faces of monogenic diabetes. *Journal of Diabetes Investigation, 5*(2), 121–133. https://doi.org/10.1111/jdi.12197

Seaberg, R. M., Smukler, S. R., Kieffer, T. J., Enikolopov, G., Asghar, Z., Wheeler, M. B., Korbutt, G., & van der Kooy, D. (2004). Clonal identification of multipotent precursors from adult mouse pancreas that generate neural and pancreatic lineages. *Nature Biotechnology, 22*(9), 1115–1124. https://doi.org/10.1038/nbt1004

Sellick, G. S., Barker, K. T., Stolte-Dijkstra, I., Fleischmann, C., Coleman, R. J., Garrett, C., Gloyn, A. L., Edghill, E. L., Hattersley, A. T., Wellauer, P. K., Goodwin, G., & Houlston, R. S. (2004). Mutations in PTF1A cause pancreatic and cerebellar agenesis. *Nature Genetics, 36*(12), 1301–1305. https://doi.org/10.1038/ng1475

Seymour, P. A., Freude, K. K., Tran, M. N., Mayes, E. E., Jensen, J., Kist, R., Scherer, G., & Sander, M. (2007). SOX9 is required for maintenance of the pancreatic progenitor cell pool. *Proceedings of National Academy of Science USA, 104*(6), 1865–1870. https://doi.org/10.1073/pnas.0609217104

Seymour, P. A., Shih, H. P., Patel, N. A., Freude, K. K., Xie, R., Lim, C. J., & Sander, M. (2012). A Sox9/Fgf feed-forward loop maintains pancreatic organ identity. *Development, 139*(18), 3363–3372. https://doi.org/10.1242/dev.078733

Shahjalal, H. M., Abdal Dayem, A., Lim, K. M., Jeon, T. I., & Cho, S. G. (2018). Generation of pancreatic β-cells for treatment of diabetes: Advances and challenges. *Stem Cell Research and Therapy, 9*(1), 355. https://doi.org/10.1186/s13287-018-1099-3

Shen, C. N., Slack, J. M., & Tosh, D. (2000). Molecular basis of transdifferentiation of pancreas to liver. *Nature Cell Biology, 2*(12), 879–887. https://doi.org/10.1038/35046522

Shu, L., Zien, K., Gutjahr, G., Oberholzer, J., Pattou, F., Kerr-Conte, J., & Maedler, K. (2012). TCF7L2 promotes beta cell regeneration in human and mouse pancreas. *Diabetologia, 55*(12), 3296–3307. https://doi.org/10.1007/s00125-012-2693-z

Singal, D. P., & Blajchman, M. A. (1973). Histocompatibility (HL-A) antigens, lymphocytotoxic antibodies and tissue antibodies in patients with diabetes mellitus. *Diabetes, 22*(6), 429–432. https://doi.org/10.2337/diab.22.6.429

Smith, S. B., Qu, H. Q., Taleb, N., Kishimoto, N. Y., Scheel, D. W., Lu, Y., Patch, A. M., Grabs, R., Wang, J., Lynn, F. C., Miyatsuka, T., Mitchell, J., Seerke, R., Désir, J., Vanden Eijnden, S., Abramowicz, M., Kacet, N., Weill, J., Renard, M. E., … German, M. S. (2010). Rfx6 directs islet formation and insulin production in mice and humans. *Nature, 463*(7282), 775–780. https://doi.org/10.1038/nature08748

Sneddon, J. B., Tang, Q., Stock, P., Bluestone, J. A., Roy, S., Desai, T., & Hebrok, M. (2018). Stem cell therapies for treating diabetes: Progress and remaining challenges. *Cell Stem Cell, 22*(6), 810–823. https://doi.org/10.1016/j.stem.2018.05.016

Solis, M. A., Moreno Velásquez, I., Correa, R., & Huang, L. L. H. (2019). Stem cells as a potential therapy for diabetes mellitus: A call-to-action in Latin America. *Diabetology and Metabolic Syndrome, 11*, 20. https://doi.org/10.1186/s13098-019-0415-0

Solomon, B. D., Pineda-Alvarez, D. E., Balog, J. Z., Hadley, D., Gropman, A. L., Nandagopal, R., Han, J. C., Hahn, J. S., Blain, D., Brooks, B., & Muenke, M. (2009). Compound heterozygosity for mutations in PAX6 in a patient with complex brain anomaly, neonatal diabetes mellitus, and microophthalmia. *American Journal of Medical Genetics. Part A, 149A*(11), 2543–2546. https://doi.org/10.1002/ajmg.a.33081

Sorenson, R. L., & Brelje, T. C. (1997). Adaptation of islets of Langerhans to pregnancy: β-cell growth, enhanced insulin secretion and the role of lactogenic hormones. *Hormone and Metabolic Research, 29*(6), 301–307. https://doi.org/10.1055/s-2007-979040

Soria, B., Roche, E., Berná, G., León-Quinto, T., Reig, J. A., & Martín, F. (2000). Insulin-secreting cells derived from embryonic stem cells normalize glycemia in streptozotocin-induced diabetic mice. *Diabetes, 49*(2), 157–162. https://doi.org/10.2337/diabetes.49.2.157

Sosa-Pineda, B., Chowdhury, K., Torres, M., Oliver, G., & Gruss, P. (1997). The Pax4 gene is essential for differentiation of insulin-producing beta cells in the mammalian pancreas. *Nature, 386*(6623), 399–402. https://doi.org/10.1038/386399a0

Spence, J. R., Lange, A. W., Lin, S. C., Kaestner, K. H., Lowy, A. M., Kim, I., Whitsett, J. A., & Wells, J. M. (2009). Sox17 regulates organ lineage segregation of ventral foregut progenitor cells. *Developmental Cell, 17*(1), 62–74. https://doi.org/10.1016/j.devcel.2009.05.012

Steward, M. C., Ishiguro, H., & Case, R. M. (2005). Mechanisms of bicarbonate secretion in the pancreatic duct. *Annual Review of Physiology, 67*, 377–409. https://doi.org/10.1146/annurev.physiol.67.031103.153247

Stoffers, D. A., Zinkin, N. T., Stanojevic, V., Clarke, W. L., & Habener, J. F. (1997). Pancreatic agenesis attributable to a single nucleotide deletion in the human IPF1 gene coding sequence. *Nature Genetics, 15*(1), 106–110. https://doi.org/10.1038/ng0197-106

Sussel, L., Kalamaras, J., Hartigan-O'Connor, D. J., Meneses, J. J., Pedersen, R. A., Rubenstein, J. L., & German, M. S. (1998). Mice lacking the homeodomain transcription factor Nkx2.2 have diabetes due to arrested differentiation of pancreatic beta cells. *Development, 125*(12), 2213–2221.

Takahashi, K., & Yamanaka, S. (2006). Induction of pluripotent stem cells from mouse embryonic and adult fibroblast cultures by defined factors. *Cell, 126*(4), 663–676. https://doi.org/10.1016/j.cell.2006.07.024

Takahashi, K., Tanabe, K., Ohnuki, M., Narita, M., Ichisaka, T., Tomoda, K., & Yamanaka, S. (2007). Induction of pluripotent stem cells from adult human fibroblasts by defined factors. *Cell, 131*(5), 861–872. https://doi.org/10.1016/j.cell.2007.11.019

Takahashi, Y., Takebe, T., & Taniguchi, H. (2016). Engineering pancreatic tissues from stem cells towards therapy. *Regeneration Theraphy, 3*, 15–23. https://doi.org/10.1016/j.reth.2016.01.002

Tateishi, K., He, J., Taranova, O., Liang, G., D'Alessio, A. C., & Zhang, Y. (2008). Generation of insulin-secreting islet-like clusters from human skin fibroblasts. *Journal of Biological Chemistry, 283*(46), 31601–31607. https://doi.org/10.1074/jbc.M806597200

Teta, M., Rankin, M. M., Long, S. Y., Stein, G. M., & Kushner, J. A. (2007). Growth and regeneration of adult beta cells does not involve specialized progenitors. *Developmental Cell, 12*(5), 817–826. https://doi.org/10.1016/j.devcel.2007.04.011

Thakur, G., Lee, H. J., Jeon, R. H., Lee, S. L., & Rho, G. J. (2020). Small molecule-induced pancreatic β-like cell development: Mechanistic approaches and available strategies. *International Journal of Molecullar of Science, 21*(7). https://doi.org/10.3390/ijms21072388

Tögel, F., Weiss, K., Yang, Y., Hu, Z., Zhang, P., & Westenfelder, C. (2007). Vasculotropic, paracrine actions of infused mesenchymal stem cells are important to the recovery from acute kidney injury. *American Journal of Physiology. Renal Physiology, 292*(5), F1626-1635. https://doi.org/10.1152/ajprenal.00339.2006

Tosh, D., Shen, C. N., & Slack, J. M. (2002). Differentiated properties of hepatocytes induced from pancreatic cells. *Hepatology, 36*(3), 534–543. https://doi.org/10.1053/jhep.2002.35060

Tse, H. M., Kozlovskaya, V., Kharlampieva, E., & Hunter, C. S. (2015). Minireview: Directed differentiation and encapsulation of islet β-cells-recent advances and future considerations. *Molecular Endocrinology, 29*(10), 1388–1399. https://doi.org/10.1210/me.2015-1085

Turque, N., Plaza, S., Radvanyi, F., Carriere, C., & Saule, S. (1994). Pax-QNR/Pax-6, a paired box- and homeobox-containing gene expressed in neurons, is also expressed in pancreatic endocrine cells. *Molecular Endocrinology, 8*(7), 929–938. https://doi.org/10.1210/mend.8.7.7984154

Ueyama, H., Horibe, T., Hinotsu, S., Tanaka, T., Inoue, T., Urushihara, H., Kitagawa, A., & Kawakami, K. (2012). Chromosomal variability of human mesenchymal stem cells cultured under hypoxic conditions. *Journal of Cellular and Molecular Medicine, 16*(1), 72–82. https://doi.org/10.1111/j.1582-4934.2011.01303.x

Van Assche, F. A., Aerts, L., & De Prins, F. (1978). A morphological study of the endocrine pancreas in human pregnancy. *British Journal of Obstetrics and Gynaecology, 85*(11), 818–820. https://doi.org/10.1111/j.1471-0528.1978.tb15835.x

Voltarelli, J. C., Couri, C. E., Stracieri, A. B., Oliveira, M. C., Moraes, D. A., Pieroni, F., Coutinho, M., Malmegrim, K. C., Foss-Freitas, M. C., Simões, B. P., Foss, M. C., Squiers, E., & Burt, R. K. (2007). Autologous nonmyeloablative hematopoietic stem cell transplantation in newly diagnosed type 1 diabetes mellitus. *JAMA, 297*(14), 1568–1576. https://doi.org/10.1001/jama.297.14.1568

Walther, C., Guenet, J. L., Simon, D., Deutsch, U., Jostes, B., Goulding, M. D., Plachov, D., Balling, R., & Gruss, P. (1991). Pax: A murine multigene family of paired box-containing genes. *Genomics, 11*(2), 424–434. https://doi.org/10.1016/0888-7543(91)90151-4

Wang, S., Yan, J., Anderson, D. A., Xu, Y., Kanal, M. C., Cao, Z., Wright, C. V., & Gu, G. (2010). Neurog3 gene dosage regulates allocation of endocrine and exocrine cell fates in the developing mouse pancreas. *Developmental Biology, 339*(1), 26–37. https://doi.org/10.1016/j.ydbio.2009.12.009

Wang, L., Zhao, S., Mao, H., Zhou, L., Wang, Z. J., & Wang, H. X. (2011). Autologous bone marrow stem cell transplantation for the treatment of type 2 diabetes mellitus. *Chinese Medical Journal (engl), 124*(22), 3622–3628.

Wang, X., He, C., & Hu, X. (2014). LIM homeobox transcription factors, a novel subfamily which plays an important role in cancer (review). *Oncology Reports, 31*(5), 1975–1985. https://doi.org/10.3892/or.2014.3112

Wang, W., Jin, S., & Ye, K. (2017). Development of islet organoids from H9 human embryonic stem cells in biomimetic 3D scaffolds. *Stem Cells and Development, 26*(6), 394–404. https://doi.org/10.1089/scd.2016.0115

Wang, M., Yuan, Q., & Xie, L. (2018). Mesenchymal stem cell-based immunomodulation: Properties and clinical application. *Stem Cells and International, 2018*, 3057624. https://doi.org/10.1155/2018/3057624

Wang, C., Feng, N., Chang, F., Wang, J., Yuan, B., Cheng, Y., Liu, H., Yu, J., Zou, J., Ding, J., & Chen, X. (2019). Injectable cholesterol-enhanced stereocomplex polylactide thermogel loading chondrocytes for optimized cartilage regeneration. *Advance Healthcare Mater, 8*(14), e1900312. https://doi.org/10.1002/adhm.201900312

Watt, A. J., Zhao, R., Li, J., & Duncan, S. A. (2007). Development of the mammalian liver and ventral pancreas is dependent on GATA4. *BMC Developmental Biology, 7*, 37. https://doi.org/10.1186/1471-213X-7-37

Weiss, M. L., Anderson, C., Medicetty, S., Seshareddy, K. B., Weiss, R. J., VanderWerff, I., Troyer, D., & McIntosh, K. R. (2008). Immune properties of human umbilical cord Wharton's jelly-derived cells. *Stem Cells, 26*(11), 2865–2874. https://doi.org/10.1634/stemcells.2007-1028

Williams, J. A. (2001). Intracellular signaling mechanisms activated by cholecystokinin-regulating synthesis and secretion of digestive enzymes in pancreatic acinar cells. *Annual Review of Physiology, 63*, 77–97. https://doi.org/10.1146/annurev.physiol.63.1.77

Wu, X. H., Liu, C. P., Xu, K. F., Mao, X. D., Zhu, J., Jiang, J. J., Cui, D., Zhang, M., Xu, Y., & Liu, C. (2007). Reversal of hyperglycemia in diabetic rats by portal vein transplantation of islet-like cells generated from bone marrow mesenchymal stem cells. *World Journal of Gastroenterology, 13*(24), 3342–3349. https://doi.org/10.3748/wjg.v13.i24.3342

Xie, H., Wang, Y., Zhang, H., Qi, H., Zhou, H., & Li, F. R. (2013). Role of injured pancreatic extract promotes bone marrow-derived mesenchymal stem cells efficiently differentiate into insulin-producing cells. *PLoS ONE, 8*(9), e76056. https://doi.org/10.1371/journal.pone.0076056

Yoshitomi, H., & Zaret, K. S. (2004). Endothelial cell interactions initiate dorsal pancreas development by selectively inducing the transcription factor Ptf1a. *Development, 131*(4), 807–817. https://doi.org/10.1242/dev.00960

Youos, J. G. (2011). The role of α-, δ- and F cells in insulin secretion and action. *Diabetes Research and Clinical Practice, 93*(Suppl 1), S25-26. https://doi.org/10.1016/S0168-8227(11)70009-2

Zhang, H., Zhang, J., Pope, C. F., Crawford, L. A., Vasavada, R. C., Jagasia, S. M., & Gannon, M. (2010). Gestational diabetes mellitus resulting from impaired beta-cell compensation in the absence of FoxM1, a novel downstream effector of placental lactogen. *Diabetes, 59*(1), 143–152. https://doi.org/10.2337/db09-0050

Zhang, X., Ye, L., Hu, J., Tang, W., Liu, R., Yang, M., Hong, J., Wang, W., Ning, G., & Gu, W. (2012). Acute response of peripheral blood cell to autologous hematopoietic stem cell transplantation in type 1 diabetic patient. *PLoS ONE, 7*(2), e31887. https://doi.org/10.1371/journal.pone.0031887

Zhou, Q., & Melton, D. A. (2008). Extreme makeover: Converting one cell into another. *Cell Stem Cell, 3*(4), 382–388. https://doi.org/10.1016/j.stem.2008.09.015

Zhou, Q., & Melton, D. A. (2018). Author correction: Pancreas regeneration. *Nature, 560*(7720), E34. https://doi.org/10.1038/s41586-018-0294-9

Zhou, Q., Law, A. C., Rajagopal, J., Anderson, W. J., Gray, P. A., & Melton, D. A. (2007). A multipotent progenitor domain guides pancreatic organogenesis. *Developmental Cell, 13*(1), 103–114. https://doi.org/10.1016/j.devcel.2007.06.001

Zorn, A. M., & Wells, J. M. (2009). Vertebrate endoderm development and organ formation. *Annual Review of Cell and Developmental Biology, 25*, 221–251. https://doi.org/10.1146/annurev.cellbio.042308.113344

Zulewski, H., Abraham, E. J., Gerlach, M. J., Daniel, P. B., Moritz, W., Müller, B., Vallejo, M., Thomas, M. K., & Habener, J. F. (2001). Multipotential nestin-positive stem cells isolated from adult pancreatic islets differentiate ex vivo into pancreatic endocrine, exocrine, and hepatic phenotypes. *Diabetes, 50*(3), 521–533. https://doi.org/10.2337/diabetes.50.3.521

Chapter 8
Application of Stem Cells in Treatment of Bone Diseases: Pre-clinical and Clinical Perspectives

Mir Sadat-Ali

Abstract The word "Stem Cell" first appeared in the scientific literature in 1868. Stem cells are cells which have the ability to self-renew and give rise to differentiated cells. In 1960, McCullough and till reported that the living tissues came from stem cells and with the concept of self-renewal. In the twentieth century with the discovery of hESC, it was believed that stem cells will give potential therapies for the chronic human diseases. There was a flood of research in every field including orthopaedic surgery. As the Mesenchymal stem cells are able to develop into tissues including bone, cartilage, muscle, tendon, and ligament. Trials were instituted to treat non-unions, long-bone defects, spinal cord Injury, osteonecrosis of head of femur, spinal cord injury, osteochondral defects, osteoarthritis, rotator cuff injuries, and tendon and ligament ruptures. Stem cell therapy requires a clear comprehension of the orthopaedic disease process before clinicians embark on the new strategies to treat old diseases. It is also imperative that practicing clinicians to have a knowledge of different cell sources like autologous, allogeneic and iPSC, and the culture methods and their limitations.

It is also strongly recommended that orthopaedic surgeons should not give up the well-known recommended treatment modalities of treatment until stem cell therapy is proved safe, efficacious, and cost effective.

Keywords Orthopaedic surgery · Autologous · Allogeneic · Stem cells · Mesenchymal stem cells · Osteoblasts · Chondrocytes · Neurocytes

Introduction

In the last 2 decades, there has been enormous interests to treat chronic diseases through cellular therapy and tissue engineering. Two aspects were considered important visa vi the limit the cost of the care of the skeletal system in the aging population

M. Sadat-Ali (✉)
Department of Orthopaedic Surgery, College of Medicine, Imam Abdulrahman Bin Faisal University, Dammam, Saudi Arabia
e-mail: smali@iau.edu.sa

F. A. Khan (ed.), *Advances in Application of Stem Cells: From Bench to Clinics*, Stem Cell Biology and Regenerative Medicine 69,
https://doi.org/10.1007/978-3-030-78101-9_8

of the world and secondly to improve the quality of life. The process in the development took place in stages from 1800's to remove the diseased organs to 1960's replace the affected joints and ligaments, in 1980's the era of repair of the skeletal system, and from 2000's clinicians and researchers took the pathway to regenerate tissues. It was expected that by 2020, reproduction of the organs will take place which has not crossed the line of bench to the bedside. There are few specializations which are rapidly advancing in the field to utilize cellular therapy and orthopaedic and trauma surgery stands clearly ahead. This chapter will deal with the common chronic conditions of the musculoskeletal system highlighting the animal works to the transfer of technology to treatment human patients.

Short History of Stem Cell Development

Even though it was 1957, the first bone marrow transplant was performed to protect patients from after effects of radiation and chemotherapy (Thomas et al., 1957), but it was Friedenstein and associates (Friedenstein et al., 1966) and Tavassoli and Crosby (1968), who reported the osteogenic material in the bone marrow. The studies of Owen and Macpherson taught us. That osteoblasts precursors lined the inner layer of the periosteum (Owen, 1970; Owen & Macpherson, 1963). Recognizing and accepting the bone was a living organ, and it was shown that the role of bone marrow in the maintenance of hematopoiesis and hematopoietic stem cells (Dexter et al., 1973) showing that bone apart from being part of skeletal system giving attachment to muscles and ligaments for the body to move and protect and support vital organs had a major function of providing hematopoiesis and hematopoietic stem cells.

Caplan (1991) renamed the bone marrow Stroma as the bone marrow cells of the yester years to Mesenchymal stem cells which was well accepted by the scientific fraternity. It was brought to light that the MSCs had the potential to form osteoblasts the bone forming cells, chondrocytes, and adipocytes (Dominici et al., 2006). The reported work of Thomson and his colleagues in 1998 (Thomson et al., 1998) of achieving the isolation of human embryonic stem cells (hESC) pushed the research to treat diseases in a high gear, and many clinicians and general public believed that the treatment of many chronic untreatable diseases is around the corner. But the issues of the human embryos to provide the cells was still under cloud due to the ethical issues associated with hESCs when the major breakthrough came with the success of work of Takahashi and Yamanaka (2006) in creating induced pluripotent cells (iPSC) that can be generated directly from a somatic cells with the fast development in the field of stem cell and its potential use in chronic diseases has met with assumed cures and misuse of stem cell which failed to withstand serious scientific scrutiny.

Osteonecrosis of Head of Femur

Introduction

Avascular necrosis of the head of femur (ANFH) or osteonecrosis as it is termed is a common condition and type of osteonecrosis which occurs due to disruption of blood supply to the head of femur, and there are multiple causes which could be secondary to the trauma or diseases. In USA alone, yearly, there are 10,000–20,000 new cases are added with ANFH (Petek et al., 2019). Traumatic causes are due to the fracture neck of femur and hip dislocations, whereas diseases form the main bulk of ANFH. The common conditions which can be listed are chronic steroid use, alcohol consumption, and other risk factors include radiation therapy, chemotherapy HIV infection, Caisson disease, Gaucher's disease and in sickle cell hemeglobinopathy, and last but not the least idiopathic. Recovery of the ANFH without treatment does not occur and many patients go through with total hip arthroplasty (THA) for the relief of pain. Even though ANFH was first described in 1785, still we do not have a definite treatment for complete recovery of the head of femur (Tetik et al., 2011). Many methods have been tried to maintain the femoral head so that the final surgery of total hip arthroplasty can be avoided. Some of the procedures include bone grafting (Mont et al., 2007; Seyler et al., 2008), core decompression (Ficat et al., 1971; Lieberman, 2004; Lieberman et al., 2004; Mont et al., 2004), and electrical stimulation (Steinberg et al., 1989; Trancik et al., 1990). Extracorporeal shock wave therapy (ESWT) had initial success which reduced osteoblast apoptosis (Wang et al., 2005). But the results of these procedure were inconsistent. Long-term results of hip arthroplasty in young patients are below the norms due to the durability of implants. The results of the THR are usually unpredictable in this age group (Ince et al., 2006; Kim et al., 2011).

Pre-clinical Studies

In the recent past cell-based therapies, particularly mesenchymal stromal cells (MSC) for repair of damaged cartilage and relief of pain have been tried in experimental animals with excellent results. Abudusaimi et al. (2011) showed that direct transplantation of autologous adipose derived MSCs into an avascular area of the femoral head of the rabbit made new vessels to grow and new bone formation. Sun et al. (2011) reported that the results of forty rabbits which showed the core decompression stem cells injection in the avascular head of femur gave excellent results on histologic and histomorphometric analyses. They concluded that local transplantation of stem cells may prove an effective treatment option for steroid-induced osteonecrosis of the femur. The other reported animal studies had similar results (Aimaiti et al., 2011; Wen et al., 2012; Xie et al., 2012).

Clinical Studies

The concept of using stem cells via bone graft was attempted as early as 1987. Hernigou et al. (1997) treated a patient with SCD in with bone marrow concentrate in the affected area of the head of humerus and obtained with good results. Other researchers used bone marrow concentrate (BMC), bone marrow stromal cells (BMSCs), and MSCs were used (Lee et al., 2003; Li et al., 2011; Sen et al., 2012; Yoshioka et al., 2011). Hernigou and Beaujean (2002) used standard core decompression and BMC which produced excellent results at 60 months follow up and only 6.2% hips requiring total hip arthroplasty (THA). In 2012, Zhao et al. (2012) conducted a randomized control trail involving one hundred patients with early stage ANFH were recruited and randomly assigned to bone marrow derived mesenchymal stromal cells (BMMSC) treatment or core decompression (CD) treatment. At 5 years after the initial surgery, only 4% bone marrow derived MSC-treated hips progressed requiring further treatment. In CD group, 23% hips progressed and underwent further definite surgery. There were no complications in either groups. This intervention is safe and effective in preventing femoral head collapse, which require THA. In an extensive review on the subject Houdek et al. (2014) suggested that, CD combined with MSCs can provide significant pain relief, improvement in function, and ultimately halt the progression of AVN of the femoral head. Using this procedure, patients in young age can return to normal activities of daily living and avoid early hip arthroplasty. Piuzzi et al. (Persiani, 2015) performed an a systematic review with a level-of- evidence of III or higher evidence and reported that the avascular lesions in 24.5% after cellular therapy progressed compared with 40% in the controls. Ninety percent of studies that reported failure rates showed a lower THA conversion rate in the cell therapy group sixteen percent compared with the control group 21%. Sadat-Ali et al. (2017) reported in a small study in which they used osteoblasts injection derived from MSCs and found to me more effective as it healed all their patients who had grade 11 and III Ficat ANFH. In an extended study, the analysis of the 63 patients who had 5 million of osteoblasts injected at the site of the avascular lesion in patients <30 years with majority of female patients with of 49.05 ± 12.9 (range 24–60) months. The VAS and modified Harris hip score improved significantly ($p < 0.0001$). The Azam-Sadat score (ASS) for quality of life score for chronic diseases also significantly improved from 2.76 ± 0.49 preoperatively to 7.92 ± 0.09 ($p < 0.0001$) at 24 months. Overall 93.6% were satisfied with improved quality of life and only 4 (6.4%) the disease progressed and had to have total hip arthroplasty.

Conclusion

Reports of pre-clinical trials are few, and majority of the published data has concentrated on the clinical trials, and the literature gives a clear and more persuasive evidence to use cellular therapy in the ANFH. The type of cells used were the MSCs and osteoblast injection in the healing of the avascular lesions and postponement of the THA. Since randomized and comparative studies are available for review, it is safe to say that cellular therapy is the treatment of choice in young patients with ANFH so that they live a more normal life.

Stem Cell Therapy in Fracture Healing

Introduction

Fracture healing is a very complex process which involves local and general factors. The reported incidence of impaired healing leading to non-union is 5–10% (Mills et al., 2017; Zimmermann & Moghaddam, 2010). The cost treating a non-union ranges in different countries differently. In USA, the hospital costs for each non-union is $25,556, and in Great Britain, it costs £16,330 GBP (Antonova et al., 2013; Kanakaris & Giannoudis, 2007). Delayed healing is usually due to failure of the local cellular structures to react to the stimulation of the growth factors which are released at the site of the fractures. In the last 40 years, surgeons got a boost to heal fractures way of rigid internal fixation (Allen et al., 1968; Allgower & Speigel, 1979), but only realized later that with adequate fixations fractures also failed to unite. The second method developed to heal fractures was mechanical stimulation (Claes et al., 1998; Hadjiargyrou et al., 1998; Ryaby, 1998). In such cases, there is a growing need to find ways to regenerate the fracture site so that adequate healing occurs in time and MSCs has been suggested a promising option. MSCs are part of the bone marrow cells which are present in the cavity of the bone and are known to give rise to cells like osteoblasts, chondrocytes, and endothelial cells which take active part in bone deposition (Bruder et al., 1994; Granero-Molto et al., 2009; Muguruma et al., 2006).

Pre-clinical Studies

In the pre-clinical field, initial studies in smaller animals like rats were quite successful. It was demonstrated as early as 2009 that MSCs can induce a fracture healing in animals with increase in the callus formation and contributed in the enhancement of all the stages of fracture healing (Undale et al., 2011). Bruder et al. (1994) have shown that the healing of the fracture depends on the quantity of cells is also an important factor, hence, injecting large number of MSCs become imperative in the healing process. Undale et al. (2011) used human MSCs to heal ununited fractures in rats. Their results indicated that both type of cells one hESC-derived MSCs and hBM-MSCs, healed the fractures good and in better time. Other studies as well reported similar results (Connolly et al., 1991; Goel et al., 2005).

Sadat-Ali et al. (Kassem, 2013) used osteogeneic differentiated cells from the MSCs instead of eESC or MSCs and used in experimentally created non-union in rats and achieved better results.

Clinical Studies

The early studies of use of MSCs in the non-union of fractures came from utilizing bone marrow injections. Connolly et al. (1991) injected autologous marrow in patients non-union of tibial fractures and achieved union in 80% of patients. This study paved the way to use bone marrow injection without much realization that MSCs were being injected to heal the fractures. Many studies reported similar results of success in healing of the non-unions using bone marrow aspirate injections (Guimarães et al., 2014; Sugaya et al., 2014).

Bajada et al. (2007) treated patient with a recalcitrant a nine year tibial non-union using autologous MSCs with calcium sulfate pellets which healed the fracture, and Kim et al. (2009) performed a multicenter, randomized clinical study of two groups of patients with non-union of the fractures; one control and had osteoblast injections. Patients with osteoblasts injection united with good results.

Recently Senthilkumar et al. (2018) compared three groups of patients one with MSCs and bone marrow aspirate and a control group. In the MSCs group 92.3% of fractures united, in the bone marrow aspirate group 40% of fractures united indicating the MSCs are more potent in uniting the non-unions. In a recent meta-analysis on human studies, Palombella et al. (2019) reported the data on 347 patients who were treated with different modalities of the stem cells and found that within a year of follow up 81–100% union took place. They concluded that bone marrow concentrate and bone marrow derived mesenchymal stromal cells (BMSCs) with scaffolds could be considered as treatment choice to treat non-unions.

Conclusion

Reports of clinical trials which are available in the literature does not give a clear and more convincing guide to use the cellular therapy in the non-unions as the reported studies used different cells with and without scaffolds. We need more structured and prospective randomized studies to recommend routine use. In the absence of other definite and successful treatments, the use of stem cells have demonstrated potential in the healing of fractures and non-unions where natural healing mechanisms are inadequate, and large number of Stem cells are needed for the fracture unions. Autologous stem cells in the form of MSCs and osteoblasts does play a role in providing a safe, non-immunogenic cells which can heal the non-unions.

Stem Cell Therapy in Osteoarthritis of Knee

Introduction

Osteoarthitis of Knee (OAK) is mostly due to aging process and was suggested many factors influence the severity of the disease. It has been extrapolated that in 2020, globally incidence of OAK annually is 86.7 million individuals >20 years and older (Jordan & Croft, 2005). In 2013, it was assessed that medical costs for treating osteoarthritis in USA was $140 billion (Cui et al., 2020) in direct costs and each year 6000 die in USA each year are due to NSAID-related complications and costing additional $2 billion (Ledingham & Snowden, 2017; Jawad & Irving, 2007; Brabant & Stichtenoth, 2005). The management of the OAK has been by non-steroidal anti-inflammatory drugs (NSAIDs), physical therapy, and nutritional supplements (Bellamy et al., 2006; Bruyere & Reginster, 2007; Clouet et al., 2009; Peat et al., 2001; Quinn et al., 2018; Schuh et al., 2007). Even though the prevalence has increased but the no new treatments have been added with all the treatments, available OAK progresses slowly till the joint is destroyed, and quality life is severely affected. On the other side, there are some patients who show rapid deterioration even after adequate treatment and end up having joint replacement. In OA, the knee is the joint most commonly affected (Chevalier, 2010). Some patients only require non-steroidal anti-inflammatory drugs (NSAIDs) and physical therapy and certain group of patients the disease progresses leading to severe disability. Many patients do not respond to the conservative therapies and require steroid and hyaluronic acid injections, arthroscopic joint washout with varying degrees of pain relief (Brittberg et al., 1996; Caminal et al., 2014; Moseley, 2009).

Pre-clinical Studies

In the recent past cell-based therapies, particularly mesenchymal stromal cells (MSC) for repair of damaged cartilage and relief of pain have been tried in experimental animals with excellent results. Cells (MSC) for repair of damaged cartilage and relief of pain in rabbits and sheep (Chiang, 2005; Grigolo et al., 2009; Im et al., 2001; Rahfoth et al., 1998; Shah et al., 2018). MSCs transplantation was shown to grow cartilage similar to hyaline cartilage and a high type II collagen presence. The efficacy of mesenchymal in a porcine model showed regeneration of hyaline cartilage in 180 days (El-Tookhy et al., 2008). Shah et al. (2018) studied over 200 dogs diagnosed with degenerative arthritis with severe chronic pain and limited activity. Allogenic adipose derived MSCs were harvested and given either intra-articular or intravenous. In this report, over 85% of dogs improved significantly in the physical activity. The study in healthy dogs, OA was created by partial-thickness cartilage defect. The effect of intra-articular injection of autologous derived chondrocytes was compared with allogenic derived chondrocytes indicated that recovery of the damaged cartilage regenerated when compared with control groups (Goshima et al., 1991; Miki et al., 2015; Wakitani et al., 1994; Zhang et al., 2018).

Clinical Studies

Osteoarthritis of the knee (OAK) a very common degenerative disease for which there is no definite treatment for cure as the articular cartilage which is damaged could repair itself. In an aging knee, the chondrocytes behave in a different way; hence, complete repair does not take place. It was reported that MSCs from the bone marrow could replace the cartilage and bone, and this lead to the pre-clinical and clinical studies to treat OAK.[27–29] Wakitani et al. (2002) treated 24 patients with OAK used bone marrow aspirate and injected MSCs in the affected knees. They performed clinical and arthroscopy assessments. Their final conclusion was that autologous bone marrow derived MSCs have the ability to the repair osteoarthritic cartilage defects due to OAK in humans. Recently Jo et al. (2014) conducted a phase I/II A proof-of-concept clinical trial injected MSCs into the osteoarthritic knee. Post-injection analysis at 3, 6, and 12 months showed total relief of pain and better function of the knee joint with no adverse events. Soler et al. (2016) used autologous MSCs in patients with Grade II and III of Kellegren and Lawrence grading and found that an injection single intra-articular injection of the MSCs was safe and complete pain relief, improved quality of life up to 4 years and radiological signs of cartilage repair. A recent meta-analysis drew positive conclusions that MSCs could be treatment of choice to increase the function, reduce pain in knee OA. The findings of this review should be confirmed using methodologically rigorous and adequately powered clinical trials (Soler et al., 2016). Park et al. (2017) used allogenic hUCB-MSCs in patients with OAK and had a follow up for 7 years and concluded that even allogenic MSCs are safe to regenerate effected knees due to osteoarthritis. For long-term effective results, Invossa-K used allogenic chondrocytes with TGF-ß1 has been used with results by which TKR can be postponed for 5–7 years (Cho et al., 2017; Park et al., 2017). Lim et al. (2017) reported the use of "Cartistem" an allogeneic human umbilical cord blood-derived mesenchymal stem cells approved by the Korean FDA (KFDA) which also received US FDA clearance to conduct Phase III clinical trials in the USA. The Phase I and II trials reported safety and efficacy in the treatment of OAK. The final results of the follow up for 60 months are awaited. (NCT01041001).

Conclusion

Many different cellular therapies have been tried and reported, MSCs, chondrocytes, from Bone marrow, adipose tissue, autologous, allogeneic, umbilical cord blood, and different cell strength 2×10^6–5×10^7 cells per patient. Despite excellent work and results of articular cartilage regeneration under the influence of chondrocytes, there are very few FDA-approved which are undergoing extended clinical trials. The literature is full of published data which gives from excellent to very good results which convince us that cellular therapy have an important role to play in the reversal of degenerative cartilage I which should pave the way for routine treatment option in OAK.

Stem Cell Therapy in Meniscus Injuries and Ligament Injuries

Introduction

Meniscus and ligament injuries are common in the young mainly due to sports-related activities, and it was found that per year the cost range between $446 million to $1.5 billion and reaches $19.2 billion yearly (Lim et al., 2017). Ligaments and meniscus have limited ability to naturally heal, and it is this reason these injuries have poor functional outcome. Many therapies have been in trials conducted and tried to repair and enhance the healing. The treatment of such injuries is always surgical if the meniscus is removed completely as it used to happen before can accelerate joint degradation and cause secondary osteoarthritis of knee. The use of stem cell in other conditions has encouraged clinicians to use stem cells, to intensify healing close to normal of the injured meniscus and ligaments.

Pre-clinical Studies

Initially, animal studies were carried out for meniscus injuries in rats and rabbits using stem cell derived from synovial membrane proved to be detrimental in healing of the iatrogenic meniscus defects created (Hatsushika et al., 2013; Horie et al., 2009). Ruiz-Ibán et al. (2011) studied the effect of adipose derived MSCs on avascular area of the meniscus and concluded that adipose derived MSCs healed the smaller and larger lesions which were created. In another comparative study, iatrogenic tears were created microminipigs in the medial meniscus of both knees and sutured. In one knee, MSCs were injected and the other was kept as a control. The healing was evaluated for 3 months, and the results showed that the MSC's group had a significantly better healing in all the parameters examined in the injected group (Nakagawa et al., 2005) Hatsushika and his colleagues (2014) reported that in their study, they treated large defects in the porcine model, but the meniscus healed under the influence of multiple injections of the MSCs, but could not ascertained how many injections was needed. Ferris et al. (2014) studied horses with autologous bone marrow derived MSCs by intrarticular injection and assessed arthroscopically and reported that the recovery of the meniscus healing appeared in 75% of the animals. Kanaya et al. (2007) studied partial torn anterior cruciate ligament (ACL) injuries in Sprague-Dawley rats with intra-articular injection of bone marrow derived MSCs and found better healing of the ACL as compated to the control groups of rats. Similar outcomes were reported by Oe et al. (2011) used MSCs and found very similar results of excellent healing of the ACL repair under the cellular therapy. At 4 weeks, the assessment of stress tests and histological studies indicated normal findings as compared to the control group of animals.

Clinical Studies

Various clinical studies have shown that cellular therapy in the treatment of meniscus injuries shows promising results. Centeno and colleagues (2008) were the first report, where autologous mesenchymal stem cells was used to heal a torn meniscus. The result was astounding reduction of the pain, increased joint range of movements, and with healing of the meniscus. Al-Sayed et al. (2018) 16 patients with the mean of 34.8 ± 5.1 years with complete tears. The study reported that there was total reversal of pain, range of movement, and healing of the meniscus under the influence of chondrocytes.

Vangsness et al. (2014) performed a randomized study to study the effects of MSC injections into the knee after the medial meniscus was removed. A MRI done after a year showed significantly increase in the volume of the meniscus as compare to the control group. Recently couple of studies is smaller groups of patients confirmed the use of MSCs, and chondrocytes are effective in the treatment of meniscus repair. Onoi et al. (2019) arthroscopically looked pre- and post-injection of stem cells reported after six months showed improved meniscus status of repair. Sekiya et al. (2019) went one step further in confirmation of the efficacy of the stem cells in healing of the meniscus. Patients were followed up for 2 years clinically and a 3D MRI showed complete healing of the torn meniscus.

Conclusion

Reports in the English language literature of the pre-clinical studies demonstrated robustly the efficacy of the cellular therapy in the healing of meniscus and ACL. The same results were replicated in the clinical studies. Since there are different types of cells available, it is difficult to decide what to use and how much cells are to be given and how many times. One fact is proved that the autologous cellular therapy is safe and effective. More randomized control trials are needed, and based on the results of such studies, the repair of meniscus using cellular therapy can be labeled as standard of care.

Stem Cell Therapy in Management of Osteoporosis

Introduction

Osteoporosis is an ageing disease which is common all over the world and is a serious health issue as 200 million people suffer worldwide (Vijayakumar & Büsselberg, 2016). The end result of osteoporosis is fragility fractures, which increase morbidity and mortality. As of 2005, it was reported there were >2 million fractures, costing $18 billion, and it is estimated that by 2025, annual fractures and costs will increase to $30 billion (Burge et al., 2007). Osteoporosis causes fractures with a mortality of 15–30%, which is quite similar to many chronic diseases (Cooper et al., 2011). The pharmacological therapy is based on either anti-resorptives or anabolic agents (Fukumoto & Matsumoto, 2017). There are only 4–5 drugs which physicians have to control the disease and prevent fractures. The most common adverse events, particularly for oral bisphosphonates, upper esophageal causing irritation and bleeding, atrial fibrillation and renal failure, and excretion of the drug happens through the kidney patients with kidney disease cannot be used. Atypical femur fractures and bisphosphonate-related osteonecrosis of the jaw are serious complications of bisphosphonate use (Garg & Kharb, 2013; Pazianas & Abrahamsen, 2016). The anabolic agent has its own complications of cost, restricted use, and duration of use. There is need of an agent which should be inexpensive, efficacious, and free from routine complications. Millions of patients will benefit from the stem cell applications and osteoporosis is one of them, and research should be more focused on diseases like osteoporosis.

Pre-clinical Studies

Osteoporosis is one of the ten targeted diseases to be studied using stem cell therapy, but studies were slow to start on this chronic disease (Perry, 2000). Wang et al. (2006) initially showed in osteoporotic rabbits that injections of bone marrow derived MSCs can increase bone formation in the study group in comparison to the control group of animals. Ocarinao and his group (Wang et al., 2006) studied the effect of bone marrow derived MSCs (BMMSCs) in bilaterally ovariectomy induced Wistar rats. They injected 0.75 million cells in the femur and histology and histophotometric analysis revealed improved bone strength. They concluded that osteoporosis could be treated BMMSCs. Using adipose derived stromal cell therapy in rats proved that the injections prevents bone loss in ovariectomized mice (Ocarino et al., 2010). Kiernan et al. (Cho et al., 2012) reported that their study showed that transplanted MSCs led to better bone formation mice with low bone mass after a single injection of MSCs. Sadat-Ali et al. (2018) used autologous bone marrow derived osteoblasts in ovarectemized rats who had developed osteoporosis. Rats were injected osteoblasts in the tail veins and were euthanized at 8 weeks, and bone morphology was examined using high-resolution peripheral quantitative computerized tomography (HRpQCT). Results indicated that there are large quantity of the new bone in the study groups as compared to the control group of animals. Another study with direction of future research in osteoporosis compared three different cells MSCs, osteoblasts, and exosomes derived from osteoblasts in ovariectomy induced osteoporosis in rats. Results suggested that under the influence of osteoblasts, bone formation was significantly more than the other groups (Sadat-Ali et al., 2019).

Clinical Studies

Even though there is robust data available on animal studies, only two clinical trials have been instituted. The first study involving 8 patients used allogeneic mesenchymal cell from umbilical cord and assessment was made by visual analog scale, improvement in the range of motion, results of bone mass density, and improvement in patients quality of life (ClinicalTrials.gov Identifier: NCT01532076). The second ongoing trial is using autologous bone marrow derived MSCs in the range of 2–5 million cells/kg (ClinicalTrials.gov Identifier: NCT02566655).

Stem Cell Therapy in Spinal Cord Injury

Introduction

Spinal cord injury (SCI) is a major cause of paralysis in young, and the majority is due to motor vehicular accidents. Over two hundred and fifty thousand US citizens are suffering with spinal cord injury. Half of those injured become paraplegic and >40% quadriplegic. Over 80% of the injured are males, and 56% of injuries occur between the ages of 16 and 30 years (The University of Alabama National Spinal Cord Injury Statistical Center, 2002). The total direct costs for spinal cord injury are a staggering direct costs for all causes of SCI in the USA are $7.736 billion, solely for direct costs related to the injury (DeVivo, 1997). The estimated traumatic SCI occurs worldwide is annual incidence of 15–40 cases per million (Toma et al., 2005). Many preventive measures have been taken to reduce the SCI, but the incidence is increasing. To make matters difficult at present, we do not have effective therapies to reverse this disabling condition. Efforts to reverse this disabling injury have been tried with little or no success. There is plethora of studies which are ongoing in both the pre-clinical and clinical aspects and soon something positive is bound to happen.

Pre-clinical Studies

It was Koshizuka and colleagues (2004) first initiated the study on stem cells in spinal cord injured mice. They transplanted hematopoietic stem cell after a week of the injured cord. The assessment included the recovery of hind limb function, which showed a good recovery from 3 weeks of the cellular therapy. Histologically it showed that the injected HSCs from bone marrow differentiated into glial cells and neural precursors in the injured spinal cord. The mice were able to walk with partial weight on their hind legs, and in the control, mice could not walk nor could weight bear on the hind limbs. No weight bearing on their hind limbs. They concluded that the HSCs had the potential to help in the recovery of the injured spinal cord. Iwanami et al. (2005) studied primates with iatrogenic spinal cord injured and used the neural cells. After 2 months of transplantation, histologic analysis revealed that the grafted human NSPCs survived and differentiated into all the neural cells. Syková et al. (2006) injected mesenchymal stem cells (MSCs) in the treatment of spinal cord compression lesion in rats. The functional improvement was seen in MSC-treated rats. The conclusion of the study was that treatment with MSCs can improve had the behavioral outcome and histopathological assessment in rats. Following decade showed many such studies and very similar results showing the efficacy of the stem cells in the recovery of the injured spinal cord in animals (Hur et al., 2016; Kim et al., 2016; Morita et al., 2016; Watanabe et al., 2015; Yousefifard et al., 2016; Zhou et al., 2016). In another meta-analysis of 5628 animals with different experimental protocols indicate that with the use of allogeneic stem cells, the improvement after spinal cold injury is 25% (Antonic et al., 2013). Sadat-Ali et al. (2020) compared autologous bone marrow derived neurocytes versus rESC. The animals were assessed using Basso, Beattie, Brenham scoring, electromyographic studies, and histopathological analysis. The results indicated significant improvement in rats receiving rESC and autologous bone marrow derived neurocytes as compared to the control group. Comparison between autologous bone marrow derived neurocytes and rESC groups, the recovery in autologous bone marrow derived neurocytes was much better than rESC.

Clinical Studies

Based on the robust animal studies, Jeon et al. (2010) performed a clinical trial to test the efficacy of autologous MSCs therapy for spinal cord injury in humans. They assessed the recovery using electromyography, evoked potential, and magnetic resonance imaging (MRI). In 60% of the patients, there was improvement of the motor power and sensory changes. In the last 10 years, over 200 patients with acute and chronic spinal cord injured. Most of the patients received autologous MSCs, but the route of administration was from intra-thecal, epidural space, intra-lesional, and intravenous. Secondly, the quantity of the cells is ranged from 8 to 10 million MSCs.

Majority of patients improved in the functions in varying percentages, and there was no adverse reaction reported in these patients due to treatment. At present, over 10 clinical trials are ongoing in various countries in Phase I/II and III.

Conclusion

Stem cell therapy is an important modality which needs expeditious research to formulate a treatment protocol to treat spinal cord injured patients who remain paralyzed with low quality of life. The pre-clinical trials have shown great promise, but this could not yet be translated in the clinical trials. Clinical trials have shown moderate recovery and partial improvements in the life style of patients. Clinicians and researchers need to come together with clear strategy and conduct clinical trials without bias to achieve expectations of the patients. Clinical trials in healing of the injured spinal cord need to be more organized with clear cut protocols in type of cells, dosage of number of cells, route of administration, and how soon after the injury cellular therapy should be instituted.

References

Abudusaimi, A., Aihemaitijiang, Y., Wang, Y.-H., Cui, L., Maimaitiming, S., & Abulikemu, M. (2011). Adipose-derived stem cells enhance bone regeneration in vascular necrosis of the femoral head in the rabbit. *Journal of International Medical Research, 39*, 1852–1860.

Aimaiti, A., Saiwulaiti, Y., Saiyiti, M., Wang, Y.-H., Cui, L., & Yusufu, A. (2011). Therapeutic effect of osteogenically induced adipose derived stem cells on vascular deprivation-induced osteonecrosis of the femoral head in rabbits. *Chinese Journal of Traumatology, 14*, 215–220.

Allen, W. C., Pitrowski, G., Burstein, A. H., & Frankel, V. H. (1968). Biomechanical principles of intramedullary fixation. *Clinical Orthopaedics, 60*, 13–20.

Allgower, M., & Speigel, P. (1979). Internal fixation of fractures: Evolution of concepts. *Clinical Orthopaedics, 138*, 26–29.

AlSayed, H. N., Sadat-Ali, M., Uddin, F. Z., Alani, F. M., & Acharya, S. (2018). Outcome of bone marrow derived chondrocyte injection for meniscal injuries: A preliminary study. *Trends in Medicine, 18*(5), 2–4.

Antonic, A., Sena, E. S., Lees, J. S., Wills, T. E., Skeers, P., & Batchelor, P. E. et al. (2013). Stem cell transplantation in traumatic spinal cord injury: A systematic review and metaanalysis of animal studies. *PLoS Biology, 11*(12), e1001738. https://doi.org/10.1371/journal.pbio.1001738

Antonova, E., Le, T., Burge, R., & Mershon, J. (2013). Tibia shaft fractures: Costly burden of nonunions. *BMC Musculoskeletal Disorders, 14*, 42.

Assessed January 2021. https://www.pbs.org/newshour/economy/making-sense/the-steep-economic-cost-of-contact-sports-injuries

Bajada, S., Harrison, P. E., Ashton, B. A., Cassar-Pullicino, V. N., Ashammakhi, N., & Richardson, J. B. (2007). Successful treatment of refractory tibial nonunion using calcium sulphate and bone marrow stromal cell implantation. *Journal of Bone and Joint Surgery, 89*(10), 1382–1386.

Bellamy, N., Campbell, J., Robinson, V., Gee, T., Bourne, R., & Wells, G. (2006). Intra-articular corticosteroid for treatment of osteoarthritis of the knee. *Cochrane Database Systematic Review, 19*(2), CD005328.

Brabant, T., & Stichtenoth, D. (2005). Pharmacological treatment of osteoarthritis in the elderly. *Zeitschrift Fur Rheumatologie, 64*(7), 467–472.
Brittberg, M., Nilsson, A., Lindahl, A., et al. (1996). Rabbit articular cartilage defects treated with autologous cultured chondrocytes. *Clinical Orthopaedics and Related Research, 326*, 270–283.
Bruder, S. P., Fink, D. J., & Caplan, A. I. (1994). Mesenchymal stem cells in bone development, bone repair, and skeletal regeneration therapy. *Journal of Cellular Biochemistry., 56*, 283–294.
Bruyere, O., & Reginster, J. Y. (2007). Glucosamine and chondroitin sulfate as therapeutic Agents for knee and hip osteoarthritis. *Drugs and Aging, 24*(7), 573–580.
Burge, R., Dawson-Hughes, B., Solomon, D. H., Wong, J. B., King, A., & Tosteson, A. (2007). Incidence and economic burden of osteoporosis-related fractures in the United States, 2005–2025. *Journal of Bone and Mineral Research, 22*, 465–475.
Caminal, M., Fonseca, C., Peris, D., Moll, X., Rabanal, R. M., Barrachina, J., et al. (2014). Use of a chronic model of articular cartilage and meniscal injury for the assessment of long-term effects after autologous mesenchymal stromal cell treatment in sheep. *New Biotechnology, 31*, 492–498.
Caplan, A. L. (1991). Mesenchymal stem Cells. *Journal of Orthopaedic Research, 9*, 641–650.
Centeno, C. J., Busse, D., Kisiday, J., Keohan, C., Freeman, M., & Karli, D. (2008). Increased knee cartilage volume in degenerative joint disease using percutaneously implanted, autologous mesenchymal stem cells. *Pain Physician, 11*, 343–353.
Chevalier, X. (2010). Intraarticular treatments for osteoarthritis: New perspectives. *Current Drug Targets, 11*, 546–560.
Chiang, H., et al. (2005). Repair of porcine articular cartilage defect with autologous chondrocyte transplantation. *Journal of Orthopaedic Research, 23*(3), 584–593.
Cho, S. W., Sun, H. J., Yang, J. Y., Jung, J. Y., Choi, H. J., An, J. H., et al. (2012). Human adipose tissue-derived stromal cell therapy prevents bone loss in ovariectomized nude mouse. *Tissue Engineering Part A, 18*, 1067–1078.
Cho, J. J., Elmallah, R. K., Cherian, J. J., Kim, T. W., Lee, M. C., & Mont, M. A. (2015). A multi-center, single- blind, Phase IIA clinical trail to evaluate the efficacy and safety of a cell mediated gene therapy in degenerative arthritis patients. *Human Gene Therapy Clinical Development, 26*(2), 125–130. https://doi.org/10.1089/humc.2014.145 Epub 2015 Apr 17.
Cho, J., Kim, T., Kang, S., & Lee, B. (2017). A Phase III clinical results of InvossaTM: A clues for the potential disease modifying of a drug. *The J of Cell Therapy, 19*(5), S148.
Claes, L. E., Helgele, C. A., Neidlinger-Wilke, C., Kasper, D., Seidl, W., & Margevicius, K. J., et al. (1998). Effect of mechanical factors on the fracture healing process. *Clinical Orthopaedics, 355*(Suppl), 132–147.
ClinicalTrials.gov Identifier (NCT01532076). Assessed January 2021. www.clinicaltrials.gov
ClinicalTrials.gov Identifier: NCT02566655. Assessed January 2021. www.clinicaltrials.gov
Clouet, J., Vinatier, C., Merceron, C., Pot-vaucel, M., Maugars, Y., Weiss, P., et al. (2009). From oste- oarthritis treatments to future regenerative therapies for cartilage. *Drug Discovery Today, 14*, 913–925.
Connolly, J. F., Guse, R., Tiedeman, J., & Dehne, R. (1991). Autologous marrow injection as a substitute for operative grafting of tibial nonunions. *Clinical Orthopaedics and Related Research, 266*, 259–270.
Cooper, C., Cole, Z. A., Holroyd, C. R., Earl, S. C., Harvey, N. C., Dennison, E. M., et al. (2011). Secular trends in the incidence of hip and other osteoporotic fractures. *Osteoporosis International, 22*(5), 1277–1288.
Cui, A., Li, H., Wang, D., Zhong, J., Chen, Y., & Lu, H. (2020). Global, regional prevalence, incidence and risk factors of knee osteoarthritis in population-based studies. *E Clinical Medicine, 29*. https://doi.org/10.1016/j.eclinm.2020.100587. Assessed January 2020. www.cdc.gov
DeVivo, M. J. (1997). Causes and costs of spinal cord injury in the United States. *Spinal Cord, 35*, 809–813.
Dexter, T. M., Allen, T. D., lajtha, L. G., Schofield, R., & Bord, B. I. (1973). Stimulation of differentiation and pruliferation of haemopoietic cells in vitro. *Journal Cellular Physiology, 82*, 461–473.

Dominici, M., Le Blanc, K., Mueller, I., Slaper-Cortenbach, I., Marini, F., & Krause, D., et al. (2006) Minimal criteria for defining multipotent mesenchymal stromal cells. The Interna- tional Society for Cellular Therapy position statement. *Cytotherapy, 8*, 315–317.

El-Tookhy, Abou Elkheir, W., Mokbel, A., & Osman, A. (2008). Intra-articular injection of autologous mesenchymal stem cells in experimental chondral defects in dogs. *Egypt Rheumatologist, 30*, 1–10.

Ferris, D. J., Frisbie, D. D., Kisiday, J. D., Mcilwraith, C. W., Hague, B. A., Major, M. D., et al. (2014). Clinical outcome after intra-articular administration of bone marrow derived mesenchymal stem cells in 33 horses with stifle injury. *Veterinary Surgery, 43*, 255–265.

Ficat, P., Arlet, J., Vidal, R., Ricci, A., & Fournial, J. C. (1971). Therapeutic results of drill biopsy in primary osteonecrosis of the femoral head (100 cases). *Revue Du Rhumatisme Et Des Maladies Ostéo-Articulaires, 38*, 269–276.

Friedenstein, A. J., Piatetzky-Shapiro, I. I., & Petrakova, K. V. (1966). Osteogenesis in transplants of bone marrow cells. *Journal Imbryol Experimental Morphology, 16*, 881–890.

Fukumoto, S., & Matsumoto, T. (2017). Recent advances in the management of osteoporosis. *F1000Research, 6*, 625.

Garg, M. K., & Kharb, S. (2013). Dual energy X-ray absorptiometry: Pitfalls in measurement and interpretation of bone mineral density. *Indian Journal of Endocrinology and Metabolism, 17*, 203–210.

Goel, A., Sangwan, S. S., Siwach, R. C., & Ali, A. M. (2005). Percutaneous bone marrow grafting for the treatment of tibial non- union. *Injury, 36*(1), 203–206.

Goshima, J., Goldberg, V. M., & Caplan, A. I. (1991). The osteogenic potential of culture-expanded rat bone marrow mes- enchymal cells assayed in vivo in calcium phosphate ceramic blocks. *Clinical Orthopaedics, 262*, 298–311.

Granero-Molto, F., Weis, J. A., Miga, M. I., Landis, B., Timothy, J., & Myers, T. J., et al. (2009). Regenerative effects of transplanted mesenchymal stem cells in fracture healing. *Stem Cells, 27*(8), 1887–1898.

Grigolo, B., Lisignoli, G., Desando, G., Cavallo, C., Marconi, E., Tschon, M., Giavaresi, G., Fini, M., & Giardino, R. (2009). Osteoarthritis treated with mesenchymal stem cells on hyaluronan-based scaffold in rabbit. *Tissue Engineering Part c, Methods, 15*, 647–658.

Guimarães, J. A. M., Duarte, M. E. L., Fernandes, M. B. C., et al. (2014). The effect of autologous concentrated bone-marrow grafting on the healing of femoral shaft non-unions after locked intramedullary nailing. *Injury, 45*, S7–S13.

Hadjiargyrou, M., McLeod, K., Ryaby, J. P., & Rubin, C. (1998). Enhancement of fracture healing by low intensity ultrasound. *Clinical Orthopaedics, 355*(Suppl), 216–229.

Hatsushika, D., Muneta, T., Horie, M., Koga, H., Tsuji, K., & Sekiya, I. (2013). Intraarticular injection of synovial stem cells promotes meniscal regeneration in a rabbit massive meniscal defect model. *Journal of Orthopaedic Research, 31*, 1354–1359.

Hatsushika, D., Muneta, T., Nakamura, T., Horie, M., Koga, H., & Nakagawa, Y., et al. (2014). Repetitive allogeneic intraarticular injections of synovial mesenchymal stem cells promote meniscus regeneration in a porcine massive meniscus defect model. *Osteoarthritis Cartilage, 22*, 941–950.

Hernigou, P., & Beaujean, F. (2002). Treatment of osteonecrosis with autologous bone marrow grafting. *Clinical Orthopaedics and Related Research, 405*, 14–23.

Hernigou, P., Bernaudin, F., Reinert, P., Kuentz, M., & Vernant, J. P. (1997). Bone-marrow transplantation in sickle-cell disease. Effect on osteonecrosis: a case report with a four year follow-up. *Journal Bone Joint Surgery, 79*(11):1726–1730.

Horie, M., Sekiya, I., Muneta, T., Ichinose, S., Matsumoto, K., Saito, H., et al. (2009). Intra-articular injected synovial stem cells differentiate into meniscal cells directly and promote meniscal regeneration without mobilization to distant organs in rat massive meniscal defect. *Stem Cells, 27*, 878–887.

Houdek, M. T., Wyles, C. C., Martin, J. R., & Sierra, R. J. (2014). Stem cell treatment for avascular necrosis of the femoral head: current perspectives. Stem cells and cloning. *Advances and Applications, 7*, 65–70.

Hur, J. W., Cho, T. H., Park, D. H., Lee, J. B., Park, J. Y., & Chung, Y. G. (2016). Intrathecal transplantation of autologous adipose-derived mesenchymal stem cells for treating spinal cord injury: A human trial. *Journal of Spinal Cord Medicine, 39*(6), 655–664.

Im, G. I., Kim, D. Y., Shin, J. H., et al. (2001). Repair of cartilage defect in the rabbit with cultured mesenchymal stem cells from bone marrow. *Journal of Bone and Joint Surgery, 83*, 289–294.

Ince, A., Lermann, J., Göbel, S., Wollmerstedt, N., & Hendrich, C. (2006). No increased stem subsidence after arthroplasty in young patients with femoral head osteonecrosis: 41 patients followed for 1–9 years. *Acta Orthopaedica, 77*, 866–870.

Iwanami, A., Kaneko, S., Nakamura, M., Kanemura, Y., Mori, H., Kobayashi, S., et al. (2005). Transplantation of human neural stem cells for spinal cord injury in primates. *Journal of Neuroscience Research, 80*(2), 182–190.

Jawad, A. S., & Irving, K. (2007). Drug treatment modalities in patients with chronic osteoarthritis of the hip or knee. *Saudi Medical Journal, 28*(3), 375–378.

Jeon, S. R., Park, J., Lee, J. H., Kim, H. S., Sung, I. Y., Choi, G. H., et al. (2010). Treatment of spinal cord injury with bone marrow-derived, cultured autologous mesenchymal stem cells. *Tissue Engineering Regenerative Medicine, 7*, 316–322.

Jo, C. H., Lee, Y. G., Shin, W. H., Kim, H., Chai, J. W., Jeong, E. C., et al. (2014). Intra-articular injection of mesenchymal stem cells for the treatment of osteoarthritis of the knee: A proof-of-concept clinical trial. *Stem Cells, 32*(5), 1254–1266.

Jordan, K., & Croft, P. (2005). The prevalence and history of knee osteoarthritis in general practice: A case–control study. *Family Practice, 22*(1), 103–108.

Kanakaris, N., & Giannoudis, P. (2007). The health economics of long-bone non-unions. *Injury, 38*, S77–S84.

Kanaya, A., Deie, M., Adachi, N., Nishimori, M., Yanada, S., & Ochi, M. (2007). Intra-articular injection of mesenchymal stromal cells in partially torn anterior cruciate ligaments in a rat model. *Arthroscopy, 23*(6), 610–617.

Kassem, M. S. (2013). Percutaneous autogenous bone marrow injec- tion for delayed union or non-union of fractures after internal fixation. *Acta Orthopaedica Belgica, 2013*(79), 711–717.

Kim, S. J., Shin, Y. W., Yang, K. H., Kim, S. B., Yoo, M. J., Suk-Ku Han, S. K., et al. (2009). A multi-center, randomized, clinical study to compare the effect and safety of autologous cultured osteoblast (Ossron) injection to treat fractures. *BMC Musculoskeletal Disorders, 10*, 20. https://doi.org/10.1186/1471-2474-10-20

Kim, Y. H., Choi, Y., & Kim, J. S. (2011). Cementless total hip arthroplasty with alumina-on-highly cross-linked polyethylene bearing in young patients with femoral head osteonecrosis. *Journal of Arthroplasty, 26*, 218–223.

Kim, Y. C., Kim, Y. H., Kim, J. W., & Ha, K. Y. (2016). Transplantation of mesenchymal stem cells for acute spinal cord injury in rats: Comparative study between intralesional injection and scaffold based transplantation. *Journal of Korean Medical Science, 31*(9), 1373–1382.

Koshizuka, S., Okada, S., Okawa, A., Koda, M., Murasawa, M., Hashimoto, M., et al. (2004). Transplanted hematopoietic stem cells from bone marrow differentiate into neural lineage cells and promote functional recovery after spinal cord injury in mice. *Journal of Neuropathology and Experimental Neurology, 63*(1), 64–72.

Ledingham, J., & Snowden, N. (2017). Diagnosis and early management of inflammatory arthritis. *BMJ, 358*, j3248.

Lee, H. S., Huang, G. T., Chiang, H., Chiou, L. L., Chen, M. H., Hsieh, C. H., & Jiang, C. C. (2003). Multipotential mesenchymal stem cells from femoral bone marrow near the site of osteonecrosis. *Stem Cells, 21*(2), 190–199.

Li, Z. H., Liao, W., Cui, X. L., Zhao, Q., Liu, M., Chen, Y.-H., Liu, T.-S., Liu, N.-L., Wang, F., Yi, Y., & Shao, N.-S. (2011). Intravenous transplantation of allogeneic bone marrow mesenchymal stem cells and its directional migration to the necrotic femoral head. *International Journal of Medical Sciences, 8*(1), 74–83.

Lieberman, J. R. (2004). Core decompression for osteonecrosis of the hip. *Clinical Orthopaedics and Related Research, 418*, 29–33.

Lieberman, J. R., Conduah, A., & Urist, M. R. (2004). Treatment of osteonecrosis of the femoral head with core decompression and human bone morphogenetic protein. *Clinical Orthopaedics and Related Research, 429*, 139–145.

Lim, H. C., Lee, B.-G., Choi, J.-H., Jeong, H.-J., Chul-won Ha, C.-W., & Yoon, J.-R., et al. (2017). *Follow-up study of CARTISTEM® versus microfracture for the treatment of knee articular cartilage injury or defect—Full Text View—ClinicalTrials.gov*. Accessed 5 January 2021. https://clinicaltrials.gov/ct2/show/NCT01626677

Miki, S., Takao, M., Miyamoto, W., Matsushita, T., & Kawano, H. (2015). Intra-articular injection of synovium-derived mesenchymal stem cells with hyaluronic acid can repair articular cartilage defects in a canine model. *Journal Stem Cell Research Therapy, 5*, 1000314.

Mills, L. A., Aitken, S. A., & Simpson, A. (2017). The risk of non-union per fracture: Current myths and revised figures from a population of over 4 million adults. *Acta Orthopaedica, 88*, 434–439.

Mont, M. A., Ragland, P. S., & Etienne, G. (2004). Core decompression of the femoral head for osteonecrosis using percutaneous multiple small-diameter drilling. *Clinical Orthopaedics and Related Research, 429*, 131–138.

Mont, M. A., Marulanda, G. A., Seyler, T. M., Plate, J. F., & Delanois, R. E. (2007). Core decompression and nonvascularized bone grafting for the treatment of early stage osteonecrosis of the femoral head. *Instructional Course Lectures, 56*, 213–220.

Morita, T., Sasaki, M., Kataoka-Sasaki, Y., Nakazaki, M., Nagahama, H., Oka, S., et al. (2016). Intravenous infusion of mesenchymal stem cells promotes functional recovery in a model of chronic spinal cord injury. *Neuroscience, 335*, 221–231.

Moseley, B. (2009). Arthroscopic surgery did not provide additional benefit to physical and medical therapy for osteoarthritis of the knee. *Journal of Bone and Joint Surgery, 91*, 1281.

Muguruma, Y., Yahata, T., Miyatake, H., Sato, T., Uno, T., Itoh, J., et al. (2006). Reconstitution of the functional human hematopoietic microenvironment derived from human mesenchymal stem cells in the murine bone marrow compartment. *Blood, 107*, 1878–1887.

Nakagawa, Y., Muneta, T., Kondo, S., Mizuno, M., Takakuda, K., Ichinose, S., Tabuchi, T., Koga, H., Tsuji, K., & Sekiya, I. (2015). Synovial mesenchymal stem cells promote healing after meniscal repair in microminipigs. *Osteoarthritis and Cartilage, 23*, 1007–1017.

Ocarino, N. D., Boeloni, J. N., Jorgetti, V., Gomes, D. A., Goes, A. M., & Serakides, R. (2010). Intra-bone marrow injection of mesenchymal stem cells improves the femur bone mass of osteoporotic female rats. *Connective Tissue Research, 51*, 426–433.

Oe, K., Kushida, T., Okamoto, N., Umeda, M., Nakamura, T., Ikehara, S., et al. (2011). New strategies for anterior cruciate ligament partial rupture using bone marrow transplantation in rats. *Stem Cells Development, 20*(4), 671–679.

Onoi, Y., Hiranaka, T., Nishida, R., Takase, K., Fujita, M., & Hida, Y., et al. (2019). Second-look arthroscopic findings of cartilage and meniscus repair after injection of adipose-derived regenerative cells in knee osteoarthrits: Report of two cases. *Regenerative Therapy, 11*, 212–216.

Owen, M. (1970). The origin of bone cells. *International Review of Cytology, 28*, 213–238.

Owen, M., & Macpherson, S. (1963). Cell population kinetics of an osteogenic tissue. *II. J Cell Biol, 19*, 44–83.

Palombella, S., Lopa, S., Gianola, S., Zagra, L., Moretti, M., & Lovati, A. B. (2019). Bone marrow-derived cell therapies to heal long-bone nonunions: A systematic review and meta-analysis—Which is the best available treatment? *Stem Cells International*, Article ID 3715964, 12. https://doi.org/10.1155/2019/3715964

Park, Y. B., Ha, C. W., Lee, C. H., Yoon, Y. C., & Park, Y. G. (2017). Cartilage regeneration in osteoarthritic patients by a composite of allogeneic umbilical cord blood-derived mesenchymal stem cells and hyaluronate hydrogel: Results from a clinical trial for safety and proof-of-concept with 7 years of extended follow-up. *Stem Cells Translational Medicine, 6*(2), 613–621.

Pazianas, M., & Abrahamsen, B. (2016). Osteoporosis treatment: Bisphosphonates reign to continue for a few more years, at least? *Annals of the New York Acadamy of Sciences, 1376*, 5–13.

Peat, G., McCarney, R., & Croft, P. (2001). Knee pain and osteoarthritis in older adults: A review of the community burden and current use of primary health care. *Annals of the Rheumatic Diseases, 60*, 91–97.

Perry, D. (2000). Patients' voices: The powerful sound in the stem cell debate. *Science, 287*, 1423.

Persiani, P., De Cristo, C., Graci, J., Noia, G., Gurzì, M., & Villani, C. (2015). Stage-related results in treatment of hip osteonecrosis with core-decompression and autologous mesenchymal stem cells. *Acta Orthopaedica Belgica, 81*, 406–412.

Petek, D., Hannouche, D., & Suva, D. (2019). Osteonecrosis of the femoral head: Pathophysiology and current concepts of treatment. *EFORT Open Reviews, 4*(3), 85–97.

Quinn, R. H., Murray, J. N., Pezold, R., & Sevarino, K. S. (2018). Surgical management of osteoarthritis of the knee. *Journal of the American Academy Orthopaedics Surgery, 26*(9), e191–e19311; Sadat-Ali, M., Al-Habdan, I., & El-Hassan, A. Y. (2006). Is there an alternative to NSAIDs and Cox-2 inhibitors in the management of osteoarthritis of knee. *Ostetoporosis International, 17*(1), P192.

Rahfoth, B., Weisser, J., Sternkopf, F., et al. (1998). Transplantation of allograft chondrocytes embedded in agarose gel into cartilage defects of rabbits. *Osteoarthritis Cartilage, 6*, 50–65.

Ruiz-Ibán, M. N., Díaz-Heredia, J., García-Gómez, I., Gonzalez-Lizán, F., Elías-Martín, E., & Abraira, V. (2011). The effect of the addition of adipose-derived mesenchymal stem cells to a meniscal repair in the avascular zone: An experimental study in rabbits. *Arthroscopy, 27*, 1688–2196.

Ryaby, J. T. (1998). Clinical effects of electromagnetic and electric fields on fracture healing. *Clinical Orthopaedics, 355*(Suppl), 205–215.

Sadat-Ali, M., Q Azam, M. Q., Elshabouri, E. M., Tantawy, A. M., & Acharya, S. (2017). Stem cell therapy for avascular necrosis of femoral head in sickle cell disease: Report of 11 cases and review of literature. *International Journal of Stem Cells, 10*(2), 179–183.

Sadat-Ali, M., Al-Turki, H. A., Acharya, S., & Al-Dakheel, D. A. (2018). Bone Marrow-derived Osteoblasts in the Management of Ovariectomy induced Osteoporosis in Rats. *Journal of Stem Cells and Regenerative Medicine JSRM, 14*(2), 1–6.

Sadat-Ali, M., Al-Dakheel, D. A., AlMousa, S. A., AlAnii, F. M., Ebrahim, W., AlOmar, H. K., et al. (2019). Stem-cell therapy for ovariectomy-induced osteoporosis in rats: A comparison of three treatment modalities. *Stem Cells and Cloning: Advances and Applications, 12*, 27–48.

Sadat-Ali, M., Al-Dakheel, D. A., Ahmed, A., Al-Turki, H. A., Al-Omran, A. S., Acharya, S., et al. (2020). Spinal cord injury regeneration using autologous bone marrow- derived neurocytes and rat embryonic stem cells: A comparative study in rats. *World Journal of Stem Cells, 12*(12), 1591–1616.

Schuh, A., Jezussek, D., Fabijani, R., & Honle, W. (2007). Conservative therapy of knee osteoarthritis. *MMW Fortschritte Der Medizin, 149*, 31–32.

Sekiya, I., Koga, H., Otabe, K., Nakagawa, Y., Katano, H., Ozeki, N., et al. (2019). Additional use of synovial mesenchymal stem cell transplantation following surgical repair of a complex degenerative tear of the medial meniscus of the knee: A case report. *Cell Transplantation, 28*, 1445–1454.

Sen, R. K., Tripathy, S. K., Aggarwal, S., Marwaha, N., Sharma, R. R., & Khandelwal, N. (2012). Early results of core decompression and autologous bone marrow mononuclear cells instillation in femoral head osteonecrosis: A randomized control study. *Journal of Arthroplasty, 27*(5), 679–686.

Senthilkumar, V., Goel, S., & Gupta, K. K. (2018). Stem cells in fracture gap non union. *Orthopaedic Proceedings, 96*(B), No. SUPP11.

Seyler, T. M., Marker, D. R., Ulrich, S. D., Fatscher, T., & Mont, M. A. (2008). Nonvascularized bone grafting defers joint arthroplasty in hip osteonecrosis. *Clinical Orthopaedics and Related Research, 466*, 1125–1132.

Shah, K., Drury, T., Roic, I., Hansen, P., Malin, M., & Boyd, R., et al. (2018). Allogeneic adult stem cell therapy and other joint defects. *Stem Cells International*, Article ID 7309201, 7. https://doi.org/10.1155/2018/7309201

Soler, R., Orozco, L., Munar, A., Huguet, M., López, R., Vives, J., et al. (2016). Final results of a phase III trial using ex vivo expanded autologous Mesenchymal Stromal Cells for the treatment of osteoarthritis of the knee confirming safety and suggesting cartilage regeneration. *The Knee, 23*(4), 647–654.

Steinberg, M. E., Brighton, C.T., & Corces, A, et al. (1989). Osteonecrosis of the femoral head. Results of core decompression and grafting with and without electrical stimulation. *Clinical Orthopaedics Related Research, 249*, 199–208.

Sugaya, H., Mishima, H., Aoto, K., et al. (2014). Percutaneous autologous concentrated bone marrow grafting in the treatment for nonunion. *European Journal of Orthopaedic Surgery and Traumatology, 24*, 671–678.

Sun, Y., Feng, Y., Zhang, C., Cheng, X., Chen, S., Ai, Z., & Zeng, B. (2011). Beneficial effect of autologous transplantation of endothelial progenitor cells on steroid-induced femoral head osteonecrosis in rabbits. *Cell Transplantation, 20*, 233–243.

Syková, E., Jendelová, P., Urdzíková, L., Lesný, P., & Hejcl, A. (2006). Bone marrow stem cells and polymer hydrogels—Two strategies for spinal cord injury repair. *Cellular and Molecular Neurobiology, 26*, 1113–1129.

Takahashi, K., & Yamanaka, S. (2006). Induction of pluripotent stem cells from mouse embryonic and adult fibroblast cultures by defined factors. *Cell, 126*, 663–676.

Tavassoli, M., & Crosby, W. H. (1968). Transplantation of marrow to extramedullary sites. *Science, 161*, 54–56.

Tetik, C., Başar, H., Bezer, M., Erol, B., Ağir, I., & Esemenli, T. (2011). Comparison of early results of vascularized and non-vascularized fibular grafting in the treatment of osteonecrosis of the femoral head. *Acta Orthopaedica Et Traumatologica Turcica, 45*, 326–334.

The University of Alabama National Spinal Cord Injury Statistical Center. (2002).

Thomas, E. D., Lochte, H. I., Jr., Lu, W. C., & Ferrebee, J. W. (1957). Intravenous infusion of the bone marrow in patients receiving radiation and chemotherapy. *New England Journal of Medicine, 257*, 491–496.

Thomson, J. A., Itskovitz-Eldor, J., Shapiro, S. S., Waknitz, M. A., Swiergiel, J. J., Marshall, V. S., & Jones, J. M. (1998). Embryonic stem cell lines derived from human blastocysts. *Science, 282*, 1145–1147.

Toma, J. G., McKenzie, I. A., Bagli, D., et al. (2005). Isolation and characterization of multipotent skin-derived precursors from human skin. *Stem Cells, 23*, 727–737.

Trancik, T., Lunceford, E., & Strum, D. (1990). The effect of electrical stimulation on osteonecrosis of the femoral head. *Clinical Orthopaedics and Related Research, 256*, 120–124.

Undale, A., Fraser, D., Hefferan, T., Kopher, R. A., Herrick, J., Evans, G. L., et al. (2011). Induction of fracture repair by mesenchymal cells derived from human embryonic stem cells or bone marrow. *Journal of Orthopaedic Research, 29*(12), 1804–1811.

Vangsness, C. T., Farr, J., Boyd, J., Dellaero, D. T., Mills, C. R., & LeRoux-Williams, M. (2014). Adult human mesenchymal stem cells delivered via intra-articular injection to the knee following partial medial meniscectomy—A Randomized, Double-Blind, Controlled Study. *Journal of Bone and Joint Surgerys, 96*, 90–98.

Vijayakumar, R., & Büsselberg, D. (2016). Osteoporosis: An under-recognized public health problem: Local and global risk factors and its regional and worldwide prevalence. *Journal of Local and Global Health Science, 2*.

Wakitani, S., Goto, T.., Pineda, S. J., Young, R. G., Mansour, J. M., & Caplan, A. I., et al. (1994). Mesenchymal cell-based repair of large, full-thickness defects of articular cartilage. *The Journal of Bone and Joint Surgery, 76*(A), 579–592.

Wakitani, S., Imoto, K., Yamamoto, T., Saito, M., Murata, N., & Yoneda, M. (2002). Human autologous culture expanded bone marrow mesenchymal cell transplantation for repair of cartilage defects in osteoarthritic knees. *Osteoarthritis and Cartilage, 10*, 199–206.

Wang, C. J., Wang, F. S., Huang, C. C., Yang, K. D., Weng, L. H., & Huang, H. Y. (2005). Treatment for osteonecrosis of the femoral head: Comparison of extracorporeal shock waves with core decompression and bone-grafting. *Journal of Bone and Joint Surgery. American Volume, 87*, 2380–2387.

Wang, Z., Goh, J., De, S. D., Ge, Z., Ouyang, H., Chong, J. S., et al. (2006). Efficacy of bone marrow–derived stem cells in strengthening osteoporotic bone in a rabbit model. *Tissue Engineering, 12*, 1753–1761.

Watanabe, S., Uchida, K., Nakajima, H., Matsuo, H., Sugita, D., Yoshida, A., et al. (2015). Early transplantation of mesenchymal stem cells after spinal cord injury relieves pain hypersensitivity through suppression of pain- related signaling cascades and reduced inflammatory cell recruitment. *Stem Cells, 33*(6), 1902–1914.

Wen, Q., Jin, D., Zhou, C.-Y., Zhou, M.-Q., Luo, W., & Ma, L. (2012). HGF-transgenic MSCs can improve the effects of tissue self-repair in a rabbit model of traumatic osteonecrosis of the femoral head. *PLoS One, 7*, e37503. https://doi.org/10.1371/journal.pone.0037503

Xie, X.-H., Wang, X.-L., He, Y.-X., Liu, Z., Sheng, H., Zhang, G., & Qin, L. (2012). Promotion of bone repair by implantation of cryopreserved bone marrow-derived mononuclear cells in a rabbit model of steroid-associated osteonecrosis. *Arthritis and Rheumatism, 64*, 1562–1571.

Yoshioka, T., Mishima, H., Akaogi, H., Sakai, S., Li, M., & Ochiai, N. (2011). Concentrated autologous bone marrow aspirate transplantation treatment for corticosteroid-induced osteonecrosis of the femoral head in systemic lupus erythematosus. *International Orthopaedics, 35*(6), 823–829.

Yousefifard, M., Nasirinezhad, F., Shardi Manaheji, H., Janzadeh, A., Hosseini, M., & Keshavarz, M. (2016). Human bone marrow-derived and umbilical cord-derived mesenchymal stem cells for alle- viating neuropathic pain in a spinal cord injury model. *Stem Cell Research & Therapy, 7*, 36.

Zhang, B. Y., Wang, B. Y., Li, S. C., Luo, D. Z., Zhan, X., Chen, S. F., Chen, Z. S., Liu, C. Y., Ji, H. Q., Bai, Y. S., Li, D. S., & He, Y. (2018). Evaluation of the curative effect of umbilical cord mesenchymal stem cell therapy for knee arthritis in dogs using imaging technology. *Stem Cells International, 2018*, 1983025.

Zhao, D., Cui, D., Wang, B., Tian, F., Guo, L., Yang, L., Liu, B., & Yu, X. (2012). Treatment of early stage osteonecrosis of the femoral head with autologous implantation of bone marrow- derived and cultured mesenchymal stem cells. *Bone, 50*, 325–330.

Zhou, H. L., Zhang, X. J., Zhang, M. Y., Yan, Z. J., Xu, Z. M., & Xu, R. X. (2016). Transplantation of human amniotic mesenchymal stem cells promotes functional recovery in a rat model of traumatic spinal cord injury. *Neurochemical Research, 41*(10), 2708–2718.

Zimmermann, G., & Moghaddam, A. (2010). Trauma: Non-union: New trends. *European Instructional Lectures, 10*, 10–19.

Chapter 9
Stem Cells in Regenerative Medicine: Clinical Trials

Firdos Alam Khan, Razan Aldahhan, and Noor Alrushaid

Abstract Stem cells have been extensively used in tissue repair and cell tissue engineering in preclinical and clinical conditions. One of the stem cells' characteristics is to provide an ample number of cells for transplantation purposes. Although embryonic stem cells (ESCs) are considered a potential source of stem cells, due to ethical concerns, there is not much progress in using ESCs in clinical conditions. The application of adult mesenchymal stem cells (MSCs) derived from bone marrow, and the umbilical cord is extensively used to treat many degenerative diseases. Many clinical trials have been successfully done in the last few years. Moreover, many clinical trials with different phases are under progress where adult MSCs are transplanted in patients with different degenerative diseases. More than 5000 registered clinical trials are in progress as per the ClinicalTrials.gov where stem cells have been applied in disease conditions.

Keywords Stem cells therapy · Regenerative medicine · Cell therapy · Clinical trials

Introduction

Stem cell therapy, also known as regenerative medicine, promotes the repair response of diseased, dysfunctional, or injured tissue using stem cells or their derivatives. One of the main requirements for cell-based therapy is to test the cells in different phases in humans, called clinical trials. There are four different clinical trial phases, such as Phase-1, Phase-2, Phase-3, and Phase-4 trials, respectively. These clinical trials are performed in healthy persons and patients, which aim to evaluate potential benefits. Such trials' primary objectives are to find out if the cell-based therapy is safe (healthy person) and therapeutically effective (patients). Often, clinical trials are used to learn if a new cell-based therapy is more effective or has less harmful side effects. In the

F. A. Khan (✉) · R. Aldahhan · N. Alrushaid
Department of Stem Cell Biology, Institute for Research and Medical Consultations, Imam Abdulrahman Bin Faisal University, Post Box No. 1982, Dammam 31441, Saudi Arabia
e-mail: fakhan@iau.edu.sa

F. A. Khan (ed.), *Advances in Application of Stem Cells: From Bench to Clinics*, Stem Cell Biology and Regenerative Medicine 69,
https://doi.org/10.1007/978-3-030-78101-9_9

year 2014, cell-based therapies clinical trials were either sponsored by academic institutions roughly constitute 70% of total clinical trials, and 30% of clinical trials were sponsored by private companies (Bersenev, 2015). It has been observed that a combination of public and private funding was strongly encouraged in the clinical trials conducted by the Californian Institute for Regenerative Medicine, California, USA (Trounson et al., 2010). The investments for stem cell clinical trials have been heavily based on the successful and positive clinical trial results. The primary concern of any investments toward stem cell-based clinical trials is to get clear information about the trial's outcome. Over the past few years, attempts have been made to better evaluate and monitor the stem cell-based trials at different intervals. It has often been observed that investors have retracted the funding due to non-getting satisfactory trial results. The outcome of clinical trials must be shared with investors with the technical team to decide whether to continue the trials.

There have been reports on clinical trials (Ratcliffe et al., 2013; Trounson et al., 2011), where different stem cells have been used in the clinical trials. Most of the publicly available data on stem cell-based therapy trials either available at the National Institute of Health, USA, and European databases. The outcome of stem cell-based clinical trials has also been published in peer-reviewed journals, and they are available at the National Center for Biotechnology Information, USA. However, some clinical trial data are not available in the public domain. As stem cell-based therapy is a new therapy, it is essential to have the outcome of all clinical trials to be able to develop effective cell-based therapy. Among different stem cells, limbal stem cells and neural stem cells showed significant promising results in regenerative repair. The application of pluripotent stem cells and mesenchymal stem cells (MSCs) is the most widely used cell types for clinical trials. There is also interest in placental-derived stem cells in regenerative medicine. The application of stem cell-based therapy is mostly used in eye-related degenerative diseases. One of the reasons is that small numbers of cells required in the eye regeneration, and tissue is easily accessible for surgery.

Concerning the type of stem cell transplantation, autologous stem cell transplantation is most widely used as there are many benefits of using autologous transplantation such as easily accepted by the body, no immuno rejection issue, and cell survivability and functionality are better. In the case of allogeneic cell transplantation, few clinical trials are conducted, but due to high immuno rejection and low cell acceptability, allogeneic cell transplantation is less favored than autologous cell transplantation. Different types of stem cells, such as pluripotent stem cells and induced pluripotent stem cells, are extensively used in clinical trials (Cyranoski, 2014; Kushner et al., 2014; Schwartz et al., 2012, 2015).

Clinical Trials of Bone and Muscle Diseases Using Stem Cell Therapy

Stem cells have been tested as the treatment for various bone and muscle-related diseases. Many clinical trials are currently in progress or have been completed where stem cells have been used in bone and muscle disease. The summary of clinical trials is listed in the tables.

Bone Disease and Regenerative Medicine

Forty clinical trials are being performed using different types of stem cells to treat bone diseases as per the USA clinical trials portals (https://www.clinicaltrials.gov/) and (https://stemcellsportal.com/stem-cells-translational-medicine-clinical-trials-portal). The list of trial which has been conducted during the year 2009–2020 is listed in Table 9.1.

Clinical Trials in Joint Diseases and Stem Cell Therapy

As per the clinical trial record, more than 110 clinical trials are being performed using different types of stem cells to treat joint diseases as per the USA clinical trials portals (https://www.clinicaltrials.gov/) and (https://stemcellsportal.com/stem-cells-translational-medicine-clinical-trials-portal). The list of trial which has been conducted during the year 2019–2020 is listed in Table 9.2.

Musculoskeletal Disease and Stem Cell Therapy

One hundred sixty clinical trials are being conducted using different stem cells to treat musculoskeletal diseases as per the USA clinical trials portals (https://www.clinicaltrials.gov/) and (https://stemcellsportal.com/stem-cells-translational-medicine-clinical-trials-portal). The list of trial which has been conducted during the year 2020 is listed in Table 9.3.

Table 9.1 Bone disease and stem cell therapy

Title of trial	Disease	Type of stem cells	Year
Evaluation of clinical and bone density improvement after implantation of allogeneic mesenchymal stem cell from umbilical cord on osteoporosis patients	Osteoporosis	Mesenchymal stem cell	2020
Effectiveness and safety of mesenchymal stem cell (MSc) implantation on degenerative disc disease patients	Degenerative disc disease, low back pain; disc degeneration	Mesenchymal stem cell + NaCl 0.9% 2 ml	2020
Effectivity of mesenchymal stem cell on vertebral bone defect due to mycobacterium tuberculosis infection	Spinal tuberculosis	Combination product: mesenchymal stem cell + NaCl 0.9%	2020
Human umbilical cord mesenchymal stem cells for the treatment of lumbar disc degeneration disease	Lumbar disc degeneration; lumbar disc herniation	Human umbilical cord mesenchymal stem cells	2020
Efficacy in alveolar bone regeneration with autologous MSCs and biomaterial in comparison to autologous bone grafting	Alveolar bone atrophy interventions:	Advanced medicinal therapy (MSCs combined with biomaterial); procedure: autologous bone graft	2020
The study of early stage osteonecrosis of femoral head with human umbilical cord mesenchymal stem cells	Osteonecrosis of femoral head	Drug: Allogeneic umbilical cord mesenchymal stem cells	2017
Repairing the defect of intervertebral disc with autologous BMSC and gelatin sponge after microendoscopic discectomy for lumbar disc herniation	Lumbar disc herniation	Autologous bone marrow stem cell (BMSC)/gelatin sponge	2016
Interbody spacers with map3® cellular allogeneic bone graft in anterior or lateral lumbar interbody fusion	Spondylosis, degenerative disc disease	Cellular allogeneic bone graft	2015
Role of mesenchymal stem cells in fat grafting	Romberg's disease, craniofacial microsomia; lipodystrophy; mixed connective tissue disease	Fat graft enriched with ex vivo expanded stem cells, procedure: Fat graft not enriched with ex vivo expanded stem cells	2015

(continued)

Table 9.1 (continued)

Title of trial	Disease	Type of stem cells	Year
Clinical trial of intravenous infusion of fucosylated bone marrow mesenchymal cells in patients with osteoporosis	Osteoporosis, spinal fractures	Fucosylated MSC for osteoporosis	2015
Clinical trial of use of autologous bone marrow stem cells seeded on porous tricalcium phosphate matrix and demineralized bone matrix in patients with osteonecrosis	Osteonecrosis	MSC construct for osteonecrosis	2015
Autologous adipose-derived stem cell therapy for intervertebral disc degeneration	Low back pain intervention	Autologous adipose-derived mesenchymal stem cell	2015
Phase 2a study on allogeneic osteoblastic cells implantation in lumbar spinal fusion	Degenerative disc disease intervention	ALLOB® cells with ceramic scaffold	2014
Mesenchymal stem cell based therapy for the treatment of osteogenesis imperfecta	Osteogenesis imperfecta intervention	Mesenchymal stem cells	2014
Adipose cells for degenerative disc disease	Degenerative disc disease	Adipose stem cells	2014
Evaluation of mesenchymal stem cells to treat avascular necrosis of the hip	Avascular necrosis of the femoral head	Cultured autologous mesenchymal cells	2014
Autologous stem cells in achilles tendinopathy	Achilles tendinitis, degeneration	Biological: autologous mesenchymal stem cells	2014
Phase II clinical trial prospective, open, nonrandomized treatment of osteonecrosis of the femoral head by the administration of autologous mesenchymal stem cells	Osteonecrosis of the femoral head	Bone marrow aspirate	2012
Autologous adipose tissue-derived mesenchymal stem cells transplantation in patient with lumbar intervertebral disc degeneration	Lumbar intervertebral disc degeneration	Autologous adipose tissue-derived MSCs transplantation	2012

(continued)

Table 9.1 (continued)

Title of trial	Disease	Type of stem cells	Year
Mesenchymal stem cells in osteonecrosis of the femoral head	Avascular necrosis of femur head	XCEL-MT-OSTEO-ALPHA	2012
Safety study of mesenchymal stem cells and spinal fusion	Lumbar spondylolisthesis involving L4-L5, degenerative discopathy involving L4-L5	XCEL-MT-OSTEO-ALPHA	2012
Safety and efficacy study of umbilical cord/placenta-derived mesenchymal stem cells to treat ankylosing spondylitis (AS)	Ankylosing spondylitis	Human umbilical cord-derived MSCs	2011
The use of autologous bone marrow mesenchymal stem cells in the treatment of articular cartilage defects	Degenerative arthritis; chondral defects; osteochondral defects	Bone marrow mesenchymal stem cell implantation	2009

Neuromuscular Disease and Stem Cell Therapy

There are 40 clinical trials being conducted using different types of stem cells to treat neuromuscular diseases as per the USA clinical trials portals (https://www.clinicaltrials.gov/) and (https://stemcellsportal.com/stem-cells-translational-medicine-clinical-trials-portal). The trial list conducted during the year 2017–2020 is listed in Table 9.4.

Clinical Trials of Brain Diseases and Stem Cell Therapy

Stem cells have been tested as the treatment for various brain-related diseases and injuries. Many clinical trials are currently in progress or have been completed where stem cells have been used in bone and muscle disease as per the USA clinical trials portals.

Table 9.2 List of clinical trials in joint diseases

Title of trial	Disease	Type of stem cells	Year
Phase 2B clinical study of chondrogen for treatment of knee osteoarthritis	Osteoarthritis, knee	Chondrogen	2020
Autologous adipose tissue-derived mesenchymal stem cells (AdMSCs) for osteoarthritis	Osteoarthritis, knee; osteoarthritis, osteoarthritis shoulder	Biological: Celltex-AdMSCs	2020
Allogeneic BM-MSCs in patients with lumbar facet arthropathy	Facet-mediated low back pain	Single bilateral intra-articular injection of allogeneic BM-MSCs for lumbar facet joint arthropathy	2020
Bone marrow versus adipose autologous mesenchymal stem cells for the treatment of knee osteoarthritis	Osteoarthritis knee	Biological: bone marrow mesenchymal stem cells; biological: adipose Mesenchymal stem cells; biological: bone marrow and adipose mesenchymal stem cells injection	2020
Comparative effectiveness of arthroscopy and non-arthroscopy using mesenchymal stem cell therapy (MSCs) and conditioned medium for osteoarthritis	Osteoarthritis, knee	Mesenchymal stem cells with arthroscopy; biological: mesenchymal stem cells without arthroscopy	2020
Study on shoulder arthritis treatment with intra-articular injections of autologous bone marrow aspirate	Shoulder arthritis	Marrow cellution system	2020
Clinical study of intra-articular injection of catholic MASTER cell (bone marrow-derived mesenchymal stem cell) in knee osteoarthritis	Osteoarthritis, knee	Bone marrow-derived mesenchymal stem cell	2020
Adipose-derived biocellular regenerative therapy for osteoarthritis	Osteoarthritis; osteoarthritis knee; osteoarthritis shoulders; osteoarthritis of multiple joints; osteoarthritis, hip; osteoarthritis—ankle/foot	Tissue stromal vascular fraction	2020

(continued)

Table 9.2 (continued)

Title of trial	Disease	Type of stem cells	Year
Treatment of osteoarthritic knee with high tibial osteotomy and implantation of allogeneic human umbilical cord blood-derived stem cells	Osteoarthritis, knee	Umbilical cord blood stem cell implantation for osteoarthritis treatment	2020
Treatment of knee osteoarthritis with autologous adipose-derived mesenchymal stem cells	Knee osteoarthritis	Autologous adipose-derived mesenchymal stem cells	2019
Clinical study of pulp mesenchymal stem cells in the treatment of primary mild to moderate knee osteoarthritis	Knee osteoarthritis	Mesenchymal stem cell	2019
The evaluation of safety and effectiveness of intra-articular administration of autologous stromal vascular fraction of adipose tissue cells for treatment of knee joint arthritis	Knee osteoarthritis intervention	Stromal vascular fraction	2019
Mesenchymal stem cell transplantation for osteoarthritis	Knee osteoarthritis	Autologous BMSCs plus autologous PRP	2019
Treatment of early knee osteoarthritis with autologous adipose-derived mesenchymal stem cells	Knee osteoarthritis	Autologous adipose-derived mesenchymal stem cells	2019
Effectiveness of autologous adipose-derived stem cells in the treatment of knee cartilage injury	Knee osteoarthritis; cartilage degeneration	Autologous adipose-derived mesenchymal stem cell	2019
Multicenter trial of stem Cell Therapy for osteoarthritis (MILES)	Osteoarthritis	Autologous bone marrow concentrate biological: adipose-derived stromal vascular fraction (SVF); biological: umbilical cord tissue (UCT)	2019
Implantation of allogeneic mesenchymal stem cell from umbilical cord blood for osteoarthritis management	Osteoarthritis, knee	Umbilical cord mesenchymal stem cell; biological: recombinant human somatropin	2019

Table 9.3 List of clinical trials in musculoskeletal disease

Title of trial	Disease	Type of stem cells	Year
Evaluation of clinical and bone density improvement after implantation of allogeneic mesenchymal stem cell from umbilical cord on osteoporosis patients	Osteoporosis	Mesenchymal stem cell	2020
Human umbilical cord mesenchymal stem cells for the treatment of lumbar disc degeneration disease	Lumbar disc degeneration; lumbar disc herniation	Human umbilical cord mesenchymal stem cells	2020
Allogeneic BM-MSCs in patients with lumbar facet arthropathy	Facet-mediated low back pain	Single bilateral intra-articular injection of allogeneic BM-MSCs for lumbar facet joint arthropathy	2020
Bone marrow versus adipose autologous mesenchymal stem cells for the treatment of knee osteoarthritis	Condition: osteoarthritis knee injection	Bone marrow mesenchymal stem cells; biological: adipose mesenchymal stem cells; biological: bone marrow and adipose mesenchymal stem cells injection	2020
Comparative effectiveness of arthroscopy and non-arthroscopy using mesenchymal stem cell therapy (MSCs) and conditioned medium for osteoarthritis	Osteoarthritis, knee	Mesenchymal stem cells	2020
Efficacy in alveolar bone regeneration with autologous MSCs and biomaterial in comparison to autologous bone grafting	Alveolar bone atrophy	Advanced medicinal therapy (MSC combined with biomaterial); Autologous bone graft	2020
Treatment of osteoarthritic knee with high tibial osteotomy and implantation of allogeneic human umbilical cord blood-derived stem cells	Osteoarthritis, knee	Umbilical cord blood stem cell implantation for osteoarthritis treatment	2020

Table 9.4 List of clinical trials in neuromuscular diseases

Title of trial	Disease	Type of stem cells	Year
Evaluation of clinical and bone density improvement after implantation of allogeneic mesenchymal stem cell from umbilical cord on osteoporosis patients	Osteoporosis	Mesenchymal stem cell	2020
BMAC nerve allograft study	Peripheral nerve injury upper limb	Nerve graft with autologous BMAC	2019
In vivo analysis of muscle stem cells in chronic and acute lower limb ischemia (MyostemIschemia)	Conditions: artery disease; muscle disorder	Gastrocnemius muscle biopsy	2019
A study to evaluate transplantation of astrocytes derived from human embryonic stem cells, in patients with amyotrophic lateral sclerosis (ALS)	Condition: ALS (amyotrophic lateral sclerosis)	Astrocytes derived from human embryonic stem cells	2018
Intrathecal autologous adipose-derived mesenchymal stromal cells for amyotrophic lateral sclerosis (ALS)	ALS; amyotrophic lateral sclerosis	Autologous adipose-derived mesenchymal stromal cells	2017
Safety/efficacy study of 2nd cycle treatment after 6 months of 1st cycle HLA-haplo matched allogeneic bone marrow-derived stem cell treatment in ALS	Amyotrophic lateral sclerosis	Bone marrow-derived stem cell	2017
Safety assessment of intravitreal mesenchymal stem cells for acute non-arteritic anterior ischemic optic neuropathy	Non-arteritic ischemic optic neuropathy	Intravitreal mesenchymal stem cells	2017
Bone marrow-derived autologous stem cells for the treatment of Duchenne muscular dystrophy	Duchenne muscular dystrophy	Bone marrow-derived autologous stem cells	2017
Adipose stem/stromal cells in RSD, CRPS, fibromyalgia	RSD (reflex sympathetic dystrophy); CRPS—complex regional pain syndrome type I; fibromyalgia	Adipose stem/stromal cells	2017

List of Clinical Trials in Brain Injuries

Forty-six clinical trials are being conducted using different types of stem cells to treat brain injuries as per the USA clinical trials portals (https://www.clinicaltrials.gov/) and (https://stemcellsportal.com/stem-cells-translational-medicine-clinical-trials-portal). The trial list conducted during the year 2017–2020 is listed in Table 9.5.

List of Clinical Trials in Brain Diseases

One hundred ten clinical trials are being conducted using different types of stem cells to treat brain diseases as per the USA clinical trials portals (https://www.clinicaltrials.gov/) and (https://stemcellsportal.com/stem-cells-translational-medicine-clinical-tri

Table 9.5 List of clinical trials in brain injuries

Title of trial	Disease	Type of stem cells	Year
Study of the therapeutic effects of cortical autograft implantation in patients with cerebral ischemia	Condition: ischemic stroke	Autologous transplant	2020
Stem cell and conditioned medium for cerebral palsy	Condition: cerebral palsy	Umbilical cord mesenchymal stem cells	2020
Clinical effect and safety of autologous umbilical cord blood transfusion in the treatment of cerebral palsy	Cerebral palsy	Autologous umbilical cord blood transfusion	2019
Alzheimer's autism and cognitive impairment stem cell treatment study	Alzheimer's autism and cognitive impairment	Intravenous bone marrow stem cell (BMSC) fraction	2018
The treatment of premature infants with brain injury by autologous umbilical cord blood stem cells	Premature infants with brain injury	Autologous umbilical cord blood stem cells	2018
Transplantation of umbilical cord-derived mesenchymal stem cells via different routes	Cerebral palsy	Umbilical cord-derived mesenchymal stem cells	2018
Allogeneic mesenchymal stem cell-derived exosome in patients with acute ischemic stroke	Acute ischemic stroke	Allogeneic mesenchymal stem cell-derived exosome	2018
Mechanism of allogeneic UCB therapy in cerebral palsy	Cerebral palsy	Allogeneic cord blood transplantation	2017

Table 9.6 List of clinical trials in brain diseases

Title of trial	Disease	Type of stem cells	Year
Evaluate the safety and explore efficacy of umbilical cord mesenchymal stem cells in acute ischemic stroke	Acute ischemic stroke	Umbilical cord mesenchymal stem cells	2020
Stereotactic transplantation of hAESCs for Parkinson's disease	Parkinson's disease	Human amniotic epithelial stem cells	2020
The safety and the efficacy evaluation of allogeneic adipose MSC-Exos in patients with Alzheimer's disease	Alzheimer's disease	Allogeneic adipose MSC-exosomes	2020
Stem cell and conditioned medium for cerebral palsy	Cerebral palsy	Umbilical cord mesenchymal stem cells	2020
Allogeneic adipose tissue-derived mesenchymal stem cells in ischemic stroke	Ischemic stroke	Allogeneic adipose tissue-derived stem cells	2020
Mesenchymal stem cells for the treatment of acute ischemic stroke	Acute ischemic stroke	Mesenchymal stem cells	2019
Effect of different transplantation time for mesenchymal stem cells(MSCs) of cerebral infarction patients	Cerebral infarction, ischemic; acute stroke	Effect of different transplantation time for mesenchymal stem cells (MSCs) o	2019
An open-labeled phase ii study to evaluate the efficacy and safety of GXNPC-1 (Autologous adipose-derived stem cells) in patients with chronic stroke	Chronic stroke	Autologous adipose-derived stem cells	2019
Alzheimer's disease stem cells multiple infusions	Alzheimer's disease	Stem cells multiple infusions; 100 million cells allogeneic hMSC	2019
Effects of growth hormone and IGF-1 on anabolic signals and stem cell recruitment in human skeletal muscle	Growth hormone deficiency	Effects of Growth Hormone and IGF-1 + Stem Cell recruitment	2019
A study on the treatment of Parkinson's disease with autologous neural stem cells	Parkinson disease	Autologous neural stem cells	2019

(continued)

Table 9.6 (continued)

Title of trial	Disease	Type of stem cells	Year
Clinical effect and safety of autologous umbilical cord blood transfusion in the treatment of cerebral palsy	Cerebral palsy	Autologous umbilical cord blood transfusion	2019
Alzheimer's autism and cognitive impairment stem cell treatment study	Alzheimer's autism and cognitive impairment	Intranasal topical bone marrow stem cell	2018
The treatment of premature infants with brain injury by autologous umbilical cord blood stem cells	Infants with brain injury	Autologous umbilical cord blood stem cells	2018
Use of mesenchymal stem cells in Parkinson disease (PD)	Parkinson disease (PD)	Mesenchymal stem cells—umbilical cord-derived MSCs	2018
Investigation of neural stem cells in ischemic stroke	Ischemic stroke, chronic stroke	Neural stem cells	2018
Regenerative stem cell therapy for stroke in Europe 1 ()	Stroke	Stem cell therapy—adipose-derived stem cell	2018
Umbilical cord-derived mesenchymal stem cells therapy in Parkinson's disease	Parkinson's disease	Umbilical cord-derived mesenchymal stem cells therapy	2018
MultiStem® administration for stroke treatment and enhanced recovery study	Ischemic stroke	MultiStem®	2018
Study of UCB and MSCs in children with CP: ACCeNT-CP	Children with cerebral palsy	Allogeneic umbilical cord blood	2018
Transplantation of umbilical cord-derived mesenchymal stem cells via different routes	Cerebral palsy	Transplantation of umbilical cord-derived mesenchymal stem cells	2018
Allogeneic mesenchymal stem cell-derived exosome in patients with acute ischemic stroke	Acute ischemic stroke Condition: cerebrovascular disorders	Allogeneic mesenchymal stem cell-derived exosome	2017
Umbilical cord mesenchymal stem cells therapy for patients with spinocerebellar ataxia	Patients with spinocerebellar ataxia	Umbilical cord mesenchymal stem cells therapy	2017
Mesenchymal stem cells therapy in patients with recent intracerebral hemorrhage	Hemorrhagic stroke; intracerebral hemorrhage	Mesenchymal stem cells therapy	2017

(continued)

Table 9.6 (continued)

Title of trial	Disease	Type of stem cells	Year
Transplantation of neural stem cell-derived neurons for Parkinson's disease	Parkinson's disease	Transplantation of neural stem cell-derived neurons—intracerebral microinjections	2017
The safety and efficacy of human umbilical cord mesenchymal stem cells in the treatment of acute cerebral infarction	Cerebral infarction	Human umbilical cord mesenchymal stem cells	2017
Follow-up study of safety and efficacy in subjects who completed NEUROSTEM® Phase-I/IIa Clinical Trial Condition: Alzheimer's disease	Alzheimer's disease	NEUROSTEM® Phase-I/IIa clinical human umbilical cord blood-derived mesenchymal stem cells	2017
Human umbilical cord mesenchymal stem cell therapy for cerebral infarction patients in convalescent period	Cerebral infarction	Human umbilical cord mesenchymal stem cell therapy—allogeneic umbilical cord mesenchymal stem cell	2017
Safety and efficacy study of human ESC-derived neural precursor cells in the treatment of Parkinson's disease	Parkinson's disease	Human ESC-derived Neural precursor cells	2017
Stem cells in umbilical blood infusion for cerebral palsy	Cerebral palsy	Stem cells in umbilical blood infusion	2017
A study to evaluate the safety and efficacy of human neural stem cells for Parkinson's disease patient	Parkinson disease	Human neural stem cells	2017

als-portal). The list of trial which has been conducted during the year 2017–2020 is listed in Table 9.6.

List of Clinical Trials in Spinal Cord Injuries

Fifty-two clinical trials are being conducted using different types of stem cells to treat brain diseases as per the USA clinical trials portals (https://www.clinicaltrials.gov/) and (https://stemcellsportal.com/stem-cells-translational-medicine-clinical-trials-portal). The trial list conducted during the year 2016–2020 is listed in Table 9.7.

Table 9.7 List of clinical trials in spinal cord injuries

Title of trial	Disease	Type of stem cells	Year
Autologous bone marrow-derived mononuclear cells for acute spinal cord injury	Spinal cord injury	Autologous bone marrow-derived mononuclear cells	2020
Autologous adipose-derived mesenchymal stem cells for spinal cord injury patients	Spinal cord injuries; paralysis	Autologous adipose-derived mesenchymal stem cells	2020
Treatment of spinal cord injuries with (AutoBM-MSCs) versus (WJMSCs)	Spinal cord injuries	Autologous bone marrow-MSCs	2020
Safety stem cells in spinal cord injury	Spinal cord injuries	Safety stem cells	2019
Umbilical cord blood cell transplant into injured spinal cord with lithium carbonate or placebo followed by locomotor training	Spinal cord injuries	Umbilical cord blood cell transplant	2019
Clinical study of an autologous stem cell product in patients with a (Sub)acute spinal cord injury	Spinal cord injuries	Autologous stem cell	2019
Intrathecal transplantation of UC-MSC in patients with early stage of chronic spinal cord injury	Spinal cord injuries	Intrathecal transplantation of umbilical cord-MSC	2018
Intrathecal transplantation of UC-MSC in patients with sub-acute spinal cord injury	Spinal cord injury	Intrathecal transplantation of umbilical cord-MSC	2018
Intrathecal transplantation of UC-MSC in patients with late stage of chronic spinal cord injury	Spinal cord injuries	Umbilical cord mesenchymal stem cells	2018
Adipose stem cells for traumatic spinal cord injury	Spinal cord injuries; paralysis	Adipose stem cells	2018
Intrathecal administration of expanded Wharton's jelly mesenchymal stem cells in chronic traumatic spinal cord injury	Spinal cord injury, chronic	Wharton's jelly mesenchymal stem cells	2016
Transplantation of autologous bone marrow or leukapheresis-derived stem cells for treatment of spinal cord injury	Spinal cord injury	Transplantation of autologous bone marrow-derived stem cells	2016

(continued)

Table 9.7 (continued)

Title of trial	Disease	Type of stem cells	Year
NeuroRegen Scaffold™ combined with stem cells for chronic spinal cord injury repair	Spinal cord injury	NeuroRegen Scaffold™ combined with stem cells	2016
NeuroRegen Scaffold™ With bone marrow mononuclear cells transplantation versus intradural decompression and adhesiolysis in SCI	Spinal cord injury	NeuroRegen Scaffold™ with bone marrow mononuclear cells transplantation	2016

List of Clinical Trials in Peripheral Nerves Diseases

Forty-six clinical trials are being conducted using different types of stem cells to treat peripheral nerve diseases as per the USA clinical trials portals (https://www.clinicaltrials.gov/) and (https://stemcellsportal.com/stem-cells-translational-medicine-clinical-trials-portal). The list of trials conducted during the year 2016–2020 is listed in Table 9.8.

Table 9.8 List of clinical trials in peripheral nerve diseases

Title of trial	Disease	Type of stem cells	Year
BMAC nerve allograft study condition: peripheral nerve injury upper limb	Peripheral nerve injury upper limb	Bone marrow autologous cell nerve allograft	2019
A study to evaluate transplantation of astrocytes derived from human embryonic stem cells, in patients with amyotrophic lateral sclerosis (ALS)	ALS (amyotrophic lateral sclerosis)	Transplantation of astrocytes derived from human embryonic stem cells	2018
Intrathecal autologous adipose-derived mesenchymal stromal cells for amyotrophic lateral sclerosis (ALS)	Amyotrophic lateral sclerosis	Intrathecal autologous adipose-derived mesenchymal stromal cells	2017
Safety/efficacy study of 2nd cycle treatment after 6 months of 1st cycle HLA-haplo matched allogeneic bone marrow-derived stem cell treatment in ALS	Amyotrophic lateral sclerosis	HLA-haplo matched allogeneic bone marrow-derived stem cell	2017

(continued)

Table 9.8 (continued)

Title of trial	Disease	Type of stem cells	Year
Safety assessment of intravitreal mesenchymal stem cells for acute non-arteritic anterior ischemic optic neuropathy	Non-arteritic ischemic optic neuropathy	Mesenchymal stem cells	2017
Bone marrow-derived autologous stem cells for the treatment of Duchenne muscular dystrophy	Duchenne muscular dystrophy	Bone marrow-derived autologous stem cells	2017
Adipose stem/stromal cells in fibromyalgia, reflex sympathetic dystrophy, complex regional pain syndrome type i; fibromyalgia	Reflex sympathetic dystrophy); CRPS—complex regional pain syndrome type i; fibromyalgia	Adipose stem/stromal cells	2016
Study of two intrathecal doses of autologous mesenchymal stem cells for amyotrophic lateral sclerosis	Amyotrophic lateral sclerosis	Autologous mesenchymal stem cells	2016
Escalated application of mesenchymal stem cells in amyotrophic lateral sclerosis patients	Motor neuron disease	Mesenchymal stem cells	2016
Neurologic stem cell treatment study	Neurologic disorders	Intravenous bone marrow stem cells; procedure: intranasal bone marrow stem cells	2016

Clinical Trials of Heart Diseases with Stem Cell Therapy

Stem cells are tested as the treatment for various heart-related diseases and injuries. There are many clinical trials either currently in progress or have been completed where stem cells have been used in heart diseases as per the USA clinical trials portals.

List of Clinical Trials in Myocardial Infarction

Ninety-six clinical trials are being conducted using different stem cells in the treatment of myocardial infarction as per the USA clinical trials portals (https://www.clinicaltrials.gov/) and (https://stemcellsportal.com/stem-cells-translational-medicine-clinical-trials-portal). The list of trial which has been conducted during the year 2017–2020 are listed in Table 9.9.

Table 9.9 List of clinical trials in myocardial infarction

Title of trial	Disease	Type of stem cells	Year
WJMSCs anti-inflammatory therapy in coronary artery disease	Coronary artery disease	Wharton's jelly-derived mesenchymal stem Cell	2020
Stem cell in acute myocardial infarction	Acute myocardial infarction	Stem cells	2020
Evaluate the safety and explore efficacy of umbilical cord mesenchymal stem cells in acute myocardial infarction condition: acute myocardial infarction	Acute myocardial infarction	Umbilical cord mesenchymal stem cells	2020
MCRcI® stem cell treatment for diffuse coronary artery disease	Coronary artery disease	MCRcI® stem cell treatment	2019
MiSaver® stem cell treatment for heart attack (acute myocardial infarction)	Myocardial infarction	MiSaver® stem cell	2019
Serial infusions of allogeneic mesenchymal stem cells in ischemic cardiomyopathy patients with left ventricular assist device	Ischemic heart disease	Allogeneic mesenchymal stem cells	2019
UC-MSC transplantation for left ventricular dysfunction after AMI	Left ventricular dysfunction; acute myocardial infarction	Umbilical cord mesenchymal stem cell transplantation for	2019
Pericardial matrix with mesenchymal stem cells for the treatment of patients with infarcted myocardial tissue	Myocardial infarction	Pericardial matrix with mesenchymal stem cells	2019
First in humans to evaluate collagen patches with stem cells in patients with ischemic left ventricular dysfunction	Heart failure with reduced ejection fraction	Collagen patches with stem cells	2019
Evaluate the safety and explore efficacy of umbilical cord mesenchymal stem cells with acute myocardial infarction	Acute myocardial infarction	Umbilical cord mesenchymal stem cells	2018

(continued)

Table 9.9 (continued)

Title of trial	Disease	Type of stem cells	Year
Sequential treatment of extra-corporeal shock wave combined with autologous bone marrow mesenchymal stem cells on patients with ischemic heart disease	Ischemic heart disease	Treatment of extra-corporeal shock wave combined with autologous bone marrow mesenchymal stem cells	2018
ICBMC-MI. Intracoronary bone marrow mononuclear cells in myocardial infarction (MI) patients	Heart failure	Intracoronary bone marrow mononuclear cells	2017
Transplantation efficacy of autologous bone marrow mesenchymal stem cells with intensive atorvastatin in AMI patients	Myocardial infarction	Autologous bone marrow mesenchymal stem cells	2017

List of Clinical Trials in Stroke and Related Diseases

There are 66 clinical trials being conducted using different types of stem cells in stroke and related disease treatment as per the USA clinical trials portals (https://www.clinicaltrials.gov/) and (https://stemcellsportal.com/stem-cells-translational-medicine-clinical-trials-portal). The trial list conducted during the year 2017–2020 is listed in Table 9.10.

Table 9.10 List of clinical trials in stroke and related diseases

Title of trial	Disease	Type of stem cells	Year
Evaluate the safety and explore efficacy of umbilical cord mesenchymal stem cells in acute ischemic stroke	Acute stroke	Umbilical cord mesenchymal stem cells	2020
Allogeneic adipose tissue-derived mesenchymal stem cells in ischemic stroke conditions: ischemic stroke	Ischemic stroke	Allogeneic adipose tissue-derived mesenchymal stem cells	2020
Mesenchymal stem cells for the treatment of acute ischemic stroke	Acute ischemic stroke	Mesenchymal stem cells	2019
Effect of different transplantation time for mesenchymal stem cells (MSCs) of cerebral infarction patients	Infarction, middle cerebral artery	Mesenchymal stem cells (MSCs)	2019

(continued)

Table 9.10 (continued)

Title of trial	Disease	Type of stem cells	Year
An open-labeled phase ii study to evaluate the efficacy and safety of GXNPC-1 (autologous adipose-derived stem cells) in patients with chronic stroke condition: chronic stroke	Chronic stroke	Autologous adipose-derived stem cells	2019
Combination therapy of umbilical cord blood and erythropoietin for stroke patients	Stroke	Combination therapy of umbilical cord blood with erythropoietin	2019
Regenerative stem cell therapy for stroke in Europe	Stroke	Regenerative stem cell therapy	2018
MultiStem® administration for stroke treatment and enhanced recovery study	Ischemic stroke	MultiStem® administration	2018
Allogeneic mesenchymal stem cell-derived exosome in patients with acute ischemic stroke	Cerebrovascular disorders	Allogeneic mesenchymal stem cell-derived exosome	2017
Mesenchymal stem cells therapy in patients with recent intracerebral hemorrhage	Hemorrhagic stroke	Mesenchymal stem cells therapy	2017
The safety and efficacy of human umbilical cord mesenchymal stem cells in the treatment of acute cerebral infarction	Cerebral infarction	Human umbilical cord mesenchymal stem cells	2017
Human umbilical cord mesenchymal stem cell therapy for cerebral infarction patients in convalescent period	Cerebral infarction	Human umbilical cord mesenchymal stem cell therapy	2017
Role of umbilical cord milking in the management of hypoxic-ischemic encephalopathy in neonates	Role of hypoxic-ischemic encephalopathy	Umbilical cord milking	2017

List of Clinical Trials in Digestive System Diseases

There are 90 clinical trials being conducted using different types of stem cells to treat digestive system diseases as per the USA clinical trials portals (https://www.clinicaltrials.gov/) and (https://stemcellsportal.com/stem-cells-translational-medicine-clinical-trials-portal). The trial list conducted during the year 2017–2020 is listed in Table 9.11.

Table 9.11 List of clinical trials in digestive system diseases

Title of trial	Disease	Type of stem cells	Year
Study of mesenchymal stem cells for the treatment of medically refractory Crohn's colitis	Crohn's colitis	Mesenchymal stem cells	2020
Study of mesenchymal stem cells for the treatment of medically refractory ulcerative colitis (UC)	Ulcerative colitis	Mesenchymal stem cells	2020
Mesenchymal stem cells for the treatment of rectovaginal fistula in participants with Crohn's disease	Rectovaginal fistula; Crohn's disease	Mesenchymal stem cells	2020
Study of mesenchymal stem cells for the treatment of ileal pouch fistula's in participants with Crohn's disease	Ileal pouch; Crohn's disease	Mesenchymal stem cells	2020
Mesenchymal stem cells for the treatment of perianal fistulizing Crohn's Disease Conditions: perianal Crohn's disease	Perianal Crohn's disease	Mesenchymal stem cells	2020
Umbilical cord mesenchymal stem cell for liver cirrhosis patient caused by hepatitis B	Liver cirrhosis	Umbilical cord mesenchymal stem cell	2020
Angiographic delivery of AD-MSC for ulcerative colitis Condition: ulcerative colitis Intervention: drug: adipose-derived, autologous mesenchymal stem cells	Ulcerative colitis	Adipose-derived, autologous mesenchymal stem cells	2020
Combination of autologous MSC and HSC infusion in patients with decompensated cirrhosis	Cirrhosis, liver	Combination of autologous MSC and HSC Infusion	2020
A phase II open-label single-arm study to evaluate the efficacy and safety of autologous adipose-derived stem cells (ADSCs) in subjects with liver cirrhosis Condition: liver cirrhosis	Liver cirrhosis	Autologous adipose-derived stem cells	2019

(continued)

Table 9.11 (continued)

Title of trial	Disease	Type of stem cells	Year
Mesenchymal stem cells for the treatment of pouch fistulas in Crohn's	Crohn's disease	Mesenchymal stem cells	2019
Evaluation of local co-administration of autologous ADIpose-derived stromal vascular fraction with microfat for refractory perianal Crohn's Fistula	Crohn's disease	Co-administration of autologous adipose-derived stromal vascular fraction	2019
Mesenchymal stem cells treatment for decompensated liver cirrhosis Condition: Decompensated liver cirrhosis Interventions: biological: umbilical cord-derived mesenchymal stem cell;	Decompensated liver cirrhosis	Mesenchymal stem cells—umbilical cord-derived mesenchymal stem cell	2019
MSC intra-articular injection in Crohn's disease patients Condition: efficacy and safety Intervention: biological: mesenchymal stromal cells	Crohn's disease	Mesenchymal stromal cells	2019
Allogeneic ABCB5-positive stem cells for treatment of acute-on-chronic liver failure Condition: acute-on-chronic liver failure	Acute-on-chronic liver failure	Allogeneic ABCB5-positive stem cells	2019
Clinical study of human umbilical cord mesenchymal stem cells in the treatment of decompensated hepatitis b cirrhosis	Hepatitis B	Human umbilical cord mesenchymal stem cells	2019
Mesenchymal stem cell transplantation for refractory primary biliary cholangitis Condition: primary biliary cirrhosis	Primary biliary cirrhosis	Mesenchymal stem cell transplantation	2018

(continued)

Table 9.11 (continued)

Title of trial	Disease	Type of stem cells	Year
Mesenchymal stem cell transplantation for acute-on-chronic liver failure Condition: acute-on-chronic liver failure	Acute-on-chronic Liver failure	Mesenchymal stem cell transplantation	2018
Clinical trial of umbilical cord mesenchymal stem cell transfusion in decompensated liver cirrhosis	Decompensated liver cirrhosis	Umbilical cord mesenchymal stem cell	2018
Stem cells treatment of complex Crohn's anal fistula Conditions: anal fistula	Anal fistula	Stem cells treatment	2018
Safety and efficacy study of mesenchymal stem cell in treating liver fibrosis	Liver cirrhosis	Mesenchymal stem cell	2018
Adult allogeneic expanded adipose-derived stem cells (eASC) for the treatment of complex perianal fistula(s) in patients with Crohn's disease	Crohn's disease	Adult allogeneic expanded adipose-derived stem cells	2017

List of Clinical Trials in Eye Diseases

There are 90 clinical trials are being conducted using different types of stem cells in the treatment of in eye diseases as per the USA clinical trials portals (https://www.clinicaltrials.gov/) and (https://stemcellsportal.com/stem-cells-translational-medicine-clinical-trials-portal). The list of trial which has been conducted during year 2017–2020 is listed in Table 9.12.

List of Clinical Trials in Kidney Diseases

There are 47 clinical trials being conducted using different types of stem cells to treat kidney diseases as per the USA clinical trials portals (https://www.clinicaltrials.gov/) and (https://stemcellsportal.com/stem-cells-translational-medicine-clinical-trials-portal). The trial list conducted during the year 2017–2020 is listed in Table 9.13.

Table 9.12 List of clinical trials in eye diseases

Title of trial	Disease	Type of stem cells	Year
Autologous transplantation of induced pluripotent stem cell-derived retinal pigment epithelium for geographic atrophy associated with age-related macular degeneration	Age-related macular degeneration	Autologous transplantation of induced pluripotent stem cell-derived retinal pigment epithelium	2020
Effect of UMSCs-derived exosomes on dry eye in patients with cGVHD Condition: dry eye	Dry eye	Umbilical mesenchymal stem cells-derived exosomes	2020
Treatment of central retinal vein occlusion using stem cells study Condition: central retinal vein occlusion	Treatment of central retinal vein occlusion	Autologous bone marrow CD34+ stem cells	2019
Interventional study of implantation of hESC-derived RPE in patients with rp due to monogenic mutation Condition: retinitis pigmentosa Intervention: biological: human embryonic stem cell-derived retinal pigment epithelium (RPE)	Retinitis pigmentosa	Human embryonic stem cell-derived retinal pigment epithelium (RPE)	2019
Safety and efficacy of subretinal transplantation of clinical human embryonic stem cell-derived retinal pigment epitheliums in treatment of retinitis pigmentosa	Retinitis pigmentosa	Human embryonic stem cell-derived retinal pigment epitheliums	2019
Treatment with allogeneic adipose-derived mesenchymal stem cells in patients with aqueous deficient dry eye disease	Dry eye	Allogeneic adipose-derived mesenchymal stem cells	2018
Corneal epithelial stem cells and dry eye disease Conditions: dry eye syndromes; dry eye; ocular inflammation; ocular surface disease; ocular discomfort; blepharitis	Inflammation; ocular surface disease; ocular discomfort; blepharitis	Corneal epithelial stem cells	2017

(continued)

Table 9.12 (continued)

Title of trial	Disease	Type of stem cells	Year
Safety assessment of intravitreal mesenchymal stem cells for acute non-arteritic anterior ischemic optic neuropathy	Non-arthritic ischemic optic neuropathy	Mesenchymal stem cells	2017
Treatment of dry age-related macular degeneration disease	Dry age-related macular degeneration	Retinal pigment epithelium derived from clinical-grade human embryonic stem cells	2017

Table 9.13 List of clinical trials in kidney diseases

Title of trial	Disease	Type of stem cells	Year
Clinical research of UC-MSCs in the treatment of diabetic nephropathy	Diabetic nephropathy	Umbilical cord-mesenchymal stem cells	2020
Clinical study of umbilical cord mesenchymal stem cells in the treatment of type 2 diabetic nephropathy	Type 2 diabetes with renal manifestations	Umbilical cord mesenchymal stem cells	2020
Clinical trial of mesenchymal stem cells in the treatment of severe acute kidney injury	Acute kidney injury; mesenchymal stem cells	Mesenchymal stem cells	2020
Cell-based therapy for the treatment of kidney disease Condition: kidney diseases	Kidney diseases	Cell-based therapy—endothelial progenitor cell	2019
Umbilical cord mesenchymal stem cells therapy for diabetic nephropathy	Diabetic nephropathy	Umbilical cord mesenchymal stem cells	2019
Safety and efficacy study of mesenchymal stem cell in treating kidney fibrosis	Renal cirrhosis	Mesenchymal stem cell	2018
Treatment of chronic renal failure with adipose tissue-derived mesenchymal stem cells Conditions: mesenchymal stem cells; chronic kidney diseases; renal interstitial fibrosis	Chronic kidney diseases; renal interstitial fibrosis	Adipose tissue-derived mesenchymal stem cells	2018
Use of Wharton's jelly in diabetic nephropathy	Diabetic nephropathy	Wharton's jelly mesenchymal stem cells	2017
Adipose-derived stem cells (ADSCs) for moderate to severe chronic kidney disease	Moderate to severe chronic kidney disease	Adipose-derived stem cells (ADSCs)	2016

(continued)

Table 9.13 (continued)

Title of trial	Disease	Type of stem cells	Year
AMSCs in reducing hemodialysis arteriovenous fistula failure Conditions: end stage renal disease (ESRD); vascular access complication	End stage renal disease	Adipose-derived mesenchymal stem cells (AMSC)	2016

List of Clinical Trials in Skin Diseases

There are 47 clinical trials in skin diseases as per the USA clinical trials portals (https://www.clinicaltrials.gov/) and (https://stemcellsportal.com/stem-cells-translational-medicine-clinical-trials-portal). The trial list conducted during the year 2017–2020 is listed in Table 9.14.

Table 9.14 List of clinical trials in skin diseases

Title of trial	Disease	Type of stem cells	Year
Human placental mesenchymal stem cells treatment on diabetic foot ulcer	Diabetic foot ulcer	Human placental mesenchymal stem cells	2020
Subcutaneous injections of autologous ASC to heal digital ulcers in patients with scleroderma	Systemic sclerosis	Autologous adult stem cells	2020
Phase 1, open-label safety study of umbilical cord lining mesenchymal stem cells (corlicyte®) to heal chronic diabetic foot ulcers	Diabetic foot ulcer	Umbilical cord lining mesenchymal stem cells	2019
Clinical study of adipose-derived stem cells in the treatment of diabetic foot	Diabetes mellitus foot ulcer	Adipose-derived stem cells	2019
Therapy of scars and cutis laxa with autologous adipose-derived mesenchymal stem cells Conditions: skin; scar; cutis laxa; keloid; cicatrix	Skin; scar; cutis laxa; keloid; cicatrix	Autologous adipose-derived mesenchymal stem cells	2019

(continued)

Table 9.14 (continued)

Title of trial	Disease	Type of stem cells	Year
Treatment of chronic wounds in diabetic foot syndrome with autologous adipose-derived mesenchymal stem cells	Diabetic foot ulcer	Autologous adipose-derived mesenchymal stem cells	2019
Clinical research on treatment of psoriasis by human umbilical cord-derived mesenchymal stem cells	Psoriasis	Human umbilical cord-derived mesenchymal stem cells	2018
Long term effects on skin hyper pigmentation with and without mesenchymal stem cell enriched adipose tissue grafting for "contour deformities with pigmentary changes on face" Conditions: skin pigmentation over contour deformities of Fac	Skin pigmentation over contour deformities of face	Mesenchymal stem cell enriched adipose tissue grafting	2018
A randomized, positive controlled trial assess the efficacy and safety of Uc-MSC in plaque psoriasis patients	Moderate and severe plaque psoriasis	Umbilical cord-mesenchymal stem cells	2018

References

Bersenev, A. (2015). Cell therapy clinical trials—2014 report. CellTrials blog, January 22, 2015. http://celltrials.info/2015/01/22/2014-report/

Clinical trials of government of USA. https://www.clinicaltrials.gov/

Cyranoski, D. (2014). Japanese woman is first recipient of next-generation stem cells. Nature News (Nature Publishing Group), September 12, 2014. http://www.nature.com/news/japanese-woman-is-first-recipient-of-next-generation-stem-cells-1.15915

Kushner, J. A., MacDonald, P. E., & Atkinson, M. A. (2014). Stem cells to insulin secreting cells: Two steps forward and now a time to pause? *Cell Stem Cell, 15*, 535–536.

List of stem cell clinical trials. https://stemcellsportal.com/stem-cells-translational-medicine-clinical-trials-portal

Ratcliffe, E., Glen, K. E., Naing, M. W., & Williams, D. J. (2013). Current status and perspectives on stem cell-based therapies undergoing clinical trials for regenerative medicine: Case studies. *British Medical Bulletin, 108*, 73–94.

Schwartz, S. D., Hubschman, J. P., Heilwell, G., Franco-Cardenas, V., Pan, C. K., Ostrick, R. M., Mickunas, E., Gay, R., Klimanskaya, I., & Lanza, R. (2012). Embryonic stem cell trials for macular degeneration: A preliminary report. *Lancet, 379*, 713–720.

Schwartz, S. D., Regillo, C. D., Lam, B. L., Eliott, D., Rosenfeld, P. J., Gregori, N. Z., Hubschman, J. P., Davis, J. L., Heilwell, G., Spirn, M., et al. (2015). Human embryonic stem cell-derived retinal pigment epithelium in patients with age-related macular degeneration and Stargardt's macular dystrophy: Follow-up of two open-label phase 1/2 studies. *Lancet, 385*, 509.

Trounson, A., Baum, E., Gibbons, D., & Tekamp-Olson, P. (2010). Developing a case study model for successful translation of stem cell therapies. *Cell Stem Cell, 6*, 513–516.

Trounson, A., & McDonald, C. (2015). Stem Cell Therapies in Clinical Trials: Progress and Challenges. *Cell Stem Cell, 17*(1), 11–22. https://doi.org/10.1016/j.stem.2015.06.007 PMID: 26140604.

Trounson, A., Thakar, R. G., Lomax, G., & Gibbons, D. (2011). Clinical trials for stem cell therapies. *BMC Medicine, 9*, 52.

Chapter 10
Stem Cell Production: Scale Up, GMP Production, Bioreactor

Naseem A. Almezel

Abstract Stem Cell production for therapeutic applications is increasingly gaining attention both for the great potential this therapy presents and the complexity of such therapies. Safety and efficacy have improved dramatically during the past decade, while technology improvement facilitated the research and development phase of cell therapy journey as well as the clinical production phase. All this led to innovative therapies that for the first time are treating diseases that once were considered untreatable; bringing hope to a wide sector of patients and their families. In this chapter we focus on stem cell production for therapeutic applications, then discuss their advantages and limitations. We briefly discuss the stem cell therapeutic applications before we list specific good manufacturing practices requirements and how this impact the cell therapy field as a whole.

Keywords Stem cell · Bioreactor · GMP · Quality assurance · Regulations · Viral vector

Stem Cell Production

Historical Overview

2019 marked the anniversaries of two milestones. BM transplant turned 40 this year, and patients are taking advantage of this therapy that became the treatment of choice for many diseases as indicated by the uniformity of the clinical care around the globe. This therapy though had a difficult start. BM transplant experiments started nearly 10 years earlier around the mid of 1950s, but most of the patient died. The failure of those experiments led many professionals to leave the field convinced the barrier between individuals cannot be crossed.

Though, those who persisted, like Thomas ED (the father of BM transplant as he became to be known) made the history. In 1979, Fred Hutch reported the 1st

N. A. Almezel (✉)
potentCELLS, Dammam, Saudi Arabia

F. A. Khan (ed.), *Advances in Application of Stem Cells: From Bench to Clinics*, Stem Cell Biology and Regenerative Medicine 69,
https://doi.org/10.1007/978-3-030-78101-9_10

successful unrelated BM transplant case; it was a priceless breakthrough. But this successful case was not the group's 1st case, it was the 2nd. The Fred Hutch group reported that their 1st patient died soon after the transplant. The Hutch group reported their 1st case got infection that caused the death. The 1979 case infection was a CMV infection. The second anniversary that 2019 marks is the 30th anniversary of Gene Therapy. Steven Rosenberg reported their successful 1989 study where they inserted a gene to determine the traffic of the tumor infiltrating lymphocytes. As we will discuss later in this chapter, gene therapy is returning after a hard beginning.

Challenges—Difficulties and Shortcomings that Faced Researchers

"*Pluripotent stem cells can potentially switch from pluripotency to uncontrolled differentiation*" (Kropp et al., 2017). The use of stem cells for therapy was not without challenges; the self-renewal and differentiation capabilities that characterize stem cells and present them as advantageous, can be a source of failure if not controlled properly. The challenges that face this field concern both the safety and efficacy of the products. As safety is concerned, the carried risk is either product-related or process related. For the product-related risk, this risk stems from the nature of the therapeutic product. Specifically, not controlling the differentiation potential of stem cells can lead to either their differentiation too soon or losing their potency, therefore, not providing the right potent dose. Alternatively, stem cells may differentiate to a cell lineage that is harmful or not desired, which present a significant risk of delivering the wrong product, or failure of the entire process. For the right cell dose to be generated, cells would need to be manipulated for extended period which increases the risk of contamination due to extended time and more material use. The contamination risk is not limited to microbial or fungal, contaminants that originate from the material such as plastics and chemical toxicities can fail the entire process.

On the efficacy side, it becomes more important to determine some parameters that impact the efficacy of stem cell products. For instance, delivering the right dose of cells at the right site can improve the therapeutic effects by utilizing the product at its maximum capacity. For stem cell therapy to be effective, the required cell dose for a specific disease in a specific patient is important. Further, if not delivered at the impacted tissue soon enough, the therapeutic cells may lose viability, potency, or differentiation capability, therefore, becoming ineffective by the time they reach the desired site. Additionally, identifying the characteristics that improve the cellular function is important. Cells tend to upregulate/downregulate their gene expression at variable phases of their growth and maturation. Identifying the phenotypic and functional characteristics that are unique to potent cells, and then testing for these characteristics to ensure the product suitability becomes favorable.

Advances—Improvements in Stem Cell Production Over the Years

Cell therapy had made remarkable advances since its rough beginning more than 50 years ago. The initial scope of many of these advances was to improve the cell-based therapeutics safety. As the safety improved the scope started to include efficacy to deliver more potent products. The classical bone marrow transplant stands now in a very strong position, thanks to the improvements that were made over the years.

Scientists have been focusing lately on expanding the reach of cell therapy to include cells that were never used for therapeutic applications before such as the immune cells. Additionally, expanding the cell therapy reach resulted in treating diseases that were once considered untreatable, such as tumor. While the list of advances is long, we will only list few that concern stem cells as the therapeutic product.

Off-the-shelf cancer treatment

The cost of producing cell-based product is often front and center when such therapy is being considered. One way that the cost may be reduced is to move from patient-specific product to a condition-specific product. By utilizing this option, production facilities can produce a large batch that may be used to treat multiple patients as an off-the-shelf drug. In 2019 two induced pluripotent stem cells (iPSC) immune therapy products were unveiled by Century Therapeutics and Fate Therapeutics paving the way for more products to come.

Widening of gene therapy scope

Of the 9 approved gene therapies so far, two were approved during 2019. These two included treatments for rare disease (spinal muscular atrophy) by Novartis and for sickle cell disease by Bluebird bio. Although the start of gene therapy was tough, it is now becoming a reality that changes patients' lives.

Standardization of cell and gene therapy

As the field continue to expand, it becomes more important to standardize the entire supply chain from collection throughout infusion. For this reason, stakeholders started to discuss/collaborate to generate services and platforms that facilitate any future development.

Scale Up of Stem Cell Production

As cellular therapeutics became more popular, some challenges became evident and needed to be addressed. On one hand, after the completion of proof of concepts studies, it became necessary to generate stem cells at quantities that are enough to treat a patient. Those quantities varied from one patient to the other, from one condition to the other, and from one protocol to the other. Additionally, the push for off-the-shelf therapies was gaining momentum, therefore, requiring even higher quantities of stem cells to be produced.

On the other hand, as pharmaceutical and biotechnology companies started investing in cellular therapeutics, they developed their processes and standardized them in a fashion that mirrors the pharmaceuticals production lines. The standardization concept relies on removing as many variables as possible. The two major variables in cell therapy production are the starting product and the human operator. While the variability amongst the starting products is not possible to be rolled out, the human factor is. Because of these challenges, scientists started to explore production methodology that satisfies few but challenging criteria:

- Maintain the product safe, pure, and effective
- Meets the regulations
- Ease of maintenance.

Scale Up Methodology

Scale up methodology varied in capacity, in automation level, and therefore, in cost. At one end of the spectrum are production systems that increased the potential quantity of produced cells but relied on human operator. On the other end are production systems that increased the produced quantity while simultaneously reducing the human involvement. Between these two ends are some production systems that have variable degrees of high quantities and human involvement.

2D Production Systems

Stacks and Factories

Stacks presented an attractive production system because it maintained the flask methodology that most R&D laboratories utilizes while utilizing the vertical incubator space. This system made scaling up production of stem cells easier for the stacks needed no special requirements or specialized staff. But this ease of use comes with an expensive price; such production system requires more of staff time, significant risk of contamination, and extended footprint.

Despite this, stacks still provide a reasonable option for small production facilities and academic centers, who may have limited budget and facility and are only working on phase I/II clinical trials that enroll small number of participants. The cell factory resembles the stacks in several aspects such as being a 2D manual system with limited capacity. One advantage of the cell factory is its improved gas exchange that the ports provide. The ports make filling, venting, and harvest an easier process.

3D Production Systems

Bioreactor

The expanded use of cell therapy products made bioreactors even more attractive option of production. Unlike the manual production method, bioreactors ensure several benefits such as reduction of cost, control process, and quality of production. There are several bioreactor types that differ in their mode of operation or the types of cells they support, the main categories include stirred tank, fix bed, hollow fiber, and rocking platforms (Eaker et al., 2017). Majority of bioreactors have the capability of real-time monitoring of critical parameters such as dissolved oxygen and CO_2, pH level, and temperature. One important factor that impact the choice of bioreactor is the anchorage dependency of the culture. Therefore, based on the nature of the culture, some cultures require substrate to adhere to while others are suspended.

Over the years the design and engineering of the bioreactor has been adjusted to overcome obstacles and challenges related to upscale and diversity of culture applications. While most of the focus has been directed towards adjusting the design and operation of the bioreactor to match the nature of the microculture environment, some studies has been designed to adjust the nature of the culture to match or improve the scalability and applications of the bioreactor. Because of their scalability and cost effectiveness, bioreactors became a very good option for large scale productions (Oppermann et al., 2014). Similarly, patient specific production of therapeutic products became a common option for multiple disorders and the use of the bioreactor for these patients became one way to control the cost of the production and to improve efficient modality.

Stirred tank bioreactor

Is a widely used 3-dimenstional (3D) stem cell production system with two major production models, the immobilized and cell aggregates. Studies have showed that the optimum production results by optimizing and controlling individual factors to reach the optimal production model of each cell type. Factors such as seeding density, aggregate size, media type, and operation mode are all critical in determining the in-process and end-of-process quality of product. Types of cells that were successfully propagated using stirred tank bioreactor include human pluripotent stem cells (hPSC) (Kropp et al., 2017), this approach was facilitated by the use of a Rho-associated coiled-coil containing kinase (ROCK) inhibitor (Watanabe et al., 2007). Besides being commercially available, stirred tank bioreactor provides a flexible operating system and efficient gas exchange of culture. Although this system is robust, the need to detach the cells from the carrier remains a critical component of the process (Weber et al., 2010). One alternative is to utilize a system that operate as a carrier-free system such as the fixed bed bioreactor. A full list of advantages and disadvantages of suspended and immobilized bioreactor systems appear in Table 10.1 (Pörtner and Faschian, 2019).

Table 10.1 Summary of advantages and disadvantages of stirred tank and fixed-bed bioreactor (Pörtner and Faschian, 2019)

	Advantages	Disadvantages
Stirred tank/suspension	Known technology	Aeration difficult at high cell densities (relevant for aerobic cells)
	Good mass transfer	call damage by shear and aeration (e.g., Mammalian cells))
	Good mixing	Foaming (relevant for aerobic cells))
	Cell count possible	Low cell density and volumetric productivity cell retention required for perfusion culture, techniques insufficient for long-term culture
High potential for scale-up		
Fixed-bed/immobilized cells	High cell density and productivity per unit	Concentration gradients
	Easy exchange of medium	Nonhomogeneous
	High productivity over long periods of time	Cell count impossible
	Low-shear rates (relevant for mammalian cells))	

Fixed and packed bed bioreactors

Immobilized bioreactor technologies are being considered for the many advantages they provide such as reduced contamination susceptibility and protection against high shear environment (Fig. 10.1). In addition, the high productivity advantage is being attributed to the culture microenvironment provided by the carrier. In a two parts study, A and B, (Weber et al., 2010) successfully expanded Human mesenchymal stem cells (hMSC) using the fixed bed bioreactor. Although the fixed bed bioreactor offers several advantages, the low industrial models is reasoned in part to the lack of both process development tools and operation concepts (Pörtner and Faschian, 2019).

Rocking bioreactor

Rocking bioreactors utilize the rocking motion to distribute gas and nutrient via certain rocking speed and angel. Davis et al. (2018) reported the successful expansion of pluripotent stem cells using a rocking bioreactor.

Hollow fiber bioreactor

Described by Knazek (1972), the hollow fiber membrane bioreactor consists of hollow fiber membrane that separate the cells from the medium compartment.

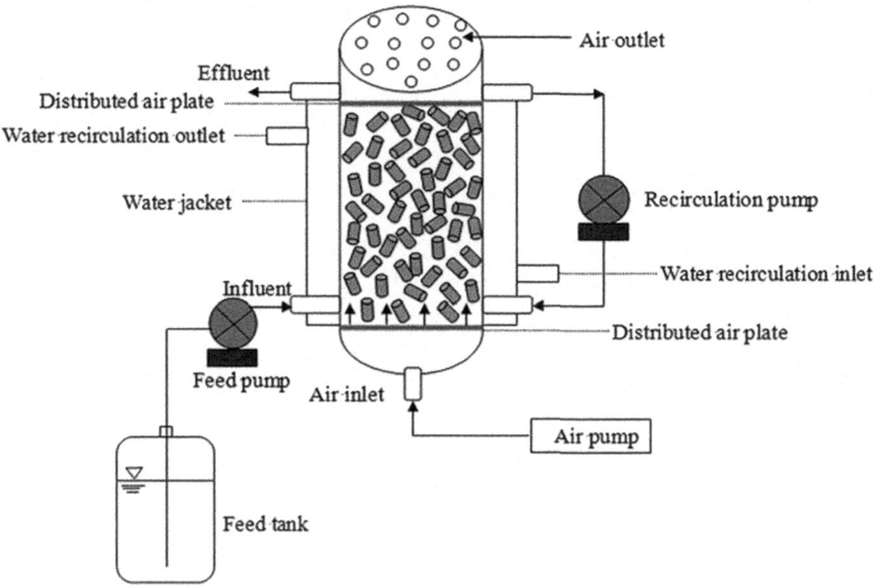

Fig. 10.1 Schematic diagram of packed-bed bioreactor

Medium flow and waste removal occurs through the membrane, which comprise a major disadvantage of this system due to the resistance of mas transfer. Several advantages are presented by this system, with the ability to grow high density cells and support of 3D growth being the most relevant to the field of cell therapy. Frank et al. (2019) in a recent study described the successful cultivation of MSC using this system, and utilizing different coating reagent. The final product met the standards as described by the international society for cell and gene therapy (ISCT).

Considerations for bioprocessing

Production facilities that are considering the use of bioreactor in their processes are utilizing a significant improvement that would lead to several advantages such as:

- Process standardization
- Eliminating /reducing man made errors
- Enhancing Product Safety /sterility
- Large scale production
- Cost reduction.

Complying with GMP requirements mandates taking reasonable measures to ensure product safety; production facilities are required to implement measures that eliminate the risks of mix-ups and cross-contamination between products. A major improvement that became available for bioprocessing utilizing facilities is the introduction of disposables bioreactor set. With this improvement, production facilities are able to eliminate the cost, labor, and risk associated with re-using a bioreactor such as:

- Purified water utility
- Validation of cleaning procedure
- Integrated pre-sterilized pH and dO_2 sensors.

Several types of bioreactors are now being offered as disposable; this facilitates the selection process as more options are now available. However, selecting the right bioreactor depends on multiple factors such as:

- Cellular growth pattern
- Scale and engineering parameters of bioreactors (e.g. flow rate/time/volume, mixing/residence times)
- Biosafety/GMP compliance
- Capital/running costs.

After phase I and II of bioreactor development, phase III marked the launch of disposable bioreactors in 1990 (Eibl et al., 2010). Currently, the available disposable bioreactors include wave-mixed, orbitally shaken or stirred bioreactors. Stirred disposable bioreactors were introduced 2006 to the market, but have since gained a major share of the technology, and by 2010, have 10 different commercially available bioreactors; some models are flexible utilizing bags while others are rigid using plastic cylinders.

One time use bioreactors are emerging as a viable option to satisfy several requirements with safety and prevention of cross-contamination being the most relevant. Hähnel et al. (2011) group evaluated one of the available one-time bioreactor models and concluded that the setup is remarkably short.

There are, though, several aspects that need to be considered when a decision to use bioreactors is made. Often, choosing a production system is done long after building and qualifying the facility, this means that there potentially are modifications that need to be made to accommodate a specific production system for a specific process; below is a list of potential changes:

Gas lines

Some bioreactors utilized 2–4 types of gas to complement the bioreactor operation. For instance, N_2 is used to reduce /adjust the oxygen concentration in cultural media. CO_2 is used to adjust the media pH to reach /maintain a set point. Utilizing gases means that there need to be a network of pipelines that delivers the gas from the closet to the production suite. Such a change to install gas pipeline requires the facility to be re-qualified. In some situations where installing pipeline is not possible, the production facility need to install the gas cylinders inside the facility utilizing all required safety measures.

Large scale production

Scaling up production requires large quantities of media, and eventually generate large volumes of waste. Moving large quantities of liquid in and out of the facility requires good planning to ensure the facility is kept in status at all times. Further, using large volumes of media requires appropriate controlled /monitored storage space for media before and during use. Similarly, using large volumes of media results in generating similar volumes of liquid waste that need to be disposed of appropriately. Additionally, facilities utilizing large volumes of media should be prepared for spill accidents with appropriate spill kits and trained staff.

Product segregation

For facilities that enroll multiple patients simultaneously, careful attention should be paid to ensure elimination of products mix up risk. Production facilities are required to make efforts and take measures that ensure proper product segregation at all stages of the production process. Particularly, stages where the product is being manipulated present the highest risk. For example, stage of bioreactor inoculation, cell transduction, expansion, and harvest.

Although the production system is expected to be closed which should lead to enhanced product safety, product mix up remains a potential risk; measures that may help reduce this risk are:

Physical segregation

Culturing different products in physically different locations, such as different suites, can significantly reduce the mix up risk. Although this model requires a proper planning during busy production times to ensure appropriate staffing in several locations, the safety enhancement is worth the cost.

Proper verification

Verification of critical steps is a GMP requirement, and should be built into the process; two technologists independently verifying the information should eliminate the mix up risk.

Identity testing

While proper labeling the product at all times of the production process is a GMP mandate, identity testing of the final product serves as the final assurance of product identity. One way for such testing is to run human leukocyte antigen (HLA) testing of the initial and final products; identical results indicate the product identity was maintained throughout the production process.

Stem Cells Types

Classification of stem cells gained a lot of attention and still does, for this field is still evolving. Comprehensive classification of stem cells has been addressed in chapter two of this book. Therefore, we will briefly classify stem cells to complement the clinical applications section of this chapter. One way of classifying stem cells is through the lens of origin and the lens of potency see Fig. 10.2. The ability of stem

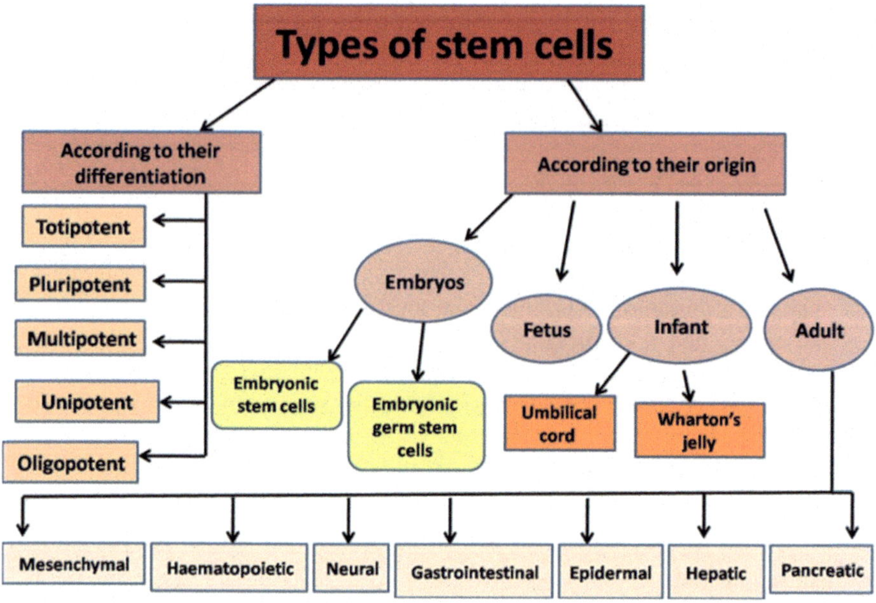

Fig. 10.2 Classification of stem cells

cells to reproduce themselves and to differentiate to any cell lineage is related to the stem cell level of potency, Good level of potency is largely dependent on the origin of stem cells. Due to this interdependency, it seems only logic to classify the stem cells based on origin and based on potency.

Stem cell classification based on origin

Embryonic stem cells ESC

Embryonic stem cells hold a unique capacity to differentiate to any cell type in the human body, which comprise a valuable source of potential therapies. The capacity to differentiate to any cell type depends on the timing of ESC isolation. ESC's isolated from embryos early after fertilization (4–5 days), or isolated primordial germline cells (PGCs) are thought to be the most potent. While a ESC isolated from the fetus organs are pluripotent and have the potential to differentiate into hematopoietic stem cells (Amira Ragab et al., 2017).

Infant stem cells

The umbilical cord has been shown to be a reliable source of potent stem cells. The most radially available source of umbilical cords stem cells (UCSC) is the umbilical cord blood containing multi potent stem cells. The other source of UCSC is umbilical cord matrix (Wharton's Jelly) Which is considered to be a source of mesenchymal stem cells.

Adult Stem Cells

Adult stem cells are isolated from mature tissues of child or adult body. Due to the stage of their development, unlike ESC's, adult stem cells have limited potential to develop into other cell types. Generally, adult stem cells are vital in repairing and regeneration of their tissue of origin, to which they are referred. Several adult stecm cells have been described, below is a brief listing:

Hematopoietic stem cells (HSC)

HSC have the potential of self-renewal and differentiation to all hematopoietic lineages. Hence, are used for transplantation, and hematologic and malignant diseases.

Neural stem cells (NSC)

NSC are established in the adult brain microenvironment and holds the potential to treat neural related disorders.

Gastrointestinal stem cells (GSC)

Residing in a niche and intestinal crypts and gastric glands, GSC nature and position is not fully established.

Epidermal stem cells

Epidermal stem cells have the capacity of self-renewal. They reside in the basal layer of the epidermis and are essential in maintaining homeostasis and wound healing.

Hepatic stem cells

Liver holds a strong regeneration capacity; therefore, liver injury gives rise of stem cell compartment who's sells later differentiate into hepatocytes.

Pancreatic stem cells

Isolated from islet cells, pancreatic stem cells are multi potent cells that can differentiate into pancreatic phenotypes.

Stem cell classification based on potency

Based on their differentiation potentials, stem cells may be classified to:

- Totipotent stem cells: have the total capacity to give rise to all self tribes and reproduce fertile offspring
- Multipotent stem cells: are capable to give rise to tissue from which they were isolated
- Unipotent stem cells: are adult stem cells that can give rise to a limited number of cell types
- Oligopotent potent stem cells: Are those cells that can differentiate into a few cell types.

Mesenchymal Stem Cells

Mesenchymal Stem Cells (MSC) are adult stem cells that can be isolated from bone marrow or cord blood. MSC potency makes them capable of differentiating to several cell lineages in the body such as neural cell, bone cells, skin cells, muscle cells, and cornea cells. MSC are thought to be able to avoid rejection by immune system and are used to treat multiple disorders due to their potency.

iPSC Stem Cells

Induced Pluripotent Stem Cells are adult cells that were isolated from skin or blood and were re-programmed to function like pluripotent stem cells. This re-programming allows the iPSC to differentiate to any cell type for therapeutic applications provided the right signals and culture micro-environment

Stem Cells Applications

The applications of stem cell therapy grew over the years to cover a wide range of conditions. While some applications have been approved by the FDA, other applications are still in their infancy. In between are applications that are progressing in clinical studies towards gaining approval.

Therapeutic Stem Cell Applications

Classical Cell Therapies

Following their success with advanced diseases, Hematopoietic Cell Transplantation (HCT) was ethically justified for malignant diseases. Transplants for patients with diseases in first remission or at early signs of relapse were largely successful. Soon after, the trials to treat non-malignant disorders were initiated. The results were not satisfying in the beginning, but had improved dramatically with changes of transplant timing, and preparations of patients. Generally, classical cell therapy includes all processes of which the product is minimally manipulated. These processes have limited impact on the product, therefore, are covered by a specific set of regulations. Despite the progress that was made in the field of cell therapy, most of these processes remained unchanged which indicate how robust these processes are. Below is a brief list of these processes:

- Volume reduction

- Red Blood Cells depletion
- Cryopreservation.

Gene Therapy

Gene therapy is a rapidly evolving field that is gaining attention from all stakeholders such as scientists, biotechnology, big pharma and patients. The field started with the intention to correct genetic diseases and was faced by several obstacles. Throughout the years, the technology has improved and the tools are now more advanced which led scientists to utilize this technology to treat diseases other than those of genetic origin. Cell-based gene therapy requires the transfer of the genetic material into the therapeutic cells. Several approaches exist to accomplish this task with each of them having advantages and disadvantages. Therefore, scientists are utilizing the mode of transfer based on the cellular nature and the disease being treated.

Viral vectors

The use of viral vectors experienced major setbacks that halted the field progress, decreased its funding, and raised a wide skepticism among the scientific community. But the persistence of gene therapist led to improvement of vector design, delivery, and safety, and eventually, regaining the trust of clinicians and scientists. Viral vectors vary with some being utilized for transient gene expression, while others being used for permanent expression. Further, the nature of cells being modified dictates the types of vectors that may be used. Some vectors can only adhere to dividing cells, while other vectors can adhere to both resting and dividing cells. Another variable is the capacity of the vector with regard to the size of the gene being inserted (Lundstrom et al., 2018).

Adenovirus

Is widely used vector that lead to temporary expression of the transferred genetic material. The 7.5 kb capacity vector initially led to strong immune reaction, but the following generations were modified to be less immunogenic.

Adeno Associated Virus (AAV)

Is a limited capacity, 4 kb, viral vector that results in long term genetic expression. Although considered of low pathogenicity and toxicity, AAV use resulted in immune response in subsequent administration, an issue that was addressed by utilizing different serotypes for subsequent administrations. Further, the limited vector capacity was increased by engineering the vector to a dual vector.

Herpes Simplix

Is considered a low toxicity long term transgenic vector, with a capacity of >30 kb of foreign DNA.

Retrovirus

Is generally a long term transgenic vector with low capacity of foreign insert; 8 kb. A major disadvantage of this vector is its inability to transduce non-dividing cells.

Lentivirus

Belongs to the retrovirus family but is capable of infecting both dividing and non-dividing cells. Lentivirus also has the integration and packaging capacity of retrovirus and can provide a long term expression. Therefore, Lentivirus vectors gained a lot of interest for cell therapy applications.

Vector Production

For cell-based therapeutic products, process components are of significant importance particularly when those components become part of the final product. Viral vectors that are used to transfer genetic material to, and permanently modify a cell product, are required to have stringent criteria for them to be approved for clinical use. Similar to clinical cell production, viral vector production for clinical applications is required to be GMP compliant. Regulatory bodies acknowledge that viral vector use in production renders the clinical product as significantly manipulated, or more than minimally manipulated per FDA's definition, and therefore apply regulations that match the category of drugs. Besides the GMP requirements, viral vector release follows a long and stringent list of criteria to demonstrate purity and safety, below is a brief list of release criteria.

- Replication competency: viral vectors are usually created from pathogenic viruses. Therefore, it is important to test the vector intended for clinical use for replication competency before final release.
- Sterility: vector products should be tested for bacterial and fungal burden, as well as for endotoxin and mycoplasma presence.
- Purity: vector product should be tested for impurities derived from the host cell system that was used to generate the vectors.
- Infectious unit/viral particles: There need to show the ratios of infectious and particle titers; this facilitate in standardizing the production process.
- Stability: vector product should be tested in temperature, volume, and container used for clinical applications to demonstrate maintenance of quality and infectivity over the expected storage time.

Regenerative Medicine

Described first by Kaiser et al. (1992), regenerative medicine (RM) implies the utilization of cells to regenerate tissues or organs and/or restore their function. The demand for unusual approaches to treat diseases has been around for decades, but being deficient in tools such as the right material, comprehensive knowledge, and appropriate facility, patients continued their efforts to manage the symptoms of untreatable diseases rather than treating them. The advances in technological tools as well as in research methodology, resulted in a shift of scientist' perception towards the possibility to treating what one day were considered untreatable diseases. These advances touched the three approaches of RM as described by Sampogna et al. (2015):

- Cell Based Therapies
- Scaffolds
- Scaffolds with cells.

Tissue or organ regeneration requires a large number of specialized cells that were differentiated to perform the physiological functions of the specific tissue or organ. In theory, the involved organ is the best source of such cells for they provide the right function, but there are rarely enough cells to aspirate especially if the organ is damaged or is malfunctioning. Therefore, there need to be innovative approaches to provide such cells in large enough quantities. To accomplish this task, the intended cells need to be capable of differentiating to the desired lineage, and capable of expanding ex-vivo to reach the required cell dose.

As we described earlier in this chapter, several stem cell types have been shown to be Pluripotent. Therefore, are capable to differentiate to the desired cell lineage. Similarly, as we described in bioreactor section, the technology has improved to allow for expansion of stem cells ex-vivo while maintaining their self-renewal and differentiation capabilities. Scientists are leveraging the late advances to guide the stem cell differentiation at the right time of the production process. Although this has been accomplished in research setting, generating a large number of appropriately differentiated cells in a GMP manner may be challenging at times. There is a need to form multidisciplinary teams of researchers and clinicians to participate in early stages of process development as well as in fine tuning late stages of clinical production.

Organoids

Organoids are three dimensional invitro tissue constructs that mimics the structure, function, and aspects of the intended organ (Natalie de Souza et al., 2018). Organoids provided the scientists with the ability to firsthand monitor the impact of therapeutic drugs on the organ tissues without affecting the patient. This ability resulted in speeding up drug discoveries while protecting humans from drug effects for the entire period of the drug development.

As the organoid technology became more robust providing the ability to generate larger and more complex organoids, attempts have started to widen the scope of organoids from only diagnostics and development to include therapeutic applications. In short, the idea of developing mini organs that can reduce the impact of a failed organ started to gain a lot of attention due to the dramatic impact of this idea on the patients and the medical field alike. Organoids generation approaches vary, hence, result in different outcomes. It is worth mentioning that the complexity of the organoid being generated is related to the potentials of the stem cells that were used to begin with; pluripotent stem cells are capable of generating complex organoids, while less capable stem cells can only differentiate into specific cell lineages.

One approach is to utilize the potentials of stem cells to differentiate to different cell lineage; this method relies on the intrinsic capabilities of stem cells to regenerate the entire organ. By providing culture environment and some development signals, scientists rely on the stem cells to do the rest of the organ generation. Another approach is to generate specific cell lineages and then fusing these cells together in an environment that cultivates the organoid generation. Unlike the first approach, this approach provides developmental signals to stem cells to ensure proper differentiation of cell stem cells to the desired cell lineages.

Organoid generation is still evolving with several obstacles that need to be addressed. Unlike the full organ, organoids lack some essential tissue components. Therefore, mimicking full organ function and structure is deficient. For example, organoid lacks vasculature, therefore, can only grow to a limited size before losing the ability to expand due to lack of nutrition. Further, lack of immune cells limits how these organoids recapitulate the organ physiological response.

Another area that scientists are working on is the maturity of organoids. Generally, organoids maturation matches the level of fetal tissue. To address this issue, scientists have used different culture environment where more control over developmental signals in stem cell differentiation is granted. This change is intended to reduce the variability between organoids being generated by unifying the level of maturity. In their recent paper, Blackford et al. (2019) reported successfully generating iPSCs-derived hepatocytes in a GMP-compliant manner. Using FDA approved scaffold material, they report generating current GMP-constructs from human pluripotent stem cells that remained viable and functional long enough after transplant for the recipient to recover from acute liver failure if product was used.

Spheroids

Spheroids are self-assembling aggregates of cells in an environment that supports 3D culture. They are generally utilized for diagnostic and experimental purposes but are also gaining increased attention in regenerative medicine for therapeutic purposes. Spheroids possess several advantages over the 2D culture such as cell–cell contact and mimicking tissue microenvironment. Therefore, are increasingly considered for applications such as tissue and organ reconstruction. For example, MSC spheroids transplant had been shown to provide advantages for organ reconstruction as well as for tissue formation (Ryu, 2019). Further, spheroids injection had shown to improve engraftment, while transplanting genetically modified spheroids led to longer periods of expression of the gene of interest.

GMP Production

The range of stem cell applications is constantly expanding, bringing hope to wider patient sectors like never before. After the proof of concept phase, scientists are constantly working to develop production methodology that simplify the process and reduce the cost while meeting the regulatory requirements. Regulations, as defined by the European medicines agency (EMA) and the Food and Drug administration (FDA), require the advanced therapy medicinal product (ATMP) and more than minimally manipulated (MMM) products to satisfy good manufacturing practice GMP requirements.

ATMP and MMM Products

Per the FDA, the minimal manipulation is "*processing that does not alter the original relevant characteristics of the tissue relating to the tissue's utility for reconstruction, repair, or replacement*". To grant an ATMP designation, EMA considers several factors including the level of manipulation. EMA defines the substantial product manipulations as "*resulting in a change of their biological characteristics, physiological functions or structural properties*". Therefore, products that do not qualify as minimally manipulated and those that qualify for the substantial manipulation designation are required to meet the GMP requirements.

GMP Regulations

GMP regulations cover the entire production process, environment, and personnel. To keep the focus on stem cell-related applications, we will briefly describe the requirements.

A. **Organization and personnel**

   Quality control

   The organization is required to define an entity that is tasked by approving or rejecting process related parts including product containers, in process materials at labeling. This entity shall also have the authority to review production records to ensure full compliance.

   Personnel qualifications

   Staff involved in any component of the production process need to have adequate education and training. The current GMP training shall be conducted on a continuous basis by qualified person. The number of the production staff should be adequate to perform and supervise the entire process. Personnel engaged in production shall wear clothing and protective apparel appropriate for the duties and necessary to protect the product from contamination. Personnel are required to practice good sanitation; their authorization to enter the restricted production areas is provided by supervisory personnel, and is contingent on being healthy and competent to protect the quality of the product.

B. **Building and facilities**

   Design and construction

The facility should be designed to facilitate proper cleaning, maintenance, and operation. Therefore, the facility size need to be appropriate for the intended operations and equipment placement. Likewise, the flow of material and personnel need to be designed to prevent mix up and contamination. Several processes need to be considered during design such as:

- Receipt/holding/storage of components
- Manufacturing and processing operations
- Quarantine/release of products
- Aseptic processing including cleanable surfaces, temperature and humidity control, - HEPA filtered air supply, environmental monitoring ventilation, air filtration, heating, and cooling.
- Adequate use of equipment is required to ensure proper ventilation and proper control over temperature, humidity, dust, pressure, and microorganisms.

C. **Equipment**
Equipment used in production areas need to be of adequate size and construction to protect the product. Likewise, equipment need to be maintained and cleaned properly, and inspected immediately before use. A routine inspection or calibration of equipment used in production is required. Computers need to be controlled, backed up, and limit their change to authorized personnel with appropriate record keeping of any change.

D. **Components, containers, and closures**
Upon receipt, components, containers, and closures (CCC) need to be inspected for damage, and where appropriate, need to be tested before being released for use for production. Each shipment of CCC needs to be identified with a unique code, quarantined until tested. CCC need to meet all approved specifications; and only those that meet all approved specifications may be released for use, otherwise, should be rejected.

E. **Production and process control**
There need to be written procedures that cover all aspects of production, such procedures should be reviewed for change and followed at all times. All components and equipment used in production need to be identified. End of process sample testing procedures need to be written and followed, approved end of process specifications need to be met.

F. **Packaging and labeling**
Similar to components, labeling and packaging materials need to meet approved specifications before being released for use, and each shipment shall be identified with a unique code. Control over label issuance is required, with an approved system to reconcile used, returned or damaged labels. Packaged and labeled products need to be examined and verified for accuracy.

G. **Laboratory control**
A written program is needed to assess the stability of products. Results of such program should be used to identify the appropriate storage conditions and expiration dates. A representative reserve sample need to be retained and stored in conditions consistent with the final product storage.

H. **Records and reports**
Production records need to be maintained. Similarly, records of components containers and closures need to be retained. Equipment use and cleaning log are needed. For each product, a master and batch production record should be described in a written procedure. Such records need to include the product name, strength, component, equipment, manufacturing instructions, specifications, sampling, in process results, identification of persons performing the process, and end dates. Production, control, and labeling records need to be reviewed by quality control unit to ensure compliance with written procedures before being approved and released. Any discrepancy or failure to meet approved specifications need to be fully investigated.

I. **Returned product**
Product that were returned after proper distribution need to be identified. If the reason for return implicate the whole batch, then the batch needs to be investigated. Return products may be used if the return condition did not impose any potential risk, provided the product was tested and met the approved release criteria. Similarly, a product may be reprocessed provided the new product meets the approved release criteria.

Cost of Clinical Production

Production of minimally manipulated cell-based therapeutics is largely standardized and the main reason being the collective experience that both, the clinical team and production teams, had built over the years. Another reason that participated in standardizing these therapeutics is the improvements of tools and materials that collectively improved the entire process. Since the first successful cases, bone marrow transplantation made wide leaps on all of its sides, transplant clinic, collection, and processing. The transplant clinic had made major improvements to the transplant process and disease management. Changes such as the timing of intervention, intensity of preparation regimen, and management of medications had improved the overall outcomes of the transplantation process. Similarly, the bone marrow collection process had improved to maximize the quality of the collected bone marrow while reducing the impact on the donor. In many cases, stem cell collection source had shifted from bone marrow to peripheral blood. This shift provided the transplant and collection teams with increased control over the quality of the product being collected and dramatically improved the donor experience. Due to this shift, the collection can now be completed in in a donor center setting instead of operating rooms. Further, product processing had utilized the higher quality material, the closed production system setting, and improved product storage and monitoring technology.

All these changes /improvements lead to shorter hospital stays, more successful collection rounds, and faster and safer processes. The improvements on the three sides of the cell therapy was reflected on the cost of the entire clinical care. The health care cost of transplant unit became largely predictable, as a result, the insurance companies started to cover the cost of the transplantation process, making a huge difference by providing this option to a wide sector of the patients.

Scientists have utilized the long experience and knowledge generated by transplantation programs around the globe to expand the scope of transplantation to include diseases that once were considered untreatable, and to include therapeutic cellular products that showed promising results in preclinical studies. Widening the scope of cell therapy have led to new set of regulations to ensure patient safety and wellbeing. Regulatory bodies then started to mandate clinical programs and production facilities to comply with these regulations.

Often, scientific breakthroughs stem from institutions of academic setting; these institutions encourage innovations and provide a supportive learning and experimenting environment. Scientists of academic centers usually provide the proof of concept, then with the help of teams who specialize in translational research, move the product from research bench to clinic. Most of the academic institutions include a teaching hospital which makes such translation from research to clinic an easier process.

Running cell-based therapeutic facility incurs fixed expenses whether the facility is being used or not (ten Ham et al., 2020). The level of this cost is related to the size of the facility, but is nonetheless a significant portion that should be considered when estimating the cost of production. Unlike the fixed cost of running the facility, operational cost varies according to the facility and to the volume of services offered. For example, utilizing platforms, such as CliniMACS, requires the use of expensive materials, but is also thought to result in reducing the cost of personnel and facility.

The major drivers of the operational cost are the specialized materials/equipment, and personnel. But this cost may be reduced by developing production modular or sharing of facility/equipment. It has been shown that such sharing can provide small-scale developers an opportunity to develop innovative therapeutics without having to make a substantial investment upfront. Considering the elevated cost of clinical GMP production, the fact that academic centers are the source of innovative therapeutic products, and due to the limited funding of academic centers research studies that depends mainly on grants from sponsors, it is essential to adapt a model that protects the patient safety while supporting academic centers to continue to produce such therapies. There is no magic solution that can create this model, but the following approaches should lay the stage for the improved model.

Scale relevant regulations

Phase I or II Clinical studies that are usually led by an academic institution tend to be of small scale with a small number of patients. Therefore, are well controlled and monitored. Applying a full-scale regulations on such studies can result in a multi-faceted burden. On one side, the GMP facility is often not available in academic institutions, and when available, would require skilled personnel to maintain it in status. On the other side, the mandated process validation (media fills) can be exhaustive to staff and management if applied in its entirety. There needs to be a set of regulation that takes the scale of production in consideration. Clinical studies that enroll few patients per year such as phase I trial, should have a relevant regulation, while phase III/IV clinical studies that enroll hundreds of patients in multicenter studies should be regulated differently.

Often, phase I/II clinical studies are led by academic institutions, then are handed over to big pharma or biotechnology companies once the feasibility studies are completed by demonstrating safety, and possibly efficacy, via Phase I/II studies. It makes logic for the proof of concept phase to be regulated accordingly. Such approach requires a close collaboration between scientists and regulators to create a set of regulations that ensure patient safety while facilitating the scientific innovations.

Semi closed system

During the process development phase, the production system is usually open such as tissue culture flasks. Proof of concept using this type of culture ware includes safety such as sterility, in addition to the efficacy. As the process progresses to be suitable for clinic, production systems are switched from open to closed, materials are changed from research level to higher quality, and production from research laboratory to stem cell laboratory. Sometimes closed production systems do not exist for a specific product purpose. In situations like this, the clinical production is carried out using open culture wear, but in a GMP facility. In other times, which is more often, a semi closed production system exists but the GMP facility does not. For a small scale clinical study, there needs to be some tolerance of facility level in exchange for an as much closed system as possible. This approach requires scientists and manufacturers working side-by-side to invent production systems that protect the product safety while being used in a clean room setting.

We have witnessed several inventions in the last decade, such as the Gas Permeable Rapid Expansion (G-Rex) that in many cases had improved the quality and quantity of the produced cells while enhancing the product safety at the same time by being semi closed (Fig. 10.3).

Process validation

Regulations require the drug manufacturer to demonstrate process safety by a simulating the production process, also known as aseptic process validation, but using

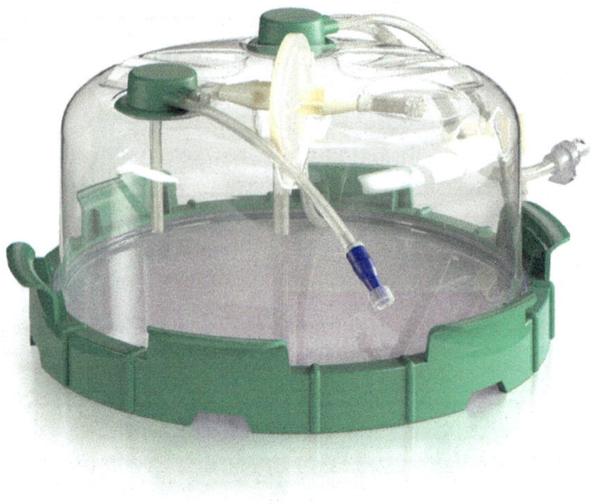

Fig. 10.3 G-Rex 500M

culture media. Following this mandate in its entirety means running simulation processes that simulate all aspects of production including:

- Number of involved staff
- Process duration
- Process interruptions
- Process materials.

While it is understandable how simulation would demonstrate the process safety in its most comprehensive manner, mandating such simulation from small scale production facility can be determinantal. Therefore, there needs to be consideration for the limited capacity of small cell therapy programs.

An ongoing process validation rather than aseptic process simulation might be the required adjustment that small production facility need to thrive while maintaining the minimum safety level. Such ongoing process validation utilizes the production data as they are being generated to demonstrate the process safety/efficacy. The results of process validation should be shared with regulators, and should be used to decide if the process need any type of adjustments. Acknowledging that small scale production facility need an appropriate set of regulations, and clarifying the mandates in terms of process material / process validation should lead to more innovative production systems. Taking all this together, the clinical production cost of cell based therapeutics should not be a hurdle that prevent cell therapy programs from offering new therapies, and should not prevent patients from taking advantage of such therapies.

Quality Assurance

To meet the regulatory standards, and one's own commitment, organizations need to develop a quality plan that serves as a road map to demonstrate how this organization will achieve the committed quality. Such a plan need to be written and controlled document, need to be accepted and supported by management, and need to be unique for specific organization.

Translating the quality plan into actionable procedures is usually achieved via developing standard operating procedures to ensure the quality plan is fully integrated in the day to day operations. Organizations are required to have a quality control unit that have adequate facilities available to them for the testing and approval of all process components such as containers, processing and packaging materials, and final products. Such unit need to have the authority to undertake their responsibility to approve or reject a product or any of its components, procedures or specifications. The quality control responsibilities need to be clearly identified in writing. Accordingly, procedures need to be developed to ensure quality control responsibilities are attained.

Facility

Cell therapy facility and process depends on each other; your process defines the facility specifications that you may utilize, similarly, your facility dictates the type of processes that you may do in that facility.

Minimally manipulated products

Because the product processing of this category is limited, the risk of contamination and mix up is therefore limited. For such products, a small dedicated, or even shared, lab space is sufficient (Leemhuis et al., 2014). However, several factors need to be considered with design and location. Starting a cell processing lab (CPL) is usually intended to support a small autologous cell therapy program at an academic center. Hence, the CPL is expected to be located on campus with close proximity to patients.

For example, sharing lab space or equipment with microbiology or radioactive isotope utilizing laboratory need to be avoided. Standard electrical supply is sufficient, but it is better to connect product storage equipment to uninterrupted power source. Access to the lab need to be restricted, and all biohazardous waste should be handled according to the appropriate hospital procedures.

Lab cleaning procedure should be clearly described and validated. Further, measures should be taken to limit the introduction of contaminants to the processing areas. For example, restrict access to processing staff, limit the delivery of materials directly to processing areas, and control of temperature and humidity to improve the material storage conditions and limit growth of contaminants. CPL requirements of equipment is limited; critical equipment need to regularly be maintained and calibrated. Further, backup for critical equipment should be identified.

Substantially manipulated products

ATMP and MMM product are considered sterile drug product that need to be processed in a GMP compliant manner. GMP facility need to be designed, constructed and maintained to provide protection against cross contamination, buildup of dirt, and any adverse effect to product quality. GMP facility need to be qualified before it may be used for production. Similarly, the cleaning protocol used at these facilities need to be validated and approved. Access to GMP facility need to be limited to qualified personnel members who work in those facilities. Likewise, personnel are required to have a periodic GMP training. Environmental monitoring (EM) is required for the facility, such monitoring is based on a previous classification of the clean areas; it need to be validated and trended. EM is expected to monitor viable and non-viable particles during both at rest and an operation status. Further, air pressure between GMP areas need to be maintained to ensure maintenance of classification.

Equipment that come in contact with products are required to not impact the product quality. Such equipment need to be installed, cleaned, validated, and maintained. Cleaning of GMP facility is a critical part of the facility maintenance. Cleaning materials need to be validated, and measures to prevent development of resistant strains such as use of more than one decent factor should be considered.

References

Amira Ragab, E. L., Barky, E. M., Ali, M., & Mohamed, T. M. (2017). Stem cells, classifications and their clinical applications . *American Journal of Pharmacology & Therapeutics, 1*(1), 001–007.

Blackford, S. J. I., Ng, S. S., Segal, J. M., King, A. J. F., Austin, A. L., Kent, D., Moore, J., Sheldon, M., Ilic, D., Anil Dhawan, B., Mitry, R. R., & Tamir Rashid, S. (2019). Validation of current good manufacturing practice compliant human pluripotent stem cell-derived hepatocytes for cell-based therapy. *Cells Translational Medicine, 8*, 124–137.

Davis, B. M., Loghin, E. R., Conway, K. R., & Zhang, X. (2018). Automated closed-system expansion of pluripotent stem cell aggregates in a rocking-motion bioreactor. *SLAS Technology, 23*(4), 364–373.

Eaker, S., Abraham, E., Allickson, J., Brieva, T. A., Baksh, D., Heathman, T. R. J., Mistry, B., & Zhang, N. (2017). Bioreactors for cell therapies: Current status and future advances. *Cytotherapy, 19*, 9–18.

Eibl, R., Kaiser, S., Lombriser, R., & Eibl, D. (2010) Disposable bioreactors: the current state-of-the-art and recommended applications in biotechnology. *Applied Microbiology and Biotechnology, 86*(1), 41–49.

Frank, N. D., Jones, M. E., Vang, B., & Coeshott, C. (2019). Evaluation of reagents used to coat the hollow-fiber bioreactor membrane of the Quantum Cell Expansion System for the culture of human mesenchymal stem cells. *Materials Science & Engineering C, 96*, 77–85.

Hähnel, A., Pütz, B., Iding, K., Niediek, T., Gudermann, F., & Lütkemeyer, D. (2011). *Evaluation of a Disposable Stirred Tank Bioreactor for Cultivation of Mammalian Cells, 5*(Suppl 8), P54.

Kaiser, L. R. (1992). The future of multihospital systems. *Topics in Health Care Financing, 18*(4), 32–45.

Knazek, R. A., Gullino, P. M., Kohler, P. O., & Dedrick, R. L. (1972). Cell culture on artificial capillaries: An approach to tissue growth in vitro. *Science, 178*(4056), 65–67.

Kropp, C., Massai, D., & Zweigerdt, R. (2017). Progress and challenges in large-scale expansion of human pluripotent stem cells. *Process Biochemistry, 59*, 244–254.

Leemhuis, T., Padley, D., Keever-Taylor, C., Niederwieser, D., Teshima, T., Lanza, F., Chabannon, C., Szabolcs, P., & Bazarbachi, A. (2014). Essential requirements for setting up a stem cell processing laboratory. *Bone Marrow Transplant, 49*(8), 1098–1105.

Lundstrom, K. (2018). Viral vectors in gene therapy. *Diseases, 6*, 42.

Manufacturing Practice Medicinal Products for Human and Veterinary Use, Annex 1, *Manufacture of Sterile Medicinal Products.*

Natalie de Souza. (2018). Organoids. *Nature Methods, 15*(1).

Oppermann, T., Leber, J., Elseberg, C., Salzig, D., & Czermak, P. (2014). *hMSC production in disposable bioreactors in compliance with cGMP guidelines and PAT*. www.americanpharaceuticalreview.com

Pharmaceutical Inspection Convention, *Guide To Good Manufacturing Practice For Medicinal Products*, PART I.

Pörtner, R., & Faschian, R. (2019). *Design and operation of fixed-bed bioreactors for immobilized bacterial culture.* Growing and Handling of Bacterial Cultures.

Ryu, N.-E., Lee, S.-H., & Park, H. (2019). Spheroid culture system methods and applications for mesenchymal stem cells. *Cells, 8*, 1620.

Sampogna, G., Guraya, S. Y., & Forgione, A. (2015). Regenerative medicine: Historical roots and potential strategies in modern medicine. *Journal of Microscopy and Ultrastructure, 3*(3), 101–107.

ten Ham, R. M. T., PharmD, A. M., Hovels, J. H., Frederix, G. W. J., Leufkens, H. G. M., Klungel, O. H., Jedema, I., Veld, S. A. J., Nikolic, T., Van Pel, M., Zwaginga, J. J., Lin, F., de Goede, A. L., Schreibelt, G., Sandy Budde, I., de Vries, J. M., Wilkie, G. M., Dolstra, H., Ovelgonne, H., … Hoefnagel, M. H. N. (2020). What does cell therapy manufacturing cost? *A Framework and Methodology to Facilitate Academic and Other Small-Scale Cell Therapy Manufacturing Costings, Cytotherapy, 000*, 1–10.

Watanabe, K., Ueno, M., Kamiya, D., Nishiyama, A., Matsumura, M., Wataya, T., Takahashi, J.B., Nishikawa, S., Nishikawa, S., Muguruma, K., & Sasai, Y. (2007). A ROCK inhibitor permits survival of dissociated human embryonic stem cells. *Nature Biotechnology, 25*(6), 681–686.

Weber, C., Freimark, D., Pörtner, R., Pino-Grace, P., Pohl, S., Wallrapp, C., Geigle, P., & Czermak, P. (2010). Expansion of human mesenchymal stem cells in a fixed-bed bioreactor system based on non-porous glass carrier – Part A: Inoculation, cultivation, and cell harvest procedures. *International Journal of Artificial Organs, 33*(8), 512–525.

Chapter 11
Stem Cell-Based Products in the Market

Alaa A. A. Aljabali, Khaled I. Seetan, Walhan Alshaer, Ejlal Abu-El-Rub, Mohammad A. Obeid, Dua Kamal, and Murtaza M. Tambuwala

Abstract There has been considerable attention paid to cutaneous stem cell in the area of regenerative medicine as a potential therapeutic goal for managing disorders of the skin and hair, possible use in quickening or supporting the process of wound healing, and numerous types of cancers related to the skin. Cutaneous stem cell plays an essential role in many processes as skin structures' renovation in case of injuries. During hemostasis, the growth of hair follicles and melanocytes is reconstructed and produced. So, obtaining applicable and valid access to skin stem cells for cutaneous interventions, which often include active molecules, is a treasured accomplishment. However, the main hindrance for drug delivery through the topical route is the active barrier represented by the skin against most exogenous molecules' penetration. Thus, this field's research is paying more attention to new strategies to avoid and by-pass this barrier efficiently. In this section, a summary of recent advancements made in stem cell and an investigation of their benefits.

A. A. A. Aljabali (✉) · M. A. Obeid
Department of Pharmaceutics and Pharmaceutical Technology-Faculty of Pharmacy, Yarmouk University, Irbid, Jordan
e-mail: alaaj@yu.edu.jo

K. I. Seetan · E. Abu-El-Rub
Department of Basic Medical Studies, Faculty of Medicine, Yarmouk University, Irbid, Jordan

W. Alshaer
Cell Therapy Center, The University of Jordan, Amman 11942, Jordan

E. Abu-El-Rub
Department of Physiology and Regenerative Medicine/School of Medicine, University of Manitoba, Winnipeg, Canada

D. Kamal
Discipline of Pharmacy, Graduate School of Health, University of Technology, Sydney, NSW 2007, Australia

M. M. Tambuwala
SAAD Centre for Pharmacy and Diabetes, School of Pharmacy and Pharmaceutical Science, Ulster University, Coleraine, UK

F. A. Khan (ed.), *Advances in Application of Stem Cells: From Bench to Clinics*, Stem Cell Biology and Regenerative Medicine 69,
https://doi.org/10.1007/978-3-030-78101-9_11

Keywords Stem cells · Regenerative medicine · Wound healing · Stem cells products

Introduction

Stem cells are broadly defined as underinflated cells to generate and differentiate into many other forms of cells. Stem cells can create every tissue in the human body and therefore have considerable potential for future use in the repair and regeneration of tissue (Aljabali et al., 2020). To be called stem Cells, a cell must have two essential characteristics. First, stem cells must produce progeny without limitations, precisely the same as the source cell. Such a feature often relates to the uncontrolled division of cancer cells, while stem cells' division is tightly supervised. Therefore, stem cells must consider the additional requirement that they may contribute to a specific cell type that is part of a healthy animal (Health, 2009). The generic "stem cell" classification covers several particular kinds of cells (embryonic or adult) (Bajada et al., 2008). There are two major classes of stem cells: pluripotent that can become any cell in the adult body, and multipotent that are restricted to becoming a more limited cell population. Pluripotent stem cells are named because they can differentiate among all kinds of cells in the body. Pluripotent stem cells are present in natural development for a short time in the embryonic stages. They are distinct from specialized multipotent stem cells, ultimately leading to the formation of the body's specialized tissue. A common fate for multipotent stem cells is that they remain quiet without dividing or differentiating, maintaining their position in the pool of stem cells. One example is the stem cells in the bone marrow waiting for body signals to be activated. The stem cells' second fate is to symmetrically self-renew the cell division creates two stem cells of the baby, much like the parent cell. This does not end in distinct progeny, but it raises the supply of stem cells required to produce specialized cells in subsequent divisions (Biehl & Russell, 2009).

Pluripotential stem cells are yet to be utilized in clinical treatment and commercialized since several of the early experimental experiments contribute to the unexpected development of unusual solid tumors or teratomas (Laflamme et al., 2007; Mehta et al., 2020). A combination of cell types from the early germ layers is produced up the teratomas. Later, successful animal studies used pluripotent cell modifications that limited this proliferative capability to a more mature phenotype. Pluripotent cells are used to cure animals effectively. For starters, insulin-producing cells that respond to glucose levels were handled for animals with diabetes (Darabi et al., 2008; Wernig et al., 2008). The FDA is now exploring the prospect of investing in medical research with commercial companies. Many animal trials to combat many disorders have been performed, such as Parkinson's, body dystrophies, and cardiac insufficiency. Companies hope that incorporating newly developed damaged cardiac myocytes into myocardial will allow cardiac operation of stem cell therapy to boost its power. Cardioactive myocyte patches derived from human embryonic stem cells can form a viable human cardiac myocardium after animal transplantation (Kehat et al., 2004; Laflamme et al., 2005). After injecting cardiac myocytes from human embryonic stem cells, the damaged rodent heart showed slightly improved cardiac function (Discher, 2010).

Stem Cells Global Market

The highest prevalence of neurological and inherited disorders has brought in greater demand for new drugs and therapy. Conventional therapies primarily target diseases' symptoms. However, there is an immediate need to identify and then treat the root causes of any disease. Stem cells are a central part of regenerative medicine and a subdivision in cell therapy, as summarized in Table 11.1. The infinite self-renewal and differentiation properties pose a wide variety of disorders that enable Stem cells to become boundaries in regenerative medicine. The worldwide demand for stem cell therapy is equally focused on the involvement of significant juggernauts versus small to medium-sized companies (Abbasalizadeh & Baharvand, 2013). The growth in federal grants and many governments and private institutions, pushing extensive stem cell work, adds to an increasing emphasis on stem cells. Throughout the worldwide stem cell therapy industry, the growing recognition and proven efficacy of stem cell therapy products are essential drivers. Thus, other stem cell therapy drugs are anticipated to be approved and introduced on the market to contribute to a shift in the healthcare sector with further research and development activities (Foley & Whitaker, 2012).

Cell therapy is by far the most quickly evolving and the leading sector in alternative medicine. The global stem cell therapy industry is estimated to be valued at USD 7.342.0 million in 2018. According to Coherent Business Insights, it is projected to show the CAGR by 21.0% over the projection period from 2018 to 2026 (Insights, 2020). However, in 2019, the stem cell therapy market was projected to be worth US$ 1.534.55 million, and it is expected to hit US$ 5.129.66 million by 2027 (Businessinsider, 2020). The rise in financing and infrastructure investments to promote new companies' launch is projected to fuel the development of sustainable cellular care revenues in the industry. For example, Bayer invested USD 215 million in July 2019 for the launch of Century Therapeutics, a US-based biotechnology company to develop therapies for solid tumors and blood cancer.

Further funding from Versant Ventures and Fujifilm cellular dynamics has increased to USD 250 billion by 35 million US dollars (Insights, 2020). Stem cell therapy is praised as the next big breakthrough in clinical improvement. Drug companies are investing heavily in increasing their stem cell portfolios based on the reprogramming of stem cells, thereby demonstrating definite indications that stem cell reprogramming is entirely plausible in treating various life-threatening diseases.

Table 11.1 Stem cell products in the market

Stem cell product generic name®	Manufacturer	Stem cell type	Clinical use
Prochymal	Osiris therapeutics Inc.	Bone marrow MSCs (BM-MSCs)	Graft-versus-host disease (GVHD), Myocardial infarction (MI), Chronic obstructive pulmonary disease (COPD)
TEMCELL HS	JCR Pharmaceuticals	Bone marrow MSCs (BM-MSCs)	Acute graft-versus-host disease (GVHD)
Cartistem (Cartilatist)	MEDIPOST Co. Ltd.	Umbilical cord-derived MSCs (UC-MSCs)	Knee joint osteoarthritis
Darvadstrocel	TiGenix	Adipose tissues derived- MSCs (AT-MSCs)	Complex perianal fistulas in Crohn's disease
Modulatist™	Regenmed Co. Ltd.	Whole Umbilical cord-derived MSCs (UC-MSCs)	Chronic obstructive pulmonary disease (COPD)
Cupistem	Anterogen	Adipose tissues derived- MSCs (AT-MSCs)	Joint osteoarthritis, Crohn's disease fistula
Queencell	Anterogen	Adipose tissues derived- MSCs (AT-MSCs)	Subcutaneous tissue defect
Cartiform	Osiris therapeutics Inc.	Chondrocytes and chondrocytes growth factors	Cartilage repair
Grafix	Osiris therapeutics Inc.	Placenta-derived mixture of endogenous stem cells, collagen matrix, and growth factors	Acute and chronic wounds
Stravix	Osiris therapeutics Inc.	Umbilical cord-derived stem cells and extracellular matrix	Acute and Chronic wounds, diabetic ulcers, pressure ulcers, and surgical wounds
iCART	Takeda pharmaceutical company	Induced pluripotent stem (iPS) cell-derived chimeric antigen receptor (CAR) T-cell	Immunotherapy
ALLOCORD	SSM cardinal glennon children's medical center	Blood-derived hematopoietic progenitor cells (HSCs)	Disorders produced or developed through myeloablative therapy involving the hematopoietic organ

(continued)

Table 11.1 (continued)

Stem cell product generic name®	Manufacturer	Stem cell type	Clinical use
Clevecord	Cleveland cord blood center	Blood-derived hematopoietic progenitor cells (HSCs)	Disorders produced or developed through myeloablative therapy involving the hematopoietic organ
Ducord	Duke university school of medicine	Blood-derived hematopoietic progenitor cells (HSCs)	Disorders produced or developed through myeloablative therapy involving the hematopoietic organ
PROVENGE	Dendreon corporation	CD54+ cells activated with GM-CSF	Asymptomatic or minimally symptomatic metastatic hormone-refractory prostate cancer
GINTUIT	Organogenesis incorporated	Allogeneic cultured keratinocytes and Fibroblasts	Treatment of clogging symptoms of Vascular wound bed
HEMACORD	New York blood center, Inc.	Blood-derived hematopoietic progenitor cells (HSCs)	Disorders produced or developed through myeloablative therapy involving the hematopoietic organ
KYMRIAH	Novartis pharmaceuticals corporation	CD19-directed genetically modified autologous T-cell	Refractory B-cell precursor acute lymphoblastic Leukemia, refractory sizeable B-cell lymphoma
TECARTUS	Kite pharma, Inc.	CD19-directed genetically modified autologous T-cell	Refractory mantle cell lymphoma
YESCARTA	Kite pharma, Inc.	CD19-directed genetically modified autologous T-cell	Refractory large B-cell lymphoma

Stem Cells Products in Diagnostics and Disease Treatment

Currently, several products are available in the market or in the final stages of clinical trials based on stem cells. These products represent good alternatives for treating various diseases, especially in the advanced stages where the current conventional therapies become unsuccessful. These stem cell-based products were treated with several conditions, such as myocardial infarction, osteoporosis, acute or chronic wounds, diabetic ulcers, anal fistulas, and many others. For example, Prochymal®

is an adult human mesenchymal stem cell produced by Osiris Therapeutics Inc to treat myocardial infarction. The use of Prochymal® for six months resulted in a significant improvement in the ejection fraction and the disease symptoms of the treated patients. This was associated with a decrease in ventricular arrhythmias and improved overall pulmonary functions (Hare et al., 2009a). Darvadstrocel (Alofisel®) is another human allogeneic adipose-derived MSC (AT-MSCs) prepared by TiGenix and approved in the EU for the treatment of complicated perianal fistulas. Alofisel® is the first MSC-based therapy for treating perianal fistulas that are resistant to conventional therapy in patients with non-active/mildly active Crohn's disease (Scott, 2018). Stem cell-based therapies have been applied for the treatment of joints and knee disorders. For example, Cartiform, prepared by Osiris Therapeutics Inc was approved for Cartilage repair (Melugin et al., 2020). On the other hand, Custom was developed by Anterogen, which is based on AT-MSCs and approved for the treatment of Joint osteoarthritis (Syed & Evans, 2013). Other stem cell-based products that are currently approved or under clinical trials can be found in Table 11.1. Moreover, several clinical trials in various phases are presently conducted for different stem cell-based products, which will seek the required regulatory approvals soon and will be translated into developments in the market.

Stem Cells in Treatment of Neurological Disorders

Various stem cell applications have usually been treated with multiple neurological diseases with successful outcomes, and the list has expanded ever since. For example, Brain paralysis is a condition caused by brain damage during or shortly after pregnancy. Convulsions, visual impairment, speech difficulties, hearing loss, lack of communication and coordination, and cognitive delays often accompany it. Laboratory studies in animal models with test-related or traumatic strokes have shown that stem cell therapy has the potential for betterment. The possibility to perform such transplants by injecting them into the brain rather than directly into the vasculature increases the probability of prompt clinical trials in humans (Abbasalizadeh & Baharvand, 2013). Consequently, human parameters suitable for testing cell injections, such as cell type, implant period, and impact on function, ought systemically to be conducted in animal models. Studies have shown that advantages of accidents may be obtained in animals of experimental strokes, or catastrophic injury, in the expectation that these trials can be rapidly converted into reality.

The researchers under Professor Kiminobu Sugaya of the University of Central Florida found that adult human stem cells develop into brain cells after implantation in adult rat brains. Neural stem cell (NSC) implants open the option and advance of supplying a much more lasting solution than present medicaments, with an innovative strategy for treating many brain disorders, including AD. These cells can migrate to lesion areas of the brain following grafting and differentiate into the required type of cells lacking in the brain, providing the cell population required to foster recovery theoretically (Oliveira Jr & Hodges, 2005).

The embryonic stem cells of mesenchymal and ectodermal stem cells originating from 4–8 weeks of old embryo corpse organ active growth zones were used for or therapy of Multiple Sclerosis. Suspensions of 1–3 mL with cells of 0,1–100 × 105/ml were performed. 2–4 different suspensions were implemented during treatment, and intracavitary, intravenous, and subcutaneous administration was applied. Early post-transplant recovery syndrome was found after therapy in 70% of patients, with reduced fatigue, increased appetite, and mood as the major depression symptoms decreased (Kimiskidis et al., 2008). On the other hand, the authors documented the potential for extreme longevity, migration, and Parkinson's symptoms. This is specifically linked to decreased amounts of dopamine in the nigrostriatal system in undifferentiated human neural stem cells (hNSCs), which were transplanted into Parkinson's infected chimpanzees (Gavira et al., 2006b, 2010; Hata et al., 2006; Sanberg, 2007).

Stem Cells in Treatment of Cardiovascular System

Nearly two decades have elapsed after initial attempts in cardiac SC therapy were made. Today, various kinds of SCs have been tested for their cardiovascular regeneration ability on multiple development levels. A substantial number of researchers have examined thoroughly how these cells behave in diverse models of small and large animals, including rats, cats, dogs, and pigs, in the care of ischemic and non-ischemic cardiomyopathies (Fukushima et al., 2008).

Such studies have shown that myoblast skeletal can be distinguished from myotubes, decreased myocardial thrombosis, reduced remodeling of the ventricle, and enhanced myocardial efficacy. These create powerfully were likely coming into clinical studies. Many of the little experiments have shown enhancement of the functional class of left ventricular discharge (LVEF) and the New York Heart Association (NYHA) and improved provincial wall motion following skeletal myoblasts implantation (Gavira et al., 2006a; Hagège et al., 2006). It was first observed by Beltrami et al. in an adult heart that c-Kit + cells could differentiate into cardiomyocytes, endothelial cells, and muscle cells to help the recovery of damaged heart tissue and others (Bearzi et al., 2007; Ellison et al., 2013). Many other specific populations have been identified in the last decade from antigen stem cells (Sca)-1 + cell, cardiac stem cells (CSC), and cardiac progenitor cells (CPCs), such as cardio sphere-derived cells (CDC) and protein (Isl)-1 + factor insulin factor enhancer cells and cardiac side population cells (Oyama et al., 2007; Uchida et al., 2013; White et al., 2013). Numerous fate-mapping studies have shown that the established cardiovascular stem and progenitor cells will contribute to adult cardiomyogenesis (Malliaras et al., 2013). In 2011, there was no evidence of mortality or CSC-related adverse consequences of intracoronary c-kit + CSC infusion in patients with ischemic cardiomyopathy in results from the very first preclinical and clinical trial (SCIPIO, NCT00474461) (Bolli et al., 2011). MRI assessment indicated that regional and global heart functions, infarction size decrease, and viable tissue increase 4 and 12 months after stem cell injection were increased (Chugh et al., 2012). The stem cells' ability, largely focused on implicit/paracrine or direct pathways, to regenerate damaged tissues. The latter involves active heart differentiation and incorporation of implanted stem cells into the myocardium to mitigate the lack of cardiomyocytes or endothelial cells. Data from several in vitro and in vivo trials have also led to the notion that paracrine activation is the standard process for mediating stem cell operation (Chong et al., 2014; Malliaras et al., 2014).

Technological Advancement Involving Stem Cells Therapy

Following the advent of stem cells' reprogramming, scientists worldwide have been committed to creating new methods of development and construction of human embryonic stem cells since the discovery of differentiation mechanisms to create human-mediated pluripotent stem cells. Recent developments in genetic engineering have provided the ability to change cells, enabling exact and complex directed functionalities, mainly stem cells. Nevertheless, unmodified stem cells also have a high capacity for healing and still are utilized in conventional therapies. For example, the primary cause of mortality in the USA is cardiovascular disease. Stem cells have developed as a possible treatment agent for permanently damaged tissue during the past decade, and MSCs for therapy have been thoroughly studied. Preliminary results showed promising results in cardiac tissue repair and generation (Karantalis & Hare, 2015). Double-blind placebo experiment in 53 post-MI patients with comparable adverse effects in each group using intravenous human allogeneic MSCs. In the pulmonary function test group that received MSC transfer, Ambulatory EKG showed reduced ventricular tachycardia episodes and improved forced exhibition volumes in one second. The MSC group significantly improved the global symptom score and ejection fractions (Hare et al., 2009b).

Related findings were found with an injection of MSC inpatient ischemic cardiomyopathies, contributing to increased functional abilities, quality of life, and ventricular remodeling (Hare et al., 2012). Investigations have shown that treatment with genetically modified bone marrow-derived MSCs that secrete IFN-β has resulted in the introduction into the malignant tissue and locally conceals IFN-β and prevents tumor development in human xenotransplantation melanoma mouse model. The systemic administration of IFN-β does not achieve this significant effect. In an investigative purpose with a canine melanoma model, cisplatin usage increases therapeutic efficacy in IFN-β-transducted adipose tissue-derived MSCs as carriers of anti-tumor drugs, has been reported (Ahn et al., 2013). Numerous different cytokines and tumor suppressor genes are currently used in the genetic engineering of MSCs for anticancer procedures. Such therapies attempt to end the cancer cells preferentially to improve the therapy's effectiveness and reduce toxic effects. Another of these is the tumor necrosis factor-related apoptosis-inducing ligand (TRAIL) related to tumor necrosis. TRAIL is a transmembrane type II protein that activates a specific apoptosis signal in tumor cells when normal cells are avoided. Despite the potential advantages of TRAIL for cancer treatments, TRAIL-based treatments' development poses many challenges: its short half-life of about 30 min, limited bioavailability, and poor pharmacokinetics (Du et al., 2012). MSCs may be used as efficient transport mechanisms for brain tumor treatments, including IL-12 and other antineoplastic agents. Peripheral administration of human MSCs transduced in the nude model of renal cell carcinoma in mice with a recombinants adenoviral vector that expresses murine IL-12 showed that these IL-12-expressing MSCs are homing to the tumor cells. These cells naturally produced IL-12 regionally, with a slight rise in IL-12 levels, thereby demonstrating the capability of adult IL-12 MSCs to mitigate cancer growth and to prolong life (Gao et al., 2010).

Stem Cells and Vaccine Production

A routine tolerance to a host of respiratory disorders such as pneumonia, measles, and chickenpox, and many others is improved by vaccination. The vaccine induces cellular and humoral immunity against viral infection, transplanting, and carcinogen-induced in humans or animals. Embryonic stem cells (ESC) are capable of fast clonal proliferation and self-renewal and can reside and prosper in different human body environments. Similarity has long been recognized between fetal development and cancer and the potential use of ESC as a vaccine have well been established and entered clinical trials on some models by the discovery of oncofetal proteins and antigens such as α-fetoprotein (AFP) had been discovered (Chism et al., 1978; Trojan et al., 1995), human chorionic gonadotropic (HCG) (Matzuk et al., 1987), and carcinoembryonic antigen (CEA) (Haynes et al., 1985). The proteins, as mentioned above, are tumor-related protein or antigen (TAA) synthesized during embryonic development and reappeared during cancer development in adults. These proteins are known for the detection and surveillance of cancer by biomarkers (Purswani & Talwar, 2011). In the induced pluripotent stem (iPSCs), four transcription factors may be produced into adult somatic cells, which turn their transcriptions and their epigenetic state into a pluripotent model resembling the ESCs. Like ESCs, iPSCs share cancer cell genetic and transcriptomic signatures, including protein markers that can be identified via the immune system (de Almeida et al., 2014; Ghosh et al., 2011; Takahashi & Yamanaka, 2006). Over a century ago, Schöne recognized that immunization with embryonic/fetal tissue could trigger the rejection of the transplanted tumors in animals (Yang et al., 2006). Later research suggested that animal vaccination of embryonic material produced humoral and cell immunity from transplantable tumors and cancer-induced tumors and endorsed the notion that antigens that are exchanged between fetal and cancer cells can trigger immune to tumors. Anti-tumor immunity. Recent studies have shown that oncofetal cancer vaccines based on antigen may elicit strong T-cell responses (Fishman et al., 1975; Purswani & Talwar, 2011).

Stem Cells and Regenerative Medicines

The present-day dilemmas for many healthcare systems worldwide are the increasing prevalence of chronic diseases and the need to implement innovative solutions that limit organ dysfunction, prevent tissue degeneration and offer a replacement for damaged tissues (Giwa et al., 2017; Jessop et al., 2016). Conventionally prescribed medications can only manage and control the symptoms but cannot repair and regenerate the injured organs3. Regenerative medicine, a recently emerged and fast-growing branch of medical science, deals with the functional restoration of damaged tissues to end the anguish of many patients with severe injuries or chronic diseases. Regenerative medicine converges the principles of stem cells-based research and molecular biology to regenerate and reinvigorate damaged cells, tissues, and organs (Mandrycky et al., 2017; Rosenthal & Badylak, 2016).

Stem cells are known for their miraculous abilities to differentiate into various committed cell types and regenerate the damaged organs (Mahla, 2016). Based on the source and origin from where Stem cells can be obtained and their stemness and transdifferentiation potential, stem cells can be divided into four broad categories accordingly: embryonic (ESCs), extraembryonic, fetal (FSCs), and adult stem cells (ASCs, among them mesenchymal stem cells—MSCs). Embryonic cells (ESCs) are pluripotent, derived 5–6 days post-fertilization from the blastocyst's inner cell mass before the implantation of the embryo. Extraembryonic stem cells can be obtained from amniotic fluid and placenta, which have a mixture of different populations of stem cells with different potency abilities; fetal Stem cells (FSCs) are multi-potent cells located in the fetal tissues and embryonic annexes. Adult stem cells (ASCs) are multipotent stem cells in fully developed organs that reside in niches with a specialized microenvironment to maintain their stemness and self-renewal. In vitro produced type of pluripotent ESC-like stem cells, epigenetically transforming, reprogramming and reestablishing the terminally differentiated cell's endogenous pluripotence factors, is regarded as a landmark discovery in the field of stem cells as an alternate way to establish an infinite stems cell source (Mahla, 2016; Rosenthal & Badylak, 2016).

Bench to bedside translational clinical trials featuring different stem cells are increasing dramatically, with 7983 registered clinical trials (ClinicalTrials.gov) (Mahla, 2016). Mesenchymal stem cells (MSCs) are considered the topmost close to clinic stem cells with 1146 registered clinical trials (ClinicalTrials.gov) as shown in Fig. 11.1 (Mahla, 2016; Rad et al., 2019). MSCs were successfully able to treat various diseases, including diabetes, cardiovascular diseases, neurodegenerative disorders, liver diseases, and kidney diseases. MSCs prevailed over other stem cells because they can be easily isolated, maintained, and expanded with no ethical or teratoma risk (Pittenger et al., 2019). Embryonic stem cells (ESCs) scientifically are the ideal stem cells for regenerating damaged organs; however, ethical concerns and high teratoma risk limit these cells' translational application (King et al., 2014). Induced pluripotent stem cells (iPSCs) can be an excellent alternative to avoid ethical issues with similar characteristics and differentiation potential to ESCs. However,

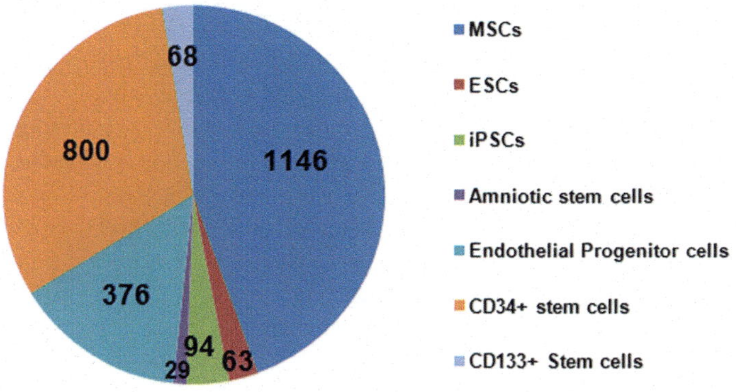

Fig. 11.1 Stem cells-based registered clinical trial

these cells still harbor the risk of teratoma, the uncontrollable pre-transplantation differentiation, and the risk of genotoxicity are the main issues that limit their clinical translation (Yoshihara et al., 2017).

Demand for launching regenerative medicine therapeutic products in the market is influenced by the need to overcome the substantial increase in the number of patients with degenerative and chronic diseases that have been exhausted many healthcare systems. The implementation of new strategies that combined advanced technologies such as nanotechnology and bioengineering with stem cells can provide and ensure efficient as well as long-term replacement and regeneration of damaged organs, tissues, and cells. It can be a promising therapeutic option for many degenerative conditions such as Parkinson's disease, Alzheimer's, stroke, and heart failure, which are poorly controlled with currently available conventional medications (Hofmann, 2014; Kwon et al., 2018). The World Regenerative Medicines Market foretold that the commercial potential of the regenerative medicine market is thriving with revenues that can reach up to $67.5 billion in 20,202. Most of the revenue is coming from bone marrow-derived products. Many governments across Europe and the USA are investing extensively in projects related to tissue engineering and regenerative medicine with endorsing new guidelines and policies that facilitate conducting a wide range of clinical trials (Rose et al., 2018).

Mesenchymal stem cells (MSCs) have been corroborated to have promising regenerative potential and ideal stem cell candidates in regenerative medicine. The regenerative molecular mechanisms of MSCs are based on their abilities to differentiate in multi-cell lineages. Further, several studies highlighted that MSCs exerted their therapeutic functions mainly by secreting large numbers of paracrine factors that have mitogenic, angiogenic, antiapoptotic, antifibrosis chemo-attractant characteristics (Linero & Chaparro, 2014). These MSCs-secreted paracrine factors can also promote the growth and the differentiation of tissue-residents stem cells following a harmful situation. The growing evidence supporting the regenerative and therapeutic role of MSCS-secreted paracrine factors has led to introducing a novel therapeutic product called "MSCs secretome," which comprises the proteins presented in MSCs culture media certain conditions (Eleuteri & Fierabracci, 2019). Paracrine factors in the MSCs-secretomes including growth factors, chemokines, cytokines, interleukins, perform many cellular and biological functions such as inhibiting the apoptosis and necrosis of cells from tolerating stressful microenvironments, inducing the proliferation of residential progenitor or stem cells, and enhancing neovascularization at the damaged site to reinvigorate affected tissues with oxygen and nutrients. MSCs immune modulation and immune suppression are other important therapeutic mechanisms that can be modulated and modified to treat a wide range of immunological diseases and to enhance the engraftment of MSCs itself (Weiss & Dahlke, 2019).

MSCs are the most suitable stem cells for bench-bedside translational applications. MSCs are superior to other stem cells because they can be isolated, maintained, and expanded quickly with no ethical or safety issues (Sugarman, 2008). More importantly, MSCs express deficient levels of significant histocompatibility factor class I (MHC- I) and do not express significant histocompatibility factor class II (MHC- II), which are immunogenic markers. Based on that, one young, healthy donor can give cells to multiple unmatched recipients "Universal donor phenomena" with minimal rejection chances8. Based on the reasons mentioned above, MSCs are the perfect stem cells for "Off-the-Shelf" products (Sheridan, 2018).

In 2012, Canada approved the first off-the-shelf product derived from mesenchymal stem cells called "Prochymal" to treat graft-versus-host disease (GVHD) (Prasad et al., 2011). This product contains MSCs from allogenic human bone marrow. In 2016, a similar product for treating (GVHD) was manufactured in Japan under the name of "Temcell HS" which also contains MSCs from bone marrow (Yamahara et al., 2019). In 2014, Korean pharmaceutical companies were developed another MSCs off-the-shelf product named "Cartistem" for treating knee osteoarthritis (Vega et al., 2015). In Europe, the Committee for Medicinal Products for Human Use (CHMP) approved (Cx601, Darvadstrocel), the first off-shelf MSC product that treats a sophisticated form of Crohn's disease (CHMP) (Meng et al., 2020). The above mentioned MSCs products are primarily based on the immune modulation abilities of MSCs.

Recently, some of the approved off-the-shelf MSCs products were used for other conditions unrelated to the immune system. In a randomized, double-blind, placebo-controlled clinical trial, Prochymal was intravenously transfused into 53 MI patients at different doses: 0.5, 1.6, and five million cells/kg. The results showed a significant improvement in all patients' ejection fraction after hMSC-treatment (Madigan & Atoui, 2018). In 2017, the first off-the-shelf cryopreserved adipose-derived stromal cells (CSCC_ASC) product from healthy donors was tested clinically for intramyocardial injection in patients with ischemic heart disease and ischemic heart failure (IHF) in Denmark (Kastrup et al., 2017). These adipose tissues derived from MSCs have been cultivated and extended in bioreactors without using animal constituents. An umbilical cord-derived MSCs were produced by Regenmedlab (Regenmed Co. Ltd) in Vietnam under the name of "Modulatist™," which has healthy immunomodulation compared to adipose tissue-derived or bone marrow-derived MSCs. Modulatist™ was clinically useful for treating and improving the symptoms in two patients suffering from a chronic obstructive pulmonary disease (COPD) (Zhou et al., 2011). Modulatist™ was administered to the COPD patients intravenously (106 cells/kg) and then evaluated at 1, 3, and 5 months post-infusion. Modulatist™ significantly improves the Patient quality of life with a significant reduction of acute COPD exacerbation. Notably, there were no adverse side effects reported with Modulatist™ at five months follow-up (Le et al., 2016). Another South Korea's Food and Drug Administration marketing product approved was Cupistem® (Anterogen). It consists of autologous adipose-derived MSCs used to reduce inflammation and regenerate damaged joint tissues and is also indicated for the treatment of Crohn's fistula (Gao et al., 2016).

With the increasing demands for MSCs off-the-shelf products and the increasing government investments in the cell manufacturing market, Pharmaceutical companies should direct efforts to manufacture a consistent product with minimal patch-patch variations to ensure the delivery of these MSCs products to the patients in a way that maximizes therapeutic efficacy. Further, they should be designing and executing successful clinical trials for post-marketing quality control and monitoring any possible side effects. The ultimate target of MSCs products -marketing companies is to assure the availability of high volume, high quality, and low-cost human MSCs on the market to ensure that many suffering patients with various incapacitating conditions can get benefit from these products (Robb et al., 2019).

Other approved stem cell products developed by Osiris Therapeutics company and are available in the market include Cartiform®, a cryopreserved osteochondral allograft composed of chondrocytes chondrocytes growth factors used for cartilage repair (Mirzayan et al., 2018). Grafix® is the only approved cryopreserved placental membrane that contains a mixture of endogenous stem cells, collagen matrix, and growth factors for managing acute and chronic wounds (Gibbons, 2015). Stravix® is another product used to treat acute and chronic wounds, diabetic ulcers, pressure ulcers, and surgical wounds (Ha et al., 2017). Stravix® contains a mixture of umbilical cord-derived stem cells and extracellular matrix, which conforms to the exposed injury site and forms an adhesion barrier.

The production of pluripotent mediated stem cells (iPSCs) offers new insights into the potential generation of successful cellular immunotherapy products. iPSCs are identical to embryonic stem cells and unrestricted in vitro development and lymphoid lineage differentiation. The generation of iPSCs cell lines allows for constant and continuous production of different types of immune cells, including T and NK lymphocytes. It offers solutions to the limited availability or expansion of primary immune cells (Bernareggi et al., 2019; Nianias & Themeli, 2019). Furthermore, iPSCs can be genetically transformed in vitro to generate modified immune cells with more therapeutic potential to overcome the difficulties associated with direct gene editing for primary immune cells. The generation of safe iPSC cell lines that are bearing genetic modifications would smooth the progress of developing "off-the-shelf" cellular immunotherapeutics for more patients. Recently, the center for iPSCs cell research and application (CiRA) at Kyoto University and Takeda pharmaceutical company limited "Takeda" announced that a novel immunotherapy product known as induced pluripotent stem (iPS) cell-derived chimeric antigen receptor (CAR) T-cell therapy (iCART) had been transferred to Takeda Company for clinical testing (Gee et al., 2020). Takeda pharmaceutical company is getting close to receiving FDA approval for commercializing the iCART product, which will be an epoch-making jump in the field of immunotherapy.

Hematopoietic stem cells (HSCs) are available in the market as FDA approves many HSCs products with different formulations and doses. Hematopoietic stem cells (HSCs) are used for patients with disorders affecting the hematopoietic system, inherited, acquired, or malignant. HSCs approved products mainly contain allogeneic cord blood-derived hematopoietic progenitor cells such as ALLOCORD, manufactured by SSM Cardinal Glennon Children's Medical Center (Ikeda et al., 2018), Clevecord which is manufactured by Cleveland Cord Blood Center and Ducord developed by Duke University School of Medicine. PROVENGE is a stem cell product generated by Dendreon Corporation to treat asymptomatic or minimally symptomatic metastatic hormone-refractory prostate cancer. PROVENGE contains autologous CD54 + cells activated with GM-CSF. The stem cells market moves forward with significant steps as more products will be available for various debilitating diseases. The launching of more stem cells derived products will hugely change the current treatment guidelines because stem cells are the standard gold therapy for regenerating and repairing damaged tissues and organs, which is impossible to be achieved with conventionally prescribed medications (Ikeda et al., 2018; Moore et al., 2020).

Stem Cell-Based Products in the Market

Allocord®

Allocord® is a biologic drug based on an allogeneic cord blood hematopoietic progenitor cell therapy (HPC, Cord Blood). This was approved Allocord®, FDA, in 2013 and manufactured by the St. Louis Cord Blood Bank (SLCBB) of the SSM Cardinal Glennon Children's Medical Center. Allocord® planned to be used in unrelated donors for the hematopoietic stem/progenitor cell transplantation procedures in patients with hematopoietic system disorders that are acquired, inherited, or resulted from myeloablative treatments. The risk–benefit assessment of Allocord® is very individualized, and therefore the decision of using Allocord® is based on hematopoietic and immunologic reconstitution, which is based on patients' characteristics including disease, risk factors, stage, and specific manifestations of the disease (Van Pham, 2016).

Hemacord®

Hemacord® is the first cord blood product approved by the US FDA in 2011 for New York Blood Center, Inc. Hemacord® is a biologic medicinal drug focused on the transplantation of HSCs (hematopoietic allogeneic stem/progenitor cells) and intended for use in the care of HSCs, including myeloablative therapies, obtained, inherited or arising from. Hemacord® is prepared by drawing the umbilical cord blood from the placenta of the newborn, followed by processing of purification, concentration, and storage. Once the Hemacord® injected into the recipients, the stem/progenitor cells migrated into the bone marrow. They started to proliferate and differentiate into mature blood cells to provide complete blood cells to restore adult blood cells (Allison, 2012).

Ducord®

Ducord® is a cord blood product approved by the US FDA in 2012 for Duke University School of Medicine, Carolinas Cord Blood Bank. Ducord® is like Allocord® and Hemacord® by means of product and clinical applications. Ducord® is a synthetic medication used for the diagnosis of blood pressure conditions (including those that have been developed, born or resulted in myeloablative treatments) dependent on transplantation by allogenic hematopoietic stem / progenitor (HSCs) (Cuende et al., 2018).

Prochymal®

Procgymal® (Remestemcel-L) is an allogeneic stem cell-based therapy developed by Osiris Therapeutics Inc in 2009. Procgymal® received conditional approval to treat pediatric steroid-refractory Graft-vs-Host Disease (GvHD) in Canada, New Zealand, and Japan. GvHD is a critical complication result from the MHC-mismatched allograft transplantation that affects 30–70% of the transplantation recipients. To date, steroids are the first-line therapy for the treatment of GvHD with 30 to 50% response rates (Locatelli et al., 2017). Therefore, patients unresponsive to steroidal therapy can benefit from second-line therapy, thereby reducing the mortality and morbidity rates that result from GvHD. Mesenchymal stem cells (MSCs) own the capability of modulating the immune system through cytokines and/or inhibition of immune cells. In 2009, a phase III clinical trial was conducted to evaluate BM-derived MSCs to treat steroid-refractory GvHD. The results showed 10 to 30% higher long-time survival chances in patients with steroid-refractory GvHD and lead to the approval of Procgymal®. Although Procgymal® approval was an outstanding achievement in stem cell-based therapy, it is still not broadly used due to strict regulations and the high production cost (Kebriaei et al., 2020).

Alofisel®

Alofisel® (Darvadstrocel) is an allogeneic stem cell therapy approved by the European Commission (EM) for TiGenix NV/Takeda Pharmaceutical Company. Alofisel® is an adipose human mesenchymal stem cell (aMSC), licensed in adult Crohn's disease patients for diagnosis and control of complicated perianal fistulas. A phase III clinical trial showed that Alofisel® provided a combined remission in more than 50% of patients, which is maintained after one year of treatment, compared to 34% of the control group. Alofisel® showed immunomodulatory and anti-inflammatory effects at the inflammation sites through impairing the proliferation of activated lymphocytes and reducing the production of the inflammatory cytokines (Panés et al., 2018). Interestingly, Alofisel® has received an orphan designation by the European Commission. Similarly, the product used by Cupistem®, which has been licensed by the Ministry of Food and Drug Security of Korea (MFDS) by Anterogen Co., Ltd, is identical to the medication used for the Alofisel® (Syed & Evans, 2013). Cupistem® showed complete remission in 82% of patients with complicated Crohn's fistula at week 8 of treatment, and around 81% of these patients showed sustained response at week 96.

Holoclar®

Haloclar® (GPLSCD01) is an expanded autologous human corneal epithelial cell containing stem cells approved in 2015 by the European union to Chiesi Farmaceutici S.p.A. to treat limbal stem cells deficiency (LSCD). LSCD is a severe and rare eye injury occurred in one or both eyes as a result of physical or chemical factors. LSCD patients lack the corneal limbal stem cells that are usually responsible for continuous regeneration of the cornea, thereby protecting the eyes from reduction or vision loss. Haloclar® treatment is focused on the creation of an ex vivo gray utilizing a biopsy of an unexposed limbus to remove autologous human corneal epituitary cells. The formed graft then can be transplanted into the injured eyes to restore the normal function of limbal stem cells thereby regenerating the corneal surface and repair patient eyes (Pellegrini et al., 2018).

Zalmoxis®

Zalmoxis® is a patient-specific cell therapy received the conditional marketing authorization by The European Medicines Agency (EMA) in the European Union (EU) as advanced therapy medicinal products to support the haploidentical hematopoietic stem cell transplant (HSCT) in patients suffering blood cancers. EMA approved Zalmoxis® in 2016 for MolMed S.p.A. chemotherapy is the primary therapeutic strategy for patients with blood cancer (Patrikoski et al., 2019). However, the repetitive cycles of chemotherapy malfunctioning the hematopoietic and immune system; thereby, patients can benefit from the transplantation of hematopoietic stem cells from donors to regenerate the functions of the hematopoietic and immune system. However, in some cases, the T lymphocytes resulted from the differentiation of donor hematopoietic stem cells can be involved in the developing GvHD and end with lethal results. Therefore, Zalmoxis® can protect the protective action of T-cells through genetic modification of the T-cells by insertion of the HSV-TK suicide gene. Therefore, the HSV-TK suicide gene allows the control of the modified T-cells once the symptoms of GvHD start to appear through the administration of ganciclovir, which in combination with HSV-TK suicide gene induce T-cell death (Mohty et al., 2016). Currently, Kiadis Pharma is developing Kiadis's Theralux platform as a photodynamic system for allodepletion of T-cells to minimize the risk of GvHD that may result from donor hematopoietic stem cells transplantation. The Allodepleted T-cell immunotherapeutics intended as an adjunct therapy to HSCT in Leukemia and thalassemia (Perruccio et al., 2008).

Clinical Trials

To date and based on ClinicalTrials.gov, there are more than 3000 trials are ongoing to investigate the therapeutic potency of different types of stem cells against different diseases Fig. 11.2a (Aly, 2020; ClinicalTrials.gov, 2020). The number of clinical trials and approved drugs is continuously growing worldwide Fig. 11.2b and c. Using stem cells in different therapeutic approaches is tricky and challenging. For example, the standardization of protocols for the isolation, characterization, and preservation of stem cells are highly demanded to provide low batch-to-batch variation and reproducing close therapeutic outcomes (Aly, 2020). Therefore, the clinical implementation of stem cell-based therapy requires the ability to bank the stem cells properly, thereby providing high quality and enough numbers of stem cells upon patient treatment and for future use. With the increased numbers of clinical trials on stem cells, regulatory guidelines are highly demanded to ensure the safety and efficacy of stem cell-based therapy.

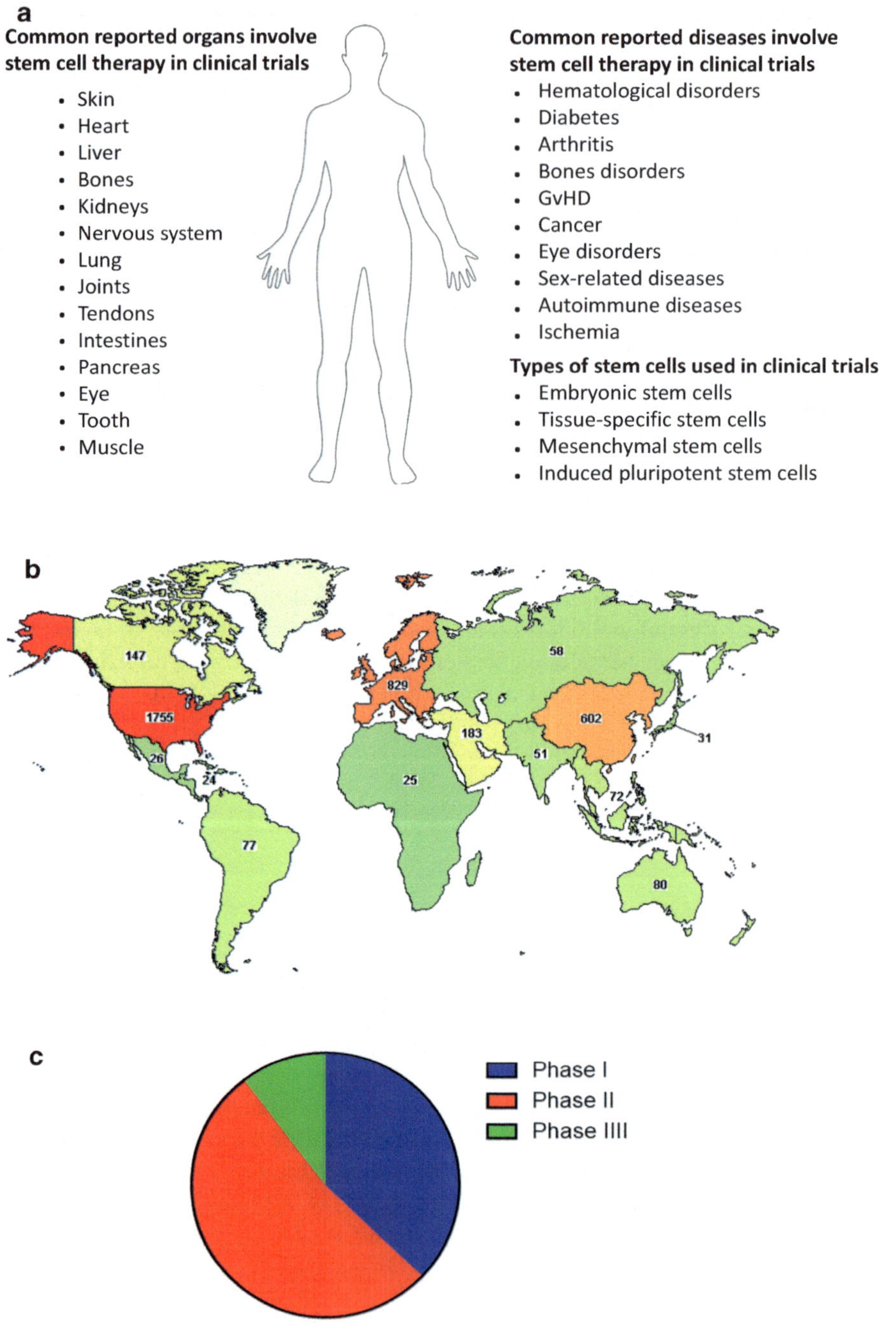
a
Common reported organs involve stem cell therapy in clinical trials
Skin
Heart
Liver
Bones
Kidneys
Nervous system
Lung
Joints
Tendons
Intestines
Pancreas
Eye
Tooth
Muscle
Common reported diseases involve stem cell therapy in clinical trials
Hematological disorders
Diabetes
Arthritis
Bones disorders
GvHD
Cancer
Eye disorders
Sex-related diseases
Autoimmune diseases
Ischemia
Types of stem cells used in clinical trials
Embryonic stem cells
Tissue-specific stem cells
Mesenchymal stem cells
Induced pluripotent stem cells
b
147
1755
829
58
602
183
51
31
26
24
25
72
77
80
c
Phase I
Phase II
Phase IIII

◂**Fig. 11.2** Clinical trials involve stem cells therapy. **a** The reported organs, diseases, and types of stem cells in clinical trials, **b** The global distribution of clinical trials involved stem cells, **c** Pie chart showing different phases of clinical trials on stem cells. The information is based on the ongoing clinical trials using stem cells as intervention that have been deposited at https://clinicaltrials.gov/ in July 2020. Moreover, these regulatory guidelines should organize and solve the controversial scientific, ethical, and legal issues in stem cell-based therapy. For example, the guidelines provided by the Food and Drug Administration (FDA) described that the stem cells planned for homogenous treatment and minimally manipulated did not require a premarket approval and subject to the efficacy in treating the disease. Finally, safety and standardization protocols that regulate the use of cellular products such as using Xeno-free culture media, growth factors, and culture supplies are highly demanded (Cyranoski, 2019; Marks et al., 2017; Volarevic et al., 2018)

Stem Cells Products in Skin Treatment and Cosmetics

Most products of stem cells available right now are plant-derived, therefore hindering the feasibility of their potential use as human stem cells (Ghieh et al., 2015). However, adipose stem cells (ASCs) are now gaining much momentum for their potential use in stem cell research and medicine (Kern et al., 2006), as ASCs are now utilized in the correction of skin defects, wound healing, facial skin rejuvenation, and scar remodeling (Kim et al., 2008). ASCs display self-renewal and developmental plasticity properties in multiple lineages and are in subcutaneous adipose tissue, specifically the stromal-vascular interface. Skin rejuvenation effects of ASCs are attributed to the properties of producing and secreting definite growth factors (Kim et al., 2007b), e.g., ASCs, in addition to the conditioned medium of ASCs (ASC-CM), enhance healing of the skin, prevent melanogenesis, correct wrinkling, and facilitate hair growth (Kim et al., 2009). ASC-derived cells or protein-based therapies are adequate for skin regeneration in the setting of the center, e.g., (PLA) cells are typically utilized for a revival of skin. PLA cells can be generally produced by a suctioned sample of adipose tissue centrifugation and separation of the high density (SVF) from a liposuction sample. ASCs, which can be obtained by culturing (PLA) cells, have been settled as a cell-based therapy.

ASCs can be utilized to be converted into adipocytes and used for soft tissue augmentation (Kim et al., 2007a). Recently, cosmetic products that contain ASC-CM have been manufactured by many companies, and their products are used in a wide range for skin and hair repair (Table 11.2) (Kim et al., 2008). ASC-CM is now attaining popularity as an effective cosmetic product. Although it has a low degree of stability and absorption by the skin, the conditioned medium that contains various growth factors has multiple points of interest over cell-based products from an industrial point of view. This review covers stem cells-based products in the market and their cosmetic cutaneous applications.

Table 11.2 List of stem cell companies and stem cell-based cosmetic products

Indications	Conditioned medium source	Product
Anti-wrinkling; anti-aging, firming, skin revitalizer	AD-SCM	Luminesce cellular rejuvenation serum
Anti-wrinkling, anti-aging, firming, hydration of skin, diminishing the appearance of age spots, evenness repair of skin tone	AD-SCM	Luminesce advance night
Dry and damaged skin repair, photoaging, skin elasticity	AD-SCM	Luminesce essential body renewal
Anti-aging, anti-wrinkling	AD-SCM	Luminesce daily moisturizing complex
Serum anti-aging	AD-SCM	Reluma skin illuminating
Cleansing	AD-SCM	Reluma skin illuminating stem cell anti-aging cleanser
Anti-aging	AD-SCM	Reluma advance stem cell facial moisturizer
For longer, thicker and darker eyelashes	AD-SCM	Reluma lash
For dry and sensitive skin, eczema, psoriasis	AD-SCM	Reluma pserene stem cell cream
Male pattern baldness	AD-SCM	Reluma hair complex for men
Female pattern baldness	AD-SCM	Reluma hair complex for women
Male and Female pattern baldness	AD-SCM	Reluma stem cell hair complex original formula
Hair growth and anti-hair loss	AD-SCM	Reluma stem cell shampoo
Cleansing	AD-SCM	Cellure restart skin cleanser
Skin hydration and skin toning	AD-SCM	Cellure recode balancing toner
Anti-aging, anti-wrinkling	AD-SCM	Cellure regenerate serum booster
Under-eye wrinkles	AD-SCM	Cellure rework eye treatment
Skincare	AD-SCM	Cellure rebuild AM day cream
Skincare	AD-SCM	Cellure recover PM night cream
Anti-aging	AD-SCM and iPS-CM	Luminesce anti-aging skin serum
Hair growth	HFB-CM, AD-SCM, and iPS-CM	Regenrxx hair serum
Anti-aging	HMSC-CM	Stemulation facial cream
Anti-aging, anti-wrinkle	HMSC-CM	Stimulation elevate eye cream

(continued)

Table 11.2 (continued)

Indications	Conditioned medium source	Product
Photo-aging	HMSC-CM	Stimulation reliance on body lotion
Anti-aging	MSC-CM	Osmosis stem factor serum
Anti-aging	HFB-CM	Regenica advance rejuvenation day repair
Anti-aging	HFB-CM	Regenica advance rejuvenation overnight repair
Anti-aging	HFB-CM	Regenica facial rejuvenation complex post-procedure
Anti-aging	PL&UC-SCM	Blue horizon stem cell special skin serum
Anti-aging	HFB-CM	AQ skin solution active serum
Haircare	HFB-CM	AQ advance hair complex
Anti-aging	BM-SCM	Cutisera
Anti-hair loss	HL AD-SCM	Dermaheal stem C'rum
Anti-aging	SC-CM	Cell revive serum complete
Skincare	SC-CM	Cell revive brightening serum
Indications	Conditioned medium source	Product
Anti-aging	BM-SCM	Dr. Mary stem cell whitening
Anti-wrinkle	BM-SCM	Dr. Mary stem cell wrinkle
Anti-wrinkle	BM-SCM	Dr. Mary stem cell renew
Anti-aging, moisturizer	HNE-SCE	Recovery night moisture serum
Anti-aging, anti-wrinkle	HNE-SCE	Defensive day moisture serum SPF 15
Anti-aging, anti-wrinkle	HFB-CM	TNS essential serum
Anti-aging, anti-wrinkle	HFB-CM	TNS recovery complex
Under-eye wrinkles	HFB-CM	TNS eye repair
Anti-wrinkle	HFB-CM	TNS line refine
Anti-wrinkle	HFB-CM	TNS Lip plump system
Under-eye wrinkles	HFB-CM	TNS illuminating eye cream
Anti-aging	BM-SCM	The prestige counter aging essence
Anti-aging	BM-SCM	The prestige counter aging cream
Skin hydration and skin cleansing	BM-SCM	Luxury cell performance toner
Under-eye wrinkles	BM-SCM	Beaucell luxury cell performance eye cream

(continued)

Table 11.2 (continued)

Indications	Conditioned medium source	Product
Anti-aging	BM-SCM	Beaucell luxury cell performance serum
Under-eye wrinkles	BM-SCM	Beaucell luxury cell performance emulsion
Anti-aging	BM-SCM	Beaucell luxury cell performance cream
Anti-wrinkle, skin whitening, UV	BM-SCM	Prestige BB cream
Anti-aging	BM-SCM	Prestige sunblock
Anti-aging	MSC-CM	Anteage serum
Anti-aging	MSC-CM	Anteage accelerator
Anti-aging	AD-SCM	U autologous adult stem cell regenerative firming serum
Under-eye wrinkles	AD-SCM	U autologous adult stem cell regenerative eye cream
Anti-aging	AD-SCM	U autologous adult stem cell regenerative moisturizer
Anti-aging	SC-CM	Carecell gold nourishing cream
Anti-aging	SC-CM	Carecell perfect skin 3 in 1 lotion & essence
Haircare	SC-CM	20% of men hair and skin cell-conditioned media

* AD-SCM-Adipose-derived stem cell-conditioned medium, BM-SCM-Bone marrow-derived stem cell-conditioned medium, iPS-CM—Induced pluripotent stem cell-conditioned medium, HFB-CM-Human Fiber

Characters of ASCs

The limited availability of human cells that can renew autonomously has made the process of tissue engineering to be delayed. ASCs provide a probable answer to this dilemma as ASCs like other MSCs with regarding surface markers, gene profile, and function. ASC-derived cell- or protein-based therapies are shown to safe and effective for the repair of damaged skin (Kim et al., 2008).

Applications of ASCs and ASC-CM in Skin

In the past, all adipose tissue-based cosmetics taken during surgery were rejected; though, other usages have been developed for adipose tissues such as microinjection with ASC-CM and PLA cell injection.

8.2.1 Regarding the clinical use of PLA cells, clinicians emphasized that fat injection has an unexpected renewal effect on facial skin (Coleman & Surgery, 2006). This rejuvenation effect is described as delicate and remarkable and is accounted to the mobilization of free ASCs during the fat injection procedure, and the different growth factors released from transplanted ASCs (PARK et al., 2008). Thus, transplantation of fat cells itself has more advantages when compared to a filling regarding the process of soft tissue augmentation. Now, ASC is easily obtained, as an automatic mechanism of gathering SVF from lipo-aspirated fat is being endorsed by many companies. Employment of suitable centrifugation is essential for maintaining cell viability, and excess centrifugation may lead to the destruction of adipocytes and ASCs. Injection of PLA cells intradermally can be employed in the management of photo-aged skin in some patients, as two consecutive injections are given two weeks apart exert a beneficial effect of improving skin texture and wrinkles within two months following the injection. Besides, adipose tissue has been utilized to repair defects in soft tissue by acting as autologous filler. Moreover, survival rates were improved, while fibrosis and steatonecrosis were dampened in transplanted fat tissue.

8.2.2 ASCs have also shown promise in the treatment of facial rhytids (PARK et al., 2008), and newly differentiated ASCs also being used in the treatment of depressed scars (Kim et al., 2007a). The potential of ASCs conversion to adipocytes in vitro plays a significant role in soft tissue augmentation (Kim et al., 2007a). Therefore, ASCs with the ability of differentiation into adipocytes is considered an excellent cell source in adipose tissue engineering, which led some companies to utilize stored lipo-aspirated fat tissue and its content of ASCs for use in the future (Miller et al., 2003). Industrial productions are now focusing on the use of stem cell banking for the use of these cells in the field of cosmetic and reconstructive surgery in the future.

8.2.3 Feasibility of skin rejuvenation therapy without using autologous adipose tissue has been demonstrated, as many cosmetic companies developed products using ASC-CM (such as AAPE, Dr Jucre, and TACS Stem Cell) for skin rejuvenation. When comparing the potential regenerative power of ASC-CM with the injection of ASC, it was found that the potential of ASC-CM is lower than ASC; despite this, ASCs possess many beneficial secretory functions compared with the cell-based therapy. Because the ability of stem cell proteins to penetrate the skin barrier is limited, microneedles and lasers are used clinically to create a hole in order to increase the absorption process. ASC-CM stimulates both collagen synthesis and migration of dermal fibroblasts.

MSCs are derived mainly from stem cells, which are considered its primary source and ASCs have a marked capability to renew damaged skin. Clinical use of PLA cells to treat aged skin and their cosmetic utility is somewhat inapplicable, as the cells are extracted from patients. Instead, ASC-CM containing cosmetics can be prepared earlier from ASC taken from healthy volunteers; and these can subsequently be accessible for use in treating aged skin. The mechanism of regeneration attributed to ASC is not well understood. Complex paracrine and building-block mechanisms may be possible methods through which ASCs exert their effects. ASCs and ASC-CM have been incorporated in clinical dermatology practice and have shown skin regenerating capabilities. Many studies have proven the potential roles of stem cells in cutaneous use through stimulation of cell proliferation, differentiation, migration, and matrix renovation to offer a new strategy for cosmetic purposes of stem cell. Stem cells represent a unique therapeutic material in the repair of soft tissue; however, more trials should be done to outline the exact mechanisms, and other relevant cosmetic distresses.

Conclusions

The conventional paradigm of stem cell research is influenced by a strategy focused on availability, where legislation discusses the standards and ideals guiding research activity and its commitment to the patient. Under this governance model, the consumer and the sales requirement are not active until the actual commodity is believed to be approved. As a hegemonic type, the paradigm retains its political supremacy and its continuity in stem cell science's political culture if customers accept that it is the only correct direction in which the development of new stem cell goods is produced. In this sense, the economic and political importance of the four models of stem cell engineering regulation is that they intermediate between market demand for stem cell therapy and therapeutic availability. Health competition for stem cell therapies demonstrates the differences between research and therapeutic advancement models by ensuring that time scales are a significant component of customer preference and, in certain instances, a significant portion. Much of this transient market aspect, most global stem cell therapy supplies, has been generated by reactive medical models, challenging economic model viability.

https://www.tandfonline.com/doi/full/10.1080/13563467.2016.1198757.

As for medicine and cosmeceuticals, stem cell therapy and technologies should be considered to produce health and cost-efficiency goods by government and private organizations. Products of conditioned medium of stem cells promise to be used widely and to transfigure the beauty industry. The usage of modern cosmetic products focused on stem cells is growing, and many customers in the cosmetic industry are looking forward to this process.

References

Abbasalizadeh, S. & Baharvand, H. J. B. A. (2013). Technological progress and challenges towards cGMP manufacturing of human pluripotent stem cells based therapeutic products for allogeneic and autologous cell therapies. *31*, 1600–1623.

Ahn, J. O., Lee, H. W., Seo, K. W., Kang, S. K., Ra, J. C. & Youn, H. Y. (2013). Anti-tumor effect of adipose tissue derived-mesenchymal stem cells expressing interferon-β and treatment with cisplatin in a xenograft mouse model for canine melanoma. *PloS ONE, 8*, e74897.

Aljabali, A. A., Obeid, M. A., Amawi, H. A., Rezigue, M. M., Hamzat, Y., Satija, S. & Tambuwala, M. M. (2020). Application of nanomaterials in the diagnosis and treatment of genetic disorders. In F. A. Khan (Ed.), *Applications of nanomaterials in human health*. Springer Singapore.

Allison, M. (2012). *Hemacord approval may foreshadow regulatory creep for HSC therapies.* Nature Publishing Group.

Aly, R. M. (2020) Current state of stem cell-based therapies: An overview. *Stem cell investigation, 7*

Bajada, S., Mazakova, I., Richardson, J. B., Ashammakhi, N. J. J. O. T. E. & Medicine, R. (2008). Updates on stem cells and their applications in regenerative medicine. *2*, 169–183.

Bearzi, C., Rota, M., Hosoda, T., Tillmanns, J., Nascimbene, A., De Angelis, A., Yasuzawa-Amano, S., Trofimova, I., Siggins, R. W. & Lecapitaine, N. J. P. O. T. N. A. O. S. (2007). Human cardiac stem cells. *104*, 14068–14073.

Bernareggi, D., Pouyanfard, S. & Kaufman, D. S. J. E. H. (2019). Development of innate immune cells from human pluripotent stem cells. *71*, 13–23.

Biehl, J. K. & Russell, B. J. T. J. O. C. N. (2009). Introduction to stem cell therapy. *24*, 98.

Bolli, R., Chugh, A. R., D'amario, D., Loughran, J. H., Stoddard, M. F., Ikram, S., Beache, G. M., Wagner, S. G., Leri, A. & Hosoda, T. J. T. L. (2011). Cardiac stem cells in patients with ischaemic cardiomyopathy (SCIPIO): initial results of a randomised phase 1 trial. *378*, 1847–1857.

Businessinsider. (2020). *Growth in the global healthcare market presents opportunities* [Online]. Available: https://markets.businessinsider.com/news/stocks/worldwide-stem-cell-therapy-industry-to-2027-growth-in-the-global-healthcare-market-presents-opportunities-1029322666. [Accessed June 03, 2020].

Chism, S. E., Burton, R. C., Warner, N. L. J. C. I. & Immunopathology. (1978). Immunogenicity of oncofetal antigens: A review. *11*, 346–373.

Chong, J. J., Yang, X., Don, C. W., Minami, E., LIU, Y.-W., Weyers, J. J., Mahoney, W. M., VAN Biber, B., Cook, S. M. & Palpant, N. J. J. N. (2014). Human embryonic-stem-cell-derived cardiomyocytes regenerate non-human primate hearts. *510*, 273–277.

Chugh, A. R., Beache, G. M., Loughran, J. H., Mewton, N., Elmore, J. B., Kajstura, J., Pappas, P., Tatooles, A., Stoddard, M. F. & Lima, J. A. J. C. (2012). Administration of cardiac stem cells in patients with ischemic cardiomyopathy: the SCIPIO trial: surgical aspects and interim analysis of myocardial function and viability by magnetic resonance. *126*, S54–S64.

https://www.clinicaltrialsregister.eu

Coleman, S. R. J. P. & Surgery, R. (2006). Structural fat grafting: More than a permanent filler. *118*, 108S–120S.

Cuende, N., Rasko, J. E., Koh, M. B., Dominici, M. & Ikonomou, L. J. C. (2018). Cell, tissue and gene products with marketing authorization in 2018 worldwide. *20*, 1401–1413.

Cyranoski, D. J. N. (2019). The potent effects of Japan's stem-cell policies. *573*, 482.

Darabi, R., Gehlbach, K., Bachoo, R. M., Kamath, S., Osawa, M., Kamm, K. E., Kyba, M. & Perlingeiro, R. C. J. N. M. (2008). Functional skeletal muscle regeneration from differentiating embryonic stem cells. *14*, 134–143.

de Almeida, P. E., Meyer, E. H., Kooreman, N. G., Diecke, S., Dey, D., Sanchez-Freire, V., Hu, S., Ebert, A., Odegaard, J., Mordwinkin, N. M., Brouwer, T. P., Lo, D., Montoro, D. T., Longaker, M. T., Negrin, R. S., & Wu, J. C. (2014). Transplanted terminally differentiated induced pluripotent stem cells are accepted by immune mechanisms similar to self-tolerance. *Nature Communications, 5*, 3903.

Discher, D. J. A. (2010). Matrix elasticity directs stem cell lineage specification. 2010, *Q7*. 003.

Du, J., Zhou, L., Chen, X., Yan, S., Ke, M., Lu, X., Wang, Z., Yu, W., & Xiang, A. P. (2012). IFN-γ-primed human bone marrow mesenchymal stem cells induce tumor cell apoptosis in vitro via tumor necrosis factor-related apoptosis-inducing ligand. *The International Journal of Biochemistry & Cell Biology, 44*, 1305–1314.

Eleuteri, S. & Fierabracci, A. J. I. J. O. M. S. (2019). Insights into the secretome of mesenchymal stem cells and its potential applications. *20*, 4597.

Ellison, G. M., Vicinanza, C., Smith, A. J., Aquila, I., Leone, A., Waring, C. D., Henning, B. J., Stirparo, G. G., Papait, R. & Scarfò, M. J. C. (2013). Adult c-kitpos cardiac stem cells are necessary and sufficient for functional cardiac regeneration and repair. *154*, 827–842.

Essop, Z. M., Al-Sabah, A., Francis, W. R. & Whitaker, I. S. J. B. M. (2016). Transforming healthcare through regenerative medicine. *14*, 115.

Fishman, W. H., Raam, S., & Stolbach, L. L. (1975). Markers for ovarian cancer: Regan isoenzyme and other glycoproteins. *Seminars in Oncology, 2*, 211–216.

Foley, L. & Whitaker, M. J. S. C. T. M. (2012). Concise review: Cell therapies: the route to widespread adoption. *1*, 438–447.

Fukushima, S., Coppen, S. R., Lee, J., Yamahara, K., Felkin, L. E., Terracciano, C. M., Barton, P. J., Yacoub, M. H. & Suzuki, K. J. P. O. (2008). Choice of cell-delivery route for skeletal myoblast transplantation for treating post-infarction chronic heart failure in rat. *3*, e3071.

Gao, P., Ding, Q., Wu, Z., Jiang, H., & Fang, Z. (2010). Therapeutic potential of human mesenchymal stem cells producing IL-12 in a mouse xenograft model of renal cell carcinoma. *Cancer Letters, 290*, 157–166.

Gao, F., Chiu, S., Motan, D., Zhang, Z., Chen, L., JI, H., Tse, H., Fu, Q. -L ., Lian, Q. J. C. D. & Disease. (2016). Mesenchymal stem cells and immunomodulation: current status and future prospects. *7*, e2062–e2062.

Gavira, J. J., Herreros, J., Perez, A., Garcia-Velloso, M. J., Barba, J., Martin-Herrero, F., Cañizo, C., Martin-Arnau, A., Martí-Climent, J. M., Hernández, M. J. T. J. O. T. & Surgery, C. (2006a). Autologous skeletal myoblast transplantation in patients with nonacute myocardial infarction: 1-year follow-up. *131*, 799–804.

Gavira, J. J., Perez-Ilzarbe, M., Abizanda, G., García-Rodríguez, A., Orbe, J., Páramo, J. A., Belzunce, M., Rábago, G., Barba, J. & Herreros, J. J. C. R. (2006b). A comparison between percutaneous and surgical transplantation of autologous skeletal myoblasts in a swine model of chronic myocardial infarction. *71*, 744–753.

Gavira, J. J., Nasarre, E., Abizanda, G., Perez-ilzarbe, M., DE Martino-Rodriguez, A., García De Jalón, J. A., Mazo, M., Macias, A., García-Bolao, I. & Pelacho, B. J. E. H. J. (2010). Repeated implantation of skeletal myoblast in a swine model of chronic myocardial infarction. *31*, 1013–1021.

Gee, P., Lung, M. S., Okuzaki, Y., Sasakawa, N., Iguchi, T., Makita, Y., Hozumi, H., Miura, Y., Yang, L. F. & Iwasaki, M. J. N. C. (2020). Extracellular nanovesicles for packaging of CRISPR-Cas9 protein and sgRNA to induce therapeutic exon skipping. *11*, 1–18.

Ghieh, F., Jurjus, R., Ibrahim, A., Geagea, A. G., Daouk, H., EL Baba, B., Chams, S., Matar, M., Zein, W. & Jurjus, A. J. B. R. I. (2015). The use of stem cells in burn wound healing: A review.

Ghosh, Z., Huang, M., Hu, S., Wilson, K. D., Dey, D., & Wu, J. C. (2011). Dissecting the oncogenic and tumorigenic potential of differentiated human induced pluripotent stem cells and human embryonic stem cells. *Cancer Research, 71*, 5030–5039.

Gibbons, G. W. J. A. I. W. C. (2015). Grafix®, a cryopreserved placental membrane, for the treatment of chronic/stalled wounds. *4*, 534–544.

Giwa, S., Lewis, J. K., Alvarez, L., Langer, R., Roth, A. E., Church, G. M., Markmann, J. F., Sachs, D. H., Chandraker, A. & Wertheim, J. A. J. N. B. (2017). The promise of organ and tissue preservation to transform medicine. *35*, 530–542.

Ha, A., Criman, E. T., Kurata, W. E., Matsumoto, K. W., Pierce, L. M. J. P. & Open, R. S. G. (2017). Evaluation of a novel hybrid viable bioprosthetic mesh in a model of mesh infection. *5*.

Hagège, A. A., Marolleau, J. -P., Vilquin, J. -T., Alhéritière, A., Peyrard, S. V., Duboc, D., Abergel, E., Messas, E., Mousseaux, E. & Schwartz, K. J. C. (2006). Skeletal myoblast transplantation in ischemic heart failure: long-term follow-up of the first phase I cohort of patients. *114*, I-108–I-113.

Hare, J. M., Traverse, J. H., Henry, T. D., Dib, N., Strumpf, R. K., Schulman, S. P., Gerstenblith, G., Demaria, A. N., Denktas, A. E., & Gammon, R. S. (2009a). A randomized, double-blind, placebo-controlled, dose-escalation study of intravenous adult human mesenchymal stem cells (prochymal) after acute myocardial infarction. *Journal of the American College of Cardiology, 54*, 2277–2286.

Hare, J. M., Traverse, J. H., Henry, T. D., Dib, N., Strumpf, R. K., Schulman, S. P., Gerstenblith, G., Demaria, A. N., Denktas, A. E., Gammon, R. S., Hermiller, J. B., Reisman, M. A., Schaer, G. L., & Sherman, W. (2009b). A Randomized, double-blind, placebo-controlled, dose-escalation

study of intravenous adult human mesenchymal stem cells (Prochymal) after acute myocardial infarction. *Journal of the American College of Cardiology, 54*, 2277–2286.
Hare, J. M., Fishman, J. E., Gerstenblith, G., Difede Velazquez, D. L., Zambrano, J. P., Suncion, V. Y., Tracy, M., Ghersin, E., Johnston, P. V., Brinker, J. A., Breton, E., Davis-Sproul, J., Byrnes, J., George, R., Lardo, A., Schulman, I. H., Mendizabal, A. M., Lowery, M. H., Rouy, D., Altman, P., Wong Po Foo, C., Ruiz, P., Amador, A., Da Silva, J., Mcniece, I. K. & Heldman, A. W. (2012). Comparison of allogeneic versus autologous bone marrow–derived mesenchymal stem cells delivered by transendocardial injection in patients with ischemic cardiomyopathy: the POSEIDON randomized trial. *JAMA,* 308, 2369–2379.
Hata, H., Matsumiya, G., Miyagawa, S., Kondoh, H., Kawaguchi, N., Matsuura, N., Shimizu, T., Okano, T., Matsuda, H., Sawa, Y. J. T. J. O. T. & Surgery, C. (2006). Grafted skeletal myoblast sheets attenuate myocardial remodeling in pacing-induced canine heart failure model. *132*, 918–924.
Haynes, W. D., Shertock, K. L., Skinner, J. M., & Whitehead, R. (1985). The ultrastructural immunohistochemistry of oncofoetal antigens in large bowel carcinomas. *Virchows Arch A Pathol Anat Histopathol, 405*, 263–275.
Health, N. I. O. (2009). National Institutes of Health guidelines on human stem cell research.
Hofmann, M. -C. (2014). Stem cells and nanomaterials. *Nanomaterial.* Springer.
Ikeda, K., Ohto, H., Okuyama, Y., Yamada-Fujiwara, M., Kanamori, H., Fujiwara, S.-I., Muroi, K., Mori, T., Kasama, K. & Iseki, T. J. T. M. R. (2018). Adverse events associated with infusion of hematopoietic stem cell products: a prospective and multicenter surveillance study. *32*, 186–194.
Insights, C. M. (2020). *Global stem cell therapy market* [Online]. Available: https://www.coherentmarketinsights.com. [Accessed June 03, 2020].
Karantalis, V. & Hare, J. M. (2015). Use of Mesenchymal stem cells for therapy of cardiac disease. *116*, 1413–1430.
Kastrup, J., Haack-Sørensen, M., Juhl, M., Harary Søndergaard, R., Follin, B., Drozd Lund, L., Mønsted Johansen, E., Ali Qayyum, A., Bruun Mathiasen, A. & Jørgensen, E. J. S. C. T. M. (2017). Cryopreserved off-the-shelf allogeneic adipose-derived stromal cells for therapy in patients with ischemic heart disease and heart failure—A safety study. *6*, 1963–1971.
Kebriaei, P., Hayes, J., Daly, A., Uberti, J., Marks, D. I., Soiffer, R., Waller, E. K., Burke, E., Skerrett, D., Shpall, E. J. B. O. B. & Transplantation, M. (2020). A phase 3 randomized study of remestemcel-L versus placebo added to second-line therapy in patients with steroid-refractory acute graft-versus-host disease. *26*, 835–844.
Kehat, I., Khimovich, L., Caspi, O., Gepstein, A., Shofti, R., Arbel, G., Huber, I., Satin, J., Itskovitz-ELDOR, J. & Gepstein, L. J. N. B. (2004). Electromechanical integration of cardiomyocytes derived from human embryonic stem cells. *22*, 1282–1289.
Kern, S., Eichler, H., Stoeve, J., Klüter, H. & Bieback, K. J. S. C. (2006). Comparative analysis of mesenchymal stem cells from bone marrow, umbilical cord blood, or adipose tissue. *24*, 1294–1301.
Kim, M., Kim, I., Kim, S. -H., Jung, M., Han, S., Lee, J., Nam, J.-S., LEE, S .-K. & Bang, S. J. C. (2007a). Cryopreserved human adipogenic-differentiated pre-adipocytes: a potential new source for adipose tissue regeneration. *9*, 468–476.
Kim, W. -S., Park, B. -S., Sung, J. -H., Yang, J. -M., Park, S. -B., Kwak, S. -J. & Park, J. -S. J. J. O. D. S. (2007b). Wound healing effect of adipose-derived stem cells: a critical role of secretory factors on human dermal fibroblasts. *48*, 15–24.
Kim, W. -S., Park, S. -H., Ahn, S. -J., Kim, H. -K., Park, J. -S., Lee, G. -Y., Kim, K. -J., Whang, K. -K., Kang, S. -H., Park, B. -S. J. B. & Bulletin, P. (2008). Whitening effect of adipose-derived stem cells: a critical role of TGF-β1. *31*, 606–610.
Kim, W. -S., Park, B. -S., Park, S. -H., Kim, H. -K. & Sung, J. -H. J. J. O. D. S. (2009). effect of adipose-derived stem cell: activation of dermal fibroblast by secretory factors. *53*, 96–102.
Kimiskidis, V., Sakellari, I., Tsimourtou, V., Kapina, V., Papagiannopoulos, S., Kazis, D., Vlaikidis, N., Anagnostopoulos, A. & Fassas, A. J. M. S. J. (2008). Autologous stem-cell transplantation in malignant multiple sclerosis: A case with a favorable long-term outcome. *14*, 278–283.

King, N. M., Perrin, J. J. S. C. R. & Therapy. (2014). Ethical issues in stem cell research and therapy. *5*, 85.

Kwon, S. G., Kwon, Y. W., Lee, T. W., Park, G. T. & Kim, J. H. J. B. R. (2018). Recent advances in stem cell therapeutics and tissue engineering strategies. *22*, 1–8.

Laflamme, M. A., Gold, J., XU, C., Hassanipour, M., Rosler, E., Police, S., Muskheli, V. & Murry, C. E. J. T. A. J. O. P. (2005). Formation of human myocardium in the rat heart from human embryonic stem cells. *167*, 663–671.

Laflamme, M. A., Chen, K. Y., Naumova, A. V., Muskheli, V., Fugate, J. A., Dupras, S. K., Reinecke, H., Xu, C., Hassanipour, M. & Police, S. J. N. B. (2007). Cardiomyocytes derived from human embryonic stem cells in pro-survival factors enhance function of infarcted rat hearts. *25*, 1015–1024.

Le, P. T.-B., Duong, T. M., Vu, N. B. & Van Pham, P. J. B. R. T. (2016). Umbilical cord derived stem cell (ModulatistTM) transplantation for severe chronic obstructive pulmonary disease: A report of two cases. *3*, 902–909.

Linero, I. & Chaparro, O. J. P. O. (2014). Paracrine effect of mesenchymal stem cells derived from human adipose tissue in bone regeneration. *9*, e107001.

Locatelli, F., Algeri, M., Trevisan, V. & Bertaina, A. J. E. R. O. C. I. (2017). Remestemcel-L for the treatment of graft versus host disease. *13*, 43–56.

Madigan, M. & Atoui, R. J. B. (2018). Therapeutic use of stem cells for myocardial infarction. *5*, 28.

Mahla, R. S. J. I. J. O. C. B. (2016). Stem cells applications in regenerative medicine and disease therapeutics.

Malliaras, K., Zhang, Y., Seinfeld, J., Galang, G., Tseliou, E., Cheng, K., Sun, B., Aminzadeh, M. & Marbán, E. J. E. M. M. (2013). Cardiomyocyte proliferation and progenitor cell recruitment underlie therapeutic regeneration after myocardial infarction in the adult mouse heart. *5*, 191–209.

Malliaras, K., Makkar, R. R., Smith, R. R., Cheng, K., Wu, E., Bonow, R. O., Marbán, L., Mendizabal, A., Cingolani, E. & Johnston, P. V. J. J. O. T. A. C. O. C. (2014). Intracoronary cardiosphere-derived cells after myocardial infarction: evidence of therapeutic regeneration in the final 1-year results of the CADUCEUS trial (CArdiosphere-Derived aUtologous stem CElls to reverse ventricUlar dySfunction). *63*, 110–122.

Mandrycky, C., Phong, K. & Zheng, Y. J. M. C. (2017). Tissue engineering toward organ-specific regeneration and disease modeling. *7*, 332–347

Marks, P. W., Witten, C. M. & Califf, R. M. J. N. E. J. M. (2017). Clarifying stem-cell therapy's benefits and risks. *376*, 1007–1009.

Matzuk, M. M., Krieger, M., Corless, C. L., & Boime, I. (1987). Effects of preventing O-glycosylation on the secretion of human chorionic gonadotropin in Chinese hamster ovary cells. *Proceedings of the National Academy of Sciences of the United States of America, 84*, 6354–6358.

Mehta, M., Dhanjal, D. S., Paudel, K. R., Singh, B., Gupta, G., Rajeshkumar, S., Thangavelu, L., Tambuwala, M. M., Bakshi, H. A. & Chellappan, D. K. J. I. (2020). Cellular signalling pathways mediating the pathogenesis of chronic inflammatory respiratory diseases: an update. 1–23.

Melugin, H. P., Ridley, T. J., Bernard, C. D., Wischmeier, D., Farr, J., Stuart, M. J., Macalena, J. A. & Krych, A. J. (2020). Prospective outcomes of cryopreserved osteochondral allograft for patellofemoral cartilage defects at minimum 2-year follow-up. *Cartilage*, 1947603520903420.

Meng, Z. W., Baumgart, D. C. J. E. R. O. G. & Hepatology. (2020). Darvadstrocel for the treatment of perianal fistulas in Crohn's disease.

Miller, M. J., Plastic, P. S. E. F. D. C. J. & Surgery, R. (2003). Cryopreservation of adult stem cells derived from lipoaspirate. *111*, 2466–2468.

Mirzayan, R., Sherman, B. & Chahla, J. J. A. T. (2018). Cryopreserved, viable osteochondral allograft for the treatment of a full-thickness cartilage defect of the glenoid. *7*, e1269–e1273.

Mohty, M., Labopin, M., Velardi, A., Van Lint, M. T., Bunjes, D., Bruno, B., Santarone, S., Tischer, J., Koc, Y. & Wu, D. (2016). *Allogeneic genetically modified T Cells (HSV-TK) as adjunctive treatment in haploidentical hematopoietic stem-cell transplantation (haplo-HSCT) of adult patients*

with high-risk hematological malignancies: a pair-matched analysis from the acute Leukemia working party of EBMT. American Society of Hematology.
Moore, T. J., Morrow, R. L., Dormuth, C. R., & Mintzes, B. (2020). US food and drug administration safety advisories and reporting to the adverse event reporting system (FAERS). *Pharmaceutical Medicine, 34*, 135–140.
Nianias, A. & Themeli, M. J. C. H. M. R. (2019). Induced pluripotent stem cell (iPSC)–derived lymphocytes for adoptive cell immunotherapy: recent advances and challenges. *14*, 261–268.
. Oliveira, Jr., A. A. & Hodges, H. M. J. C. A. R. (2005). Alzheimer's disease and neural transplantation as prospective cell therapy. *2*, 79–95.
Oyama, T., Nagai, T., Wada, H., Naito, A. T., Matsuura, K., Iwanaga, K., Takahashi, T., Goto, M., Mikami, Y. & Yasuda, N. J. T. J. O. C. B. (2007). Cardiac side population cells have a potential to migrate and differentiate into cardiomyocytes in vitro and in vivo. *176*, 329–341.
Panés, J., García-Olmo, D., Van Assche, G., Colombel, J. F., Reinisch, W., Baumgart, D. C., Dignass, A., Nachury, M., Ferrante, M. & Kazemi-Shirazi, L. J. G. (2018). Long-term efficacy and safety of stem cell therapy (Cx601) for complex perianal fistulas in patients with Crohn's disease. *154*, 1334–1342. e4.
Park, B. S., Jang, K. A., Sung, J. H., Park, J. S., Kwon, Y. H., Kim, K. J. & Kim, W. S. J. D. S. (2008). Adipose-derived stem cells and their secretory factors as a promising therapy for skin aging. *34*, 1323–1326.
Patrikoski, M., Mannerström, B. & Miettinen, S. J. S. C. I. (2019). Perspectives for clinical translation of adipose stromal/stem cells.
Pellegrini, G., Ardigò, D., Milazzo, G., Iotti, G., Guatelli, P., Pelosi, D. & De Luca, M. J. S. C. T. M. (2018). Navigating market authorization: the path holoclar took to become the first stem cell product approved in the European Union. *7*, 146–154.
Perruccio, K., Topini, F., Tosti, A., Carotti, A., Aloisi, T., Aversa, F., Martelli, M. F., Velardi, A. J. B. C., Molecules, & Diseases (2008). Photodynamic purging of alloreactive T cells for adoptive immunotherapy after haploidentical stem cell transplantation. *40*, 76–83.
Pittenger, M. F., Discher, D. E., Péault, B. M., Phinney, D. G., Hare, J. M. & Caplan, A. I. J. N. R. M. (2019). Mesenchymal stem cell perspective: cell biology to clinical progress. *4*, 1–15.
Prasad, V. K., Lucas, K. G., Kleiner, G. I., Talano, J. A. M., Jacobsohn, D., Broadwater, G., Monroy, R., Kurtzberg, J. J. B. O. B. & Transplantation, M. (2011). Efficacy and safety of ex vivo cultured adult human mesenchymal stem cells (Prochymal™) in pediatric patients with severe refractory acute graft-versus-host disease in a compassionate use study. *17*, 534–541.
Purswani, S., & Talwar, G. P. (2011). Development of a highly immunogenic recombinant candidate vaccine against human chorionic gonadotropin. *Vaccine, 29*, 2341–2348.
Rad, F., Ghorbani, M., Roushandeh, A. M. & Roudkenar, M. H. J. M. B. R. (2019). Mesenchymal stem cell-based therapy for autoimmune diseases: emerging roles of extracellular vesicles. *46*, 1533–1549.
Robb, K. P., Fitzgerald, J. C., Barry, F. & Viswanathan, S. J. C. (2019). Mesenchymal stromal cell therapy: progress in manufacturing and assessments of potency. *21*, 289–306.
Rose, L., Wolf, E., Brindle, T., Cernich, A., Dean, W., Dearth, C., Grimm, M., Kusiak, A., Nitkin, R. & Potter, K. J. N. R. M. (2018). The convergence of regenerative medicine and rehabilitation: federal perspectives. *3*, 1–7.
Rosenthal, N. & Badylak, S. J. N. R. M. (2016). Regenerative medicine: today's discoveries informing the future of medical practice. *1*, 1–3.
Sanberg, P. R. J. P. O. T. N. A. O. S. (2007). Neural stem cells for Parkinson's disease: To protect and repair. *104*, 11869–11870.
Scott, L. J. (2018). Darvadstrocel: a review in treatment-refractory complex perianal fistulas in Crohn's Disease. *BioDrugs, 32*, 627–634.
Sheridan, C. (2018). *First off-the-shelf mesenchymal stem cell therapy nears European approval*. Nature Publishing Group.
Sugarman, J. J. C. R. (2008). Ethical issues in stem cell research and treatment. *18*, S176–S176.

Syed, B. A. & Evans, J. B. (2013). Stem cell therapy market. *Nature reviews. Drug discovery, 12*(3), 185

Takahashi, K., & Yamanaka, S. (2006). Induction of pluripotent stem cells from mouse embryonic and adult fibroblast cultures by defined factors. *Cell, 126*, 663–676.

Trojan, J., Naval, X., Johnson, T., Lafarge-Frayssinet, C., Hajeri-Germond, M., Farges, O., Pan, Y., Uriel, J., Abramasky, O., & Ilan, J. (1995). Expression of serum albumin and of alphafetoprotein in murine normal and neoplastic primitive embryonic structures. *Molecular Reproduction and Development, 42*, 369–378.

Uchida, S., De Gaspari, P., Kostin, S., Jenniches, K., Kilic, A., Izumiya, Y., Shiojima, I., Grosse Kreymborg, K., Renz, H. & Walsh, K. J. S. C. R. (2013). Sca1-derived cells are a source of myocardial renewal in the murine adult heart. *1*, 397–410.

Van Pham, P. J. B. R. T. (2016). Stem cell drugs: the next generation of pharmaceutical products. *3*, 857–871.

Vega, A., Martín-Ferrero, M. A., Del Canto, F., Alberca, M., García, V., Munar, A., Orozco, L., Soler, R., Fuertes, J. J. & Huguet, M. J. T. (2015). Treatment of knee osteoarthritis with allogeneic bone marrow mesenchymal stem cells: A randomized controlled trial. *99*, 1681–1690.

Volarevic, V., Markovic, B. S., Gazdic, M., Volarevic, A., Jovicic, N., Arsenijevic, N., Armstrong, L., Djonov, V., Lako, M. & Stojkovic, M. J. I. J. O. M. S. (2018). Ethical and safety issues of stem cell-based therapy. *15*, 36.

Weiss, A. R. R. & Dahlke, M. H. J. F. I. I. (2019). Immunomodulation by mesenchymal stem cells (MSCs): mechanisms of action of living, apoptotic, and dead MSCs. *10*, 1191.

Wernig, M., Zhao, J.-P., Pruszak, J., Hedlund, E., FU, D., Soldner, F., Broccoli, V., Constantine-Paton, M., Isacson, O. & Jaenisch, R. J. P. O. T. N. A. O. S. (2008). Neurons derived from reprogrammed fibroblasts functionally integrate into the fetal brain and improve symptoms of rats with Parkinson's disease. *105*, 5856–5861.

White, A. J., Smith, R. R., Matsushita, S., Chakravarty, T., Czer, L. S., Burton, K., Schwarz, E. R., Davis, D. R., Wang, Q. & Reinsmoen, N. L. J. E. H. J. (2013). Intrinsic cardiac origin of human cardiosphere-derived cells. *34*, 68–75.

Yamahara, K., Hamada, A., Soma, T., Okamoto, R., Okada, M., Yoshihara, S., Yoshihara, K., Ikegame, K., Tamaki, H. & Kaida, K. J. B. O. (2019). Safety and efficacy of amnion-derived mesenchymal stem cells (AM01) in patients with steroid-refractory acute graft-versus-host disease after allogeneic haematopoietic stem cell transplantation: a study protocol for a phase I/II Japanese trial. *9*, e026403.

Yang, R., Cai, Z., Zhang, Y., Yutzy, W. H., Roby, K. F. & Roden, R. B. J. C. R. (2006). CD80 in immune suppression by mouse ovarian carcinoma–associated Gr-1+ CD11b+ myeloid cells. *66*, 6807–6815.

Yoshihara, M., Hayashizaki, Y., Murakawa, Y. J. S. C. R. & Reports (2017). Genomic instability of iPSCs: Challenges towards their clinical applications. *13*, 7–16.

Zhou, C., Yang, B., Tian, Y., Jiao, H., Zheng, W., Wang, J. & Guan, F. J. C. I. (2011). Immunomodulatory effect of human umbilical cord Wharton's jelly-derived mesenchymal stem cells on lymphocytes. *272*, 33–38.

Chapter 12
Commercialization, IPR, and Market of Stem Cell Products

Sumira Malik

Abstract Stem cells and their derivatives stem cell products have broad and diverse applications in regenerative medicine. Specific stem cell lines are used as potential targets for the study of several diseases in pharmaceutical applications. Recently, stem cell products combined with an engineering approach are novel targets for replacing damaged and degenerated tissues for the numerous applications of recent tissue engineering applications. The emerging cellular therapeutics manufacturing unit is also dependent on an intricate array of stem cell products. The wide range of stem cell products is dynamically burgeoning with accelerating demand in the market due to their contribution as potential therapeutic effectors. The scientists and researchers involved in biotechnology companies, pharmaceutical companies, and academic platforms leverage stem cell products for a wide range of essential and functional applications. The chapter reviews the brief introduction of stem cell products, market availability, market-based competition for various stem cell products, and funding. Besides, topics such as the commercial status of stem cell products of particular reference to the clinical therapeutic application and intellectual property rights associated with regulatory policies of stem cell products and research are also discussed.

Keywords Stem cell products · Commercialization · IPR

S. Malik (✉)
Amity Institute of Biotechnology, Amity University Jharkhand,
Ranchi, Jharkhand 834001, India
e-mail: smalik@rnc.amity.edu

F. A. Khan (ed.), *Advances in Application of Stem Cells: From Bench to Clinics*,
Stem Cell Biology and Regenerative Medicine 69,
https://doi.org/10.1007/978-3-030-78101-9_12

Introduction to Stem Cell Products

Introduction to Broad Categories of Stem Cell Research Products and Regulations Associated with Stem Cell Research Products

In the current scenario, stem cells are one of the most promising entities that can be used for medicinal and research-based purposes as they aid in the regeneration and replacement of disease and damage organs and tissues. Several life-threatening diseases such as cardiomyopathy, spinal cord injury, Parkinson's, liver disease, tumor, and myocardial infarction can be benefitted from cell-based therapy. However, the translation of stem cell research and their respective benefits for their application as therapeutics is a sequential multi-step complicated and complex procedure. Also, the stem cell-derived products are different from the pharmaceutical-based origin, and therefore, the issues of stem cell-based product efficacy, consistency, and safety require addressable attention. Therefore, it necessitates experimenting on relevant issues related to the management of therapeutic outcomes and the potential risks with the fulfillment of existing guidelines and regulations attached to stem cell product development and their marketing. In the developing countries for "stem cell-based products (SCBP)," there is an immediate and high requirement of well-structured and well-defined rules and regulatory frame work like the regulations developed countries such as European Union (EU), Japan, and USA (US) which have well-defined and functional regulatory frame work (George, 2011).

"Stem cell-based products (SCBP)" term refers to the products which are descendants and derivative of stem cells and, further, these derived and developed products need to be administered into a patient and that contain or are derived from stem cells (Halme & Kessler, 2006). Globally, the commercial availability of stem cell-based products is very high in the market in form therapeutics but, still, these products and therapeutics do not fulfill the safety guidelines as most of them do not fall under formal clinical trial inspection which may intend to the detrimental effect on the physical health of the patient and also the financial exploitation remain extremely high. The clinical test should be performed with utmost safety guidelines and attention before delivering the market product (Giuseppe et al., 2010). Thus, there is a requirement for the appropriate well-structured and well-regulated system for the commercial supplies of SCBP to safeguard public health safety and trust issues. However, this situation and requirement impose presents numerous regulatory challenges.

Currently, scientists promote the use of stem cell research products as therapeutics for a wide range of applied applications. In contrast, clinical researchers support the concept by integrating both stem cells and their application in regenerative medicine for the treatment of diseases. Further, the pharmaceutical industry plays a significant role in the regulated use of stem cell products to conduct pharmacological testing on cell-specific tissues. Furthermore, tissue engineering scientists and researchers are intensively working upon replacement and repairing damaged tissues and organs by developing new techniques to combine bioengineering techniques and stem cell-based products.

The broad categories of stem cells research products available in market for the expedition of research are explained below:

- Stem cell lines which includes different types of cell lines. Example: iPSCs, MSCs, HSCs, NSCs, and ESCs.
- Stem cell culture media with and without supplements.
- Instruments related to stem cells culture and maintenance.
- Stem cell culture reagents.
- Stem cell-specific cytokines and growth factors.
- Primary antibodies against specific stem cell antigens.
- Bead-based stem cell separation systems.
- Fluorescent-based labeling and detection.
- Stem cell protein purification and analysis tools.
- Tools for DNA and RNA-based characterization of stem cells.
- Isolation/characterization services for stem cells and cell-based specific cell lines.
- Molecular tools for stem cell gene regulation.
- Mechanisms for in vivo and in vitro stem cell tracking.
- Expansion/differentiation medium for stem cell media.
- RNAi products.

Common Types of Stem Cell Product and Supply

There are various facilities, products, and supplies provided by the stem cell products industry. Stem cells were discovered before 30 years; therefore, there is a vast availability of stem cell-derived products, related services, and their market. There is a diverse market for stem cell products because of their complexity and technologies required to maintain and supply stem cell-based products. The standard type of stem cell products which are used in stem cell research is stem cell lines, differentiated type of stem cells, stem cell cultures, stem cell maintenance products such as growth factors and cytokines, primary antibodies, molecular and analytical tools for stem cells cloning, tools for gene expression and regulation such RNA and protein purification kits, imaging and tracking systems. Such stem cells, stem cell-derived products, and their maintenance and research stem cell products can be sold as an individual product or in bundles or as complete functional kits for the characterization and research studies in the field of stem cells. Currently, therapies which include mesenchymal stem cell (MSC), iinduced pluripotent cells (iPSCs) and embryonic stem cells (ESC) are used as therapy for the treatment of human diseases and are under pre-clinical development.

These stem cell-based products are bought by different researchers in diverse fields in academics, biotechnology companies, clinical institutions, and pharmaceutical companies-based researchers to develop new formulation and therapies for different diseases. The stem cell industry has been heavily driven and boosted by developing and manufacturing stem cell therapeutics and their products. The large-scale production of stem cell-based products is mediated by the use of a wide range of bioreactors, biofermentors, and 3D manufacturing systems through different industries.

Market Competition with Perception to Stem Cell Products

As per market perspective, the major focus of investors is on stem cell products as these are highly promising entities in the treatment of a wide range of genetic diseases and the development of new artificial organs and tissues that cannot be cured by non-cell products. Currently, adult stem cells (ASCs) market, which includes ESCs and cord cells derived from fetus placenta, is the largest commissioned market with huge market potential. There are around 180 companies which are involved in the marketing of stem cell products. The developed nations as USA are one of the leading master followed by the European and Asia–Pacific regions in stem cell and stem cell products market. Stem cell research is presumed to boost rapidly in the upcoming years because of the regulatory amendments in several countries in the next few years. For developing countries like India, a market share of about \$540 million with an annual growth rate of 15% is expected. This indicates that there could be an enormous possible investment from pharmaceutical, biotechnology, and bioengineering companies of different developed and developing countries in the market to develop stem cell-based products (Korde, 2008). The process of development of stem cell therapy and its flow to the market is explained in Fig. 12.1.

Different companies in the market supply the tools for the isolation, differentiation, expansion, culture, and characterization of stem cells, along with the technologies based enabled the production of these products at a small to large scale. These companies' versatile function includes stem cell research applications, the satisfaction of relative demand for stem cell products, analysis of stem cell manufacturing technologies, analysis of market trends including opportunities and threats related to the stem cell-based products. With market competition growing increasingly fierce, leading competitors within the sector include the companies in the market involved in the global competition are listed in Table 12.1.

With the increase in demand for stem cell-based products as therapeutics, there are high possibilities for rising stem cell-based production in the market. There are various emerging market opportunities for developing new products as per requirement or demand in the market depending upon increased acceptance of stem cell technologies and recognition of regenerative medicine's potential to reduce globally accelerating healthcare costs. In the current scenario, increasing health costing

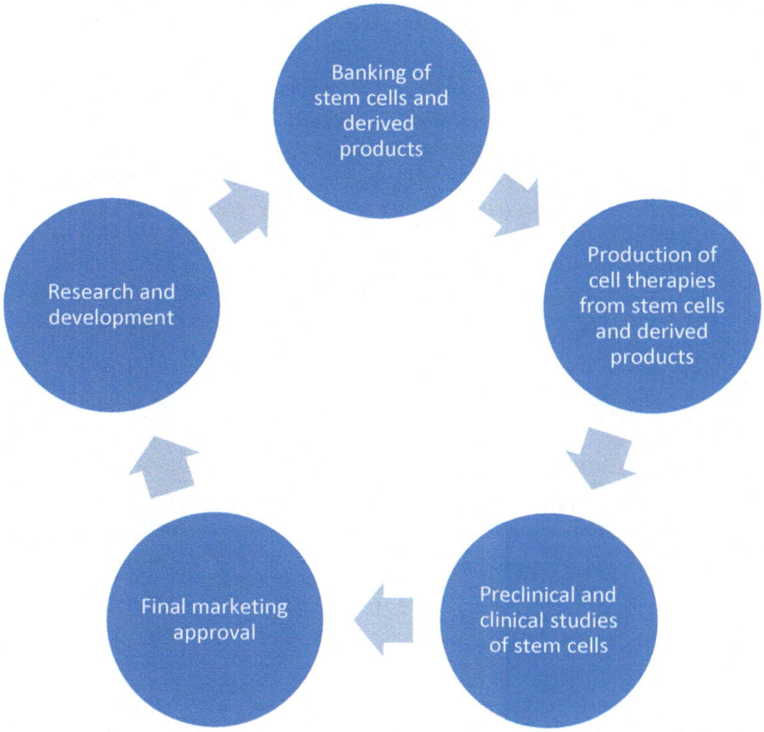

Fig. 12.1 Workflow of stem cell product therapy to market approval

services may drive essential factor costs for the investment into stem cell products and therapies on a global scale.

Funding for Stem Cell Research

The federal funding of stem cell research in developing counties such as the USA, from the National Institutes of Health (NIH), made an investment of $1.495 billion into stem cell research projects, with the pharmaceutical industry and private sources contribution of over 1.7 billion dollars annually (NIH Categorical Spending -NIH Research Portfolio Online Reporting Tools (Report). N.p., 2016. Web. 26 Jan. 2016; $40 Million to Support Future Stem Cell Scientists. CA Institute of Regenerative Medicine, Press Release. Available at: http://www.cirm.ca.gov/PressRelease_061809. Accessed Jan 24, 2016).

Table 12.1 Companies in the market involved in production and supply of stem cells and stem products at global level

S. No.	Specialization of company	Name of company	Product supply
1	Stem cell therapy	Mesoblast, Cynata Therapeutics, and Steminent Biotherapeutics	Stem cells
2	Industrial-scale production and manufacturing of stem cells and differentiated cells	Cellular Dynamics, Fujifilm CDI, Rooster Bio, ReproCELL and Ncardia	IPSCs and MSCs
3	Stem cell research tools	STEMCELL Technologie, Thermo Fisher Scientific, BD Biosciences, and Miltenyi Biotec	All stem cell-derived research products
4	Stem cell products	Corning (specializes in Matrigel®)	Products to support pluripotent stem cell culture and culture ware
5	Stem cells culture and maintenance products and instruments	Thermo Fisher Scientific, BD Biosciences, a Division of Becton Dickinson (BD), Merck KGaA, Miltenyi Biotec, Lonza Group, Takara Bio, GE Healthcare Life Sciences, Sigma Aldrich	Products to support stem cell culture and maintenance in vivo and in vitro

Current Regulatory Challenges in Stem Cell Product Development

In this section, the regulatory changes related to safety, efficacy, and quality are encountered in stem cell-based product (SCBPs) developments.

These issues are related to the preparation of cell therapies and tissue-based therapies at the clinical level as well as the commercial supply of SCBPs. The testing to ensure the safety of the products for administration in patients includes the examination and assay of product for any microbial and toxin contamination, followed by in vitro functional assays for the assessment of its clinical effectiveness and pervasiveness, including standards and controls to satisfy regulatory framework (Collins, 2009; Rayment & Williams, 2010). However, it is observed that all the model organisms used for experimental purposes during assays for pre-clinical and clinical studies have inherent limitations (Bianco & Robey, 2008). Also, the pre-clinical data needs to be studied before conducting the relevant examinations (George, 2011).

Commercialization

Stem Cell Therapy and Its Inclusion at Global Level in Clinical Research

Stem cell therapy and SCBPs are the best resources for the treatment of various diseases such cardiovascular diseases, diabetes, neurological disorders, spinal, orthopedic injuries, and regenerative medicine for the replacement of damaged tissues and organs through the clinical approval of a number of optimized techniques (Ghasroldasht et al., 2014; Lavoie & Rosu-Myles, 2013; Naderi-Meshkin et al., 2015). As previously reported till 2014, 4776, studies are registered on the US registry for clinical trials 2014. According to global research data, hematopoietic and bone marrow stem cells are top cells used in stem cell therapy, accounting 36 and 34% of total studies, respectively, followed by neural stem cells (14%), mesenchymal stem cells (11%), adipose-derived stem cells (4%), and embryonic stem cells (1%) as shown in Fig. 12.2. There was a remarkable growth discerned in one last decade with an emphasis on MSCs. In previous last ten years, a remarkable improvement has been observed in the increase in stem cell research and techniques to overcome such challenges (Naderi et al., 2011). The current challenges that the researchers still need to address in stem cell therapy are summarized in Fig. 12.3.

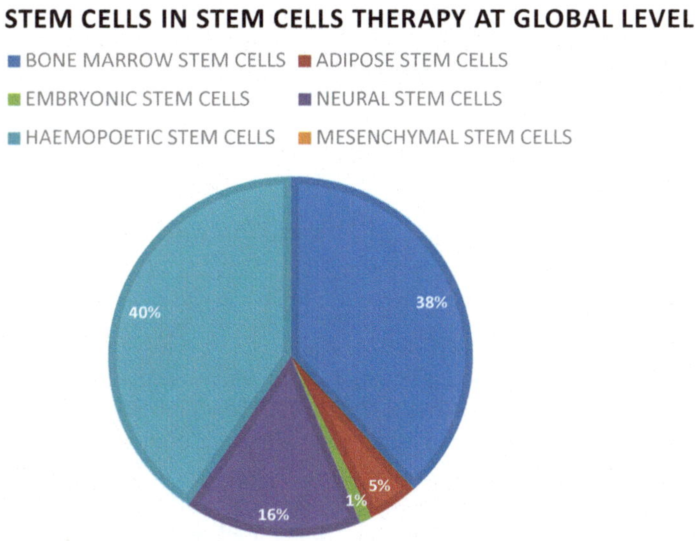

Fig. 12.2 Involvement of stem cells in stem cell therapy at global level

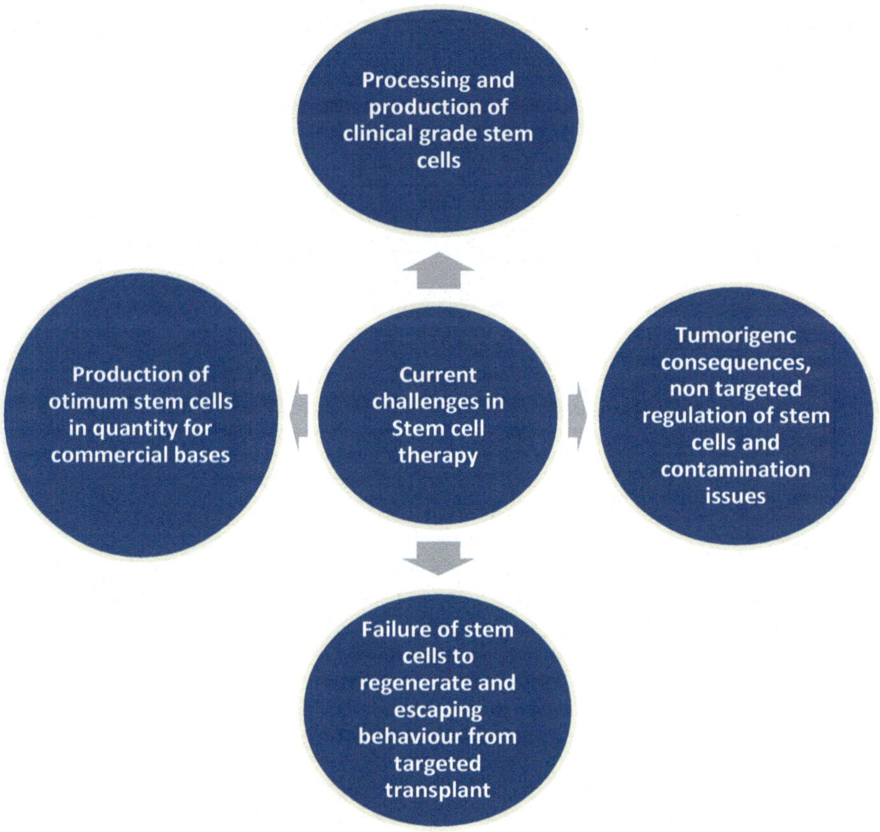

Fig. 12.3 Current challenges in stem cell therapies

Stem Cell Production at Clinical Dimension and cGMP Protocol

As per current reports, researchers are not aware of the cGMP protocol. The initial step for clinical-grade production is testing their regenerative potential as pre-clinical studies on animal models with efficacy evaluation through both in vivo and in vitro assays. Following initial assessment and promising pre-clinical data, follow-up of cGMP protocol for phase I and II monitoring is performed. Further, following the results obtained from Phase I and Phase II, the trials of phase III clinical trials, which shows 99% efficacy, can be finalized for commercial production. In case the data is conceded concerning phase I and II, results may be requested to be tried and re-produced again for trials to the researchers of the Research and Development Department of the respective company for further clinical-grade productions (Sensebe, 2008; Sensebe et al., 2011). The flow chart of cGMP protocol is summarized in Fig. 12.4.

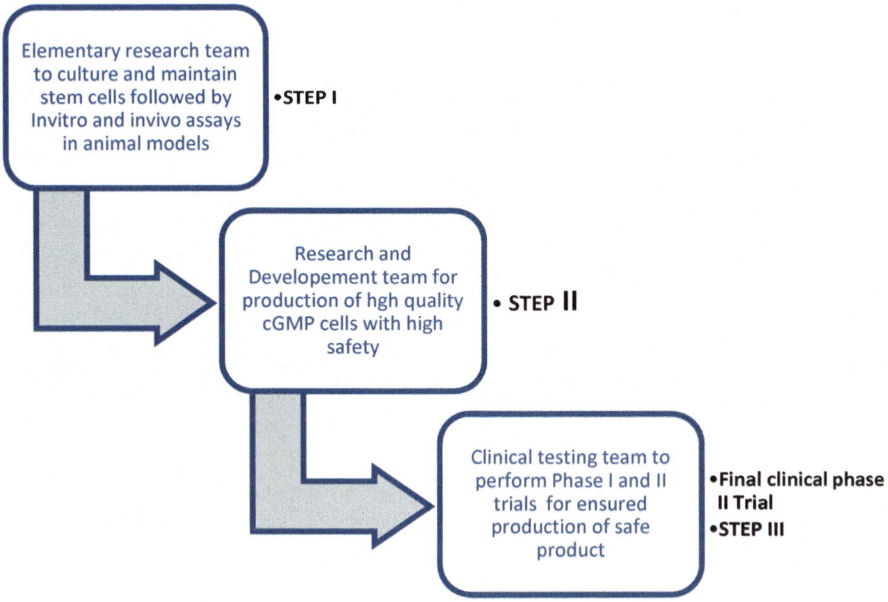

Fig. 12.4 Flow chart of cGMP protocol

Stem Cell Commercialization Establishment

Stem cell commercialization requires the transformation of stem cell research into the health industry with direct implementation into clinical studies to develop stem cell product-based therapeutics (Cuende & Izeta, 2010). It requires investment to gain skills expertise by the researchers to develop cell-based therapies with good manufacturing practice (cGMP) before applications and commercial supply of products. Robust coordination and dedication are required as several small biotech incorporations have shown their potential to commercialize cell therapeutic research (Lysaght et al., 2008). The developed nations such as Japan, the Republic of Korea, Germany, the USA, and Brazil have raised and supported public funds to perform stem cell clinical trials on a human to reduce the gap between basic and clinical research and their connectivity with the general public (Daniels et al., 2006).

- Commercialization of cell therapeutic research should maintain a strong pipeline, along with a planned growth of intellectual property (Hourd et al., 2008). The steps for the commercialization of therapeutic sciences are explained. The measurement of efficacy of pre-clinical trials in suitable and reliable validated animal models.
- Validation of safety measures for effective production of pre-clinical therapeutic products.
- Confirmation to attain FDA regulatory approval safe and effective pre-clinical products to performing prior to human clinical trials.

Commercial Perspective of Stem Cell Products

The process of transformation of research-based stem cell-based products into clinical practice requires immense and intensive coordination among academic institutions, hospitals with associations of patient and research organizations such as pharmaceutical and biotechnology-based companies followed by ethical and moral regulations for clinical commercialization. If these regulatory issues do not need to be addressed, they may account for the adverse effects on developing final products at the later stages before commercialization (Feigal, 2014). Companies require investment in the last stages of product development because of the low growth in the first initial years, which can be compensated by the following options mentioned.

(1) Institutions support in the form of grants and funds through public and private institutes.
(2) Outsourcing of research by different companies.
(3) Mutual collaborations among universities in a joint venture to exchange research facilities and ideas among researchers with expertise in respective areas.

Intellectual Property Rights and Stem Cell Products

Intellectual Property Rights and Stem Cell Products

Stem cells and their derived stem cell-based products have remarkable attributes to serve as commercial entities in the global market. Conceivably, making SCBPs readily available to the general public and common man as patients for the treatment of various diseases and organ transplantation or tissue replacement for a therapeutic application requires manufacturers' particular interest to commercialize the product and gain some profit. On a contradictory basis, if there is the least prospect for gaining profit, it is unlikely that the companies will manufacture the product. Regardless of considerable investments in research and development secured by numerous biotechnology industries to generate stem cell products, the simultaneous risk factors are very high, including zero guarantees of meeting the regulatory requirements imposed by different nations (Mummery et al., 2014).

Regulatory Intellectual Properties and Laws of Stem Cell Research in Different Developed Nations

The regulatory policies are followed by different well-developed nations such as the USA, Europe, and Canada (Zachariades, 2013). As shown in Fig. 12.5, the regulations of different nations such as USA, intellectual property (IP) regulations support the

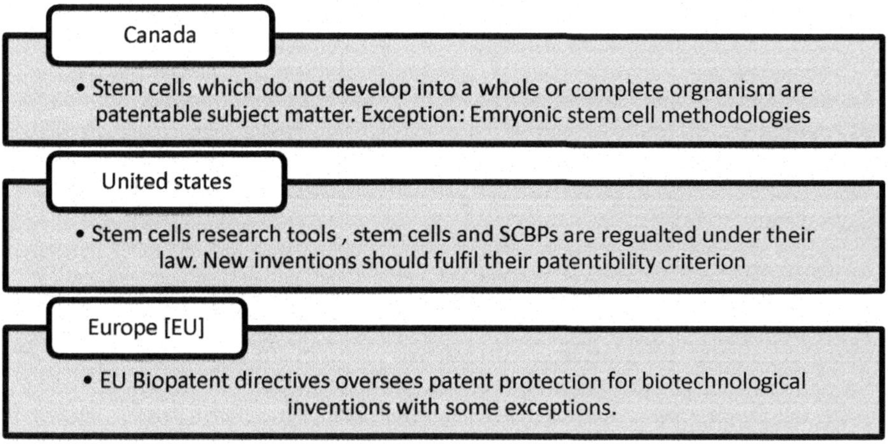

Fig. 12.5 Regulatory intellectual properties and laws of stem cell research in different developed nation

human embryonic stem cells (hESCs) and pluripotent stem cells and methods to develop such cells patentable in the USA. However, it is wholly inhibited in Europe EU. Thus, different research organizations, academic institutions, biotechnology, and pharmaceutical companies may collaborate at this point. They should reexamine their IP, regulatory, and commercial strategies based on jurisdictional laws and regulations. They must strictly comply with the current patent law regulations. The summary of intellectual properties and laws followed in different nations is explained in Fig. 12.5.

Market of Stem Cells Products

Global Stem Cell-Based Products and Stem Cell Therapy Market

The stem cell therapy and SCBPs market at a global level with its market size is predicted to reach USD 214.5 million by 2024. This market is expected to be primarily driven by enhancement of awareness among the general public about the therapeutic potency and application of stem cells, banking of stem cells and SCBPs, and their processing for future applications.

Market Segmentation

The market is segmented on the on bases of type of cells, therapeutic applications, and sources of cell. Each section is explained below.

Type of Cells

Stem cell therapies are classified into two types—autologous stem cell therapies and allogeneic stem cell therapies. Among two, the allogeneic stem cell therapies market has a larger share because of its intensive therapeutic applications, promotion in a clinical trial for developments of allogenic products, and finally cumulative commercialization of allogeneic products, which can quickly be produced through scale-up process. This suggests their rapid demand in the market in the upcoming future.

Type of Therapeutic Application

The stem cell-based products and stem cell therapy market have been various therapeutic applications in treating cardiovascular diseases, muscular diseases, skeletal diseases such as bone or joint related diseases, spinal and neural surgeries, gastrointestinal diseases, and skin diseases as burns, injuries, and wounds. It was reported that the musculoskeletal disorders category contributed to the most considerable revenue in the market for the treatment of musculoskeletal disorders and growing patient preference for effective and early treatment strategies.

Types of Cell Source

The primary cell resources in stem cell therapy and stem cell-based products are comprised of mesenchymal stem cells (MSCs), bone marrow-derived mesenchymal stem cells, placenta or cord blood cells from the fetal origin, and adipose tissue-derived mesenchymal stem cell. Among all these categories, the bone marrow-derived mesenchymal stem cells are the major market shareholders for increasing demand in the market due to diversified therapeutic applications.

Market Dynamics

Growth drivers, challenges, forecast parameters, data validation.

The industry experts should conduct the primary research in the market to understand the market dynamics, followed by valid market data validation. For market research, consumer-based surveys, which are comprised of consumer feedback and requirement, can also be conducted to understand and know consumer behavior and demand. Different growth driver regulates the market dynamics explained below:

Growth Drivers

1. **Awareness**: The emerging awareness in the general public about their health driven by the knowledge of the therapeutic potency of stem cells and its application in the future is one of the significant growth drivers for the market development. This envisages the customer or client to invest in the development of research for the promotion of advanced genome-based cell analysis techniques in the development infrastructure related to stem cell banking and processing. It eventually encourages customers to invest in the development of stem cell therapies for their upcoming future generations for longevity and healthy life. As per the World Health Organization (WHO) information, more than 50,000 transplants are carried out annually globally.
2. **Increasing risk of acute and chronic diseases**: The increasing risk of several acute and chronic diseases such as multiple sclerosis, cardiac arrest, heart failure, cerebral palsy, Parkinson's diseases, and hearing loss has promoted the interest of the general public toward the stem cell therapies-based treatments.
3. **Transplantation substituted by organ regeneration-based treatments**: Currently, there are several restrictions on traditional organ transplantation because of the dependency of patients on organ donor, transplanted organ rejection, and suppression of the immune system. These factors are also boosting the growth of the stem cell therapy market.

Challenges

1. **Technical limitations**: The limitation related to production during the scale-up, socio-ethical issues related to the use of stem cells in disease treatment. Another high possibility is in the systematic follow-up of the regulatory guidelines for product development and commercialization for the stem cell therapy market's growth.
2. **Socio-ethical issues**: This involves the religious briefs among the ordinary people in society.
3. **Economic perspective**: The investment of capital in research, poorly developed research infrastructure, and facilities for the development of stem cell therapeutics, stem cell-based product, and their preservation are also some of the challenges in the stem cell therapy market.

Forecast Parameters

The parameters which help in the identification of variables that may influence the establishment of the stem cell-based product in the market are as follows:

1. Adoption, production, import, export, and follow-up of regulatory frame work for product development.
2. Uniform of the establishment of the market according to the region.
3. Analyses of market penetration and respective opportunities according to understanding product commercialization, regional expansion.
4. Analyses and study of the historical background of the product to be launched.
5. Analyses of demand and supply trends and making alternations in industry dynamics to establish future growth.
6. Analyses of prolonged sustainability strategies abide by market partakers to determine the future course of the market.

Data Validation

Data validation is required to smooth the marketing of stem cell therapy and stem cell-based product supply in the market. The method responsible for data validation in the sustenance of the market is summarized in Fig. 12.6.

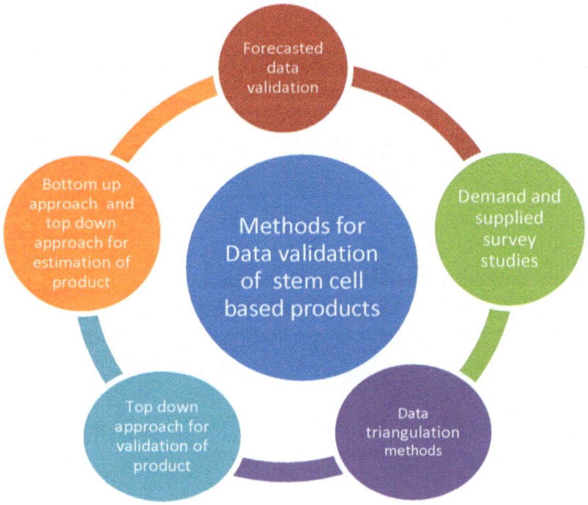

Fig. 12.6 Methods for data validation in sustenance of stem cell therapy and stem cell-based product market

References

Bianco, P. G., & Robey, S. P. J. (2008). Mesenchymal stem cells: Revisiting history, concepts, and assays. *Cell Stem Cell, 2*, 313–319.

CIRM provides $40 Million to support future stem cell scientists. CA Institute of Regenerative Medicine, Press Release. Accessed January 24, 2016. http://www.cirm.ca.gov/PressRelease_061809

Collins, N. H. (2009). Product review, release, and administration. In A. Gee (Ed.), *Cell therapy: cGMP facilities and manufacturing* (pp. 215–228). Springer.

Cuende, N., & Izeta, A. (2010). Clinical translation of stem cell therapies: A bridgeable gap. *Cell Stem Cell, 6*(6), 508–512.

Daniels, J. T., Secker, G. A., Shortt, A. J., Tuft, S. J., & Seetharaman, S. (2006). Stem cell therapy delivery: Treading the regulatory tightrope. *Regenerative Medicine, 1*(5), 715–719.

Feigal, E. G., Tsokas, K., Viswanathan, S., Zhang, J., Priest, C., Pearce, J., & Mount, N. (2014). Proceedings: International regulatory considerations on development pathways for cell therapies. *Stem Cells Translational Medicine, 3*(8), 879–887.

George, B. (2011). Regulations and guidelines governing stem cell based products: Clinical considerations. *Perspectives in Clinical Research, 2*, 94–99.

Ghasroldasht, M. M., Irfan-Maqsood, M., Matin, M. M., Bidkhori, H. R., Naderi-Meshkin, H., Moradi, A., & Bahrami, A. R. (2014). Mesenchymal stem cell based therapy for osteo-diseases. *Cell Biology International, 38*(10), 1081–1085.

Giuseppe, A., Sabrina, S., Viviana Lo, C., Francesco, S., Daniel, S., Lucia, T., et al. (2010). *American Journal of Translational Research, 2*, 285–295.

Halme, D. G., & Kessler, D. A. (2006). FDA regulation of stem-cell based therapies. *New England Journal of Medicine, 355*, 1730–1735.

Hourd, P., Chandra, A., Medcalf, N., & Williams, D. J. (2008). *Regulatory challenges for the manufacture and scale-out of autologous cell therapies*. StemBook.

Korde, J. (2008). Paving way to commercialization of mesenchymal stem cells. The business model of "off-the-shelf" products in the stem cells arena is slowly but steadily unfurling. *Biospectrum, 2008*: 38–40.

Lavoie, J. R., & Rosu-Myles, M. (2013). Uncovering the secretes of mesenchymal stem cells. *Biochimie.*

Lysaght, M. J., Jaklenec, A., & Deweerd, E. (2008). Great expectations: Private sector activity in tissue engineering, regenerative medicine, and stem cell therapeutics. *Tissue Engineering Part A, 14*(2), 305–315.

Mummery, C., van de Stolpe, A., Roelen, B. A. J., Clevers, H. (2014). *Patents, opportunities, and challenges: Legal and intellectual property issues associated with stem cells, stem cells* (2nd ed., pp. 381–395). Academic Press.

Naderi, H., Matin, M. M., & Bahrami, A. R. (2011). Review paper: Critical issues in tissue engineering: Biomaterials, cell sources, angiogenesis, and drug delivery systems. *Journal of Biomaterials Applications, 26*(4), 383–417.

Naderi-Meshkin, H., Bahrami, A. R., Bidkhori, H. R., Mirahmadi, M., & Ahmadiankia, N. (2015). Strategies to improve homing of mesenchymal stem cells for greater efficacy in stem cell therapy. *Cell Biology International, 39*(1), 23–34.

Rayment, E. A., & Williams, D. J. (2010). Concise review: Mind the gap: Challenges in characterizing and quantifying cell- and tissue-based therapies for clinical translation. *Stem Cells, 28*, 996–1004.

Report.nih.gov., NIH Categorical Spending -NIH Research Portfolio Online Reporting Tools (Report). N.p., 2016. Web. 26 Jan. 2016.

Sensebe, L. (2008). Clinical grade production of mesenchymal stem cells. *BioMedical Materials and Engineering, 18*(1 Suppl), S3-10.
Sensebe, L., Bourin, P., & Tarte, K. (2011). Good manufacturing practices production of mesenchymal stem/stromal cells. *Human Gene Therapy, 22*(1), 19–26.
Zachariades, N. A. (2013). Stem cells: Intellectual property issues in regenerative medicine. *Stem Cells and Development, 22.*

Index

F. A. Khan (ed.), *Advances in Application of Stem Cells: From Bench to Clinics*, Stem Cell Biology and Regenerative Medicine 69,
https://doi.org/10.1007/978-3-030-78101-9

Printed by Printforce, the Netherlands